BISON
BOOKS

Outward Odyssey
A People's History of Spaceflight

Series editor
Colin Burgess

HOMESTEADING SPACE
THE SKYLAB STORY

David Hitt, Owen Garriott, and Joe Kerwin

FOREWORD BY HOMER HICKAM

UNIVERSITY OF NEBRASKA PRESS • LINCOLN AND LONDON

Manufactured in the
United States of America
Photograph 48 used
courtesy David Hitt.
Photograph 50 used
courtesy Alan Bean.
All other photographs
used courtesy NASA.
∞
Library of Congress
Cataloging-in-Publication Data
Hitt, David.
Homesteading space :
the Skylab story / David Hitt,
Owen Garriott, and Joe Kerwin ;
foreword by Homer Hickam.
p. cm.
— (Outward odyssey :
a people's history of spaceflight)
Includes bibliographical references and index.
ISBN 978-0-8032-2434-6 (cloth : alk. paper)
ISBN 978-0-8032-3639-4 (paper : alk. paper)
1. Skylab Program.
2. Manned space flight—United
States—History—20th century.
3. Outer space—Exploration—
United States—History—20th century.
I. Garriott, Owen K., 1930–
II. Kerwin, Joe, 1932–
III. Title.
TL789.8.U6S5557 2008
629.45'4—dc22
2008017968
Designed and set in
Adobe Garamond
and Futura by
R. W. Boeche.

In honor of the thousands of men and women
down here
who made possible what was accomplished
up there.

While it would be impossible to name individually
in these pages the legions of people
who contributed to the Skylab program,
their dedication and hard work
are truly appreciated.

Contents

Illustrations

Photographs

Table

Foreword

Homer Hickam

The book that follows is a riveting, insightful account of the Skylab missions flown by the United States in 1973 and 1974. It is also simply a great yarn. Skylab began as an underdog, was nearly knocked out several times, staggered back to its feet, and fought on against overwhelming odds until it became a champion. In a lot of ways, it was the *Rocky* of space, and just like the story in that great film, it is an inspiration for all who know it. The difference is the remarkable saga of Skylab is all true.

For those of us who are old hands at NASA and in the space business, it is sometimes easy to forget what a great adventure it was and still is. Ultimately when all the layered explanations of why we go into space are peeled away, adventure remains at its core. But adventure aside, there are many quite practical reasons to go off our home planet. For one, the solar system is awash in energy resources such as microwaves and solar energy, and even the helium-3 isotopes that cover our moon seem perfect for futuristic fusion reactors. For another, the absence of gravity might ultimately produce wonderful new products, even life-saving medicines. And where else but space can we go to get above our light and radio-wave-polluted Earth and gain unobstructed views of our sun, the solar system, and the universe? Space is a scientific gold mine, and I believe some day it will be an economic one as well. But to be successful in the cosmos, we have to first figure out how to get there and stay. In other words, we have to learn to homestead space. This book tells how we first began to understand how to do that, through the program known as Skylab. Although often neglected by spaceflight historians, Skylab provided the key to all human space activities that followed. Quite simply, it was the series of flights that proved to the world that humans could live and work for long periods in space.

I grew up in the golden era of science fiction where all the spacemen (and spacewomen, though often scantily clad) were stalwart and brave. They were sort of ingenious, techno-savvy Davy and Polly Crocketts conquering the wild frontier while riding rockets. The robots in those tales were usually built only to help their humans through some difficulty ("Danger, Will Robinson!"), and the mightiest computer was the one every human had between his ears. If people were to explore space, they'd just have to go there themselves and have a look around. There was no other way. Not many of my favorite old-time writers guessed that by the time we were actually able to go into space, there would be a revolution in robotics combined with minimizing the size and maximizing the capabilities of computers. The reality of early spaceflight (and that's where we are now—very, *very* early) is that it is far easier, cheaper, and faster to send a robot than a human into space to explore and send back information on anything we please. But does that satisfy us? No indeed, and it shouldn't. For instance, we are also perfectly capable of purchasing a video travelogue of Paris. From the comfort of our living rooms, we can see the traffic passing beneath the Arc de Triomphe and the strollers along the Champs Élysées. But can we experience Paris with a video? No. We can only get a sense of what it is like. We can't look around a corner to see where some interesting alley might lead, or sit on a park bench and smell the aroma of fresh bread, or discover a new artist in the Louvre. It is the same for space. Ultimately to experience it, to gain from it all the riches it holds, the old sci-fi writers were correct. We humans must climb into pressurized containers and boldly rocket into the cosmic vacuum and there wrest from it with our own two hands all that it holds. In other words we still need spacefaring Meriwether Lewises and William Clarks off on bold adventures while accomplishing important scientific and economic work for the nation. The men and women who built and operated Skylab understood this and were determined to make such space accomplishments possible.

Skylab was designed to gain scientific knowledge in Earth orbit by utilizing equipment originally designed to carry men to the moon and back. It could be fairly said that Skylab was built from the spare parts of the Apollo program. Accordingly it was often neglected while the moon shots got all the energy and money, but eventually its time in the sun came, and what a grand time it was! Looking back now it's astonishing what we learned from it. During its three crewed missions, a trove of scientific knowledge was harvested that is yet unmatched by any other space facility, including the International Space Station. Skylab's huge volume, its well-constructed and considered scientific packages, its ability to generate more than adequate electrical power (after some emergency repairs!), and its focused crews made it, in my opinion, the finest comprehensive science and technological platform any country has ever sent into space. But I have to confess what I really, *really* like about Skylab is this: When it got into trouble, spacemen armed with wrenches, screwdrivers, and tin snips were sent up to fix it. No robots, no computers, no remotely controlled manipulating arms, just guys in suits carrying tools. The old sci-fi writers would have loved it!

Of course, with any space mission there is far more to the story than the spacecraft itself, or the crews. There must first be the visionaries who conceive the mission, then the politicians who must back it, followed by the armies of engineers, managers, accountants, and myriad other professionals who make it all work on the ground before the first rocket engine is lit on the pad. As this book informs us, one of Skylab's visionaries was a favorite of mine, none other than Dr. Wernher von Braun. In my memoir, *Rocket Boys/October Sky,* I told how when I was a teenager, more than anything in the world, I wanted to work for Dr. von Braun. In fact his brilliance was the distant, flickering flame for all the rocket boys and girls of that era and the reason a lot of us became engineers and scientists. Part of the fun of this book is reading how Dr. von Braun just went ahead and did things, including building the giant Neutral Buoyancy Simulator (NBS) at Marshall Space Flight Center in Huntsville, Alabama. The NBS was a big tank of water that allowed astronauts and engineers to simulate the weightless conditions of space. I am very appreciative that Dr. von Braun cut a few bureaucratic corners and built the NBS. Not only did his tank ultimately save Skylab, it also saved *me* when I suffered a bout of decompression sickness and had to be treated in its chamber. It was a great facility, although now sadly abandoned and fallen into disrepair. People ask me these days if I miss working for NASA. I do, sometimes, but mostly because I can't dive in the grand old NBS.

Although Skylab was accomplished before I became a NASA engineer, I did work on similar space missions, including training astronauts to repair the Hubble Space Telescope. That was an intricate, difficult mission but we knew we could do it because we had the example of Skylab's repair. I also worked on Spacelab, which was a science laboratory carried in the Space Shuttle's cargo bay. The Spacelab program, which proved to be a wonderful set of science missions, was profoundly affected by Skylab. Many times while working on a Spacelab situation, I heard, "Well, when I worked on Skylab, something like this happened and we . . ." Invariably the information given solved the problem we were working. One might suspect we Spacelabbers resented help from the old Skylab hands but not so. When there's work to be done in the space business, listening to veterans who've already done it is a smart thing to do. I'm proud to say that's what we did, at least on Spacelab and the Hubble Space Telescope repair missions.

I count as a good friend one of the authors of this book, astronaut Owen Garriott. With our friends and family, he and I have explored the Galapagos Islands and also hunted in Montana for dinosaur bones. It is fascinating to read this book and see a somewhat younger Owen aboard Skylab. Actually, from this account, he hasn't changed much. He's still a detailed observer of his surroundings and an amazing fount of scientific knowledge. He is also quite competitive and intensely focused. In other words he's challenging to be around and, therefore, the kind of friend we should all cultivate. Over the years I've also met all the other astronauts who flew on Skylab, plus backup Rusty Schweickart and Capcom (and future first Shuttle pilot) Bob Crippen. When *October Sky* the movie came out, I invited Pete Conrad to attend. I was gratified when he showed up for the premiere, and it didn't take long before we were deep in conversation, mostly about Skylab and our mutual experiences in the NBS. While my agent kept tugging at my elbow ("Homer, Steven Spielberg wants to say hello!"), I kept fending him off. Finally, I turned and barked, "Look, don't you understand? I'm talking to *Pete Conrad!*" My agent slunk off, and Pete and I finished our talk, one I still savor. I also once had Dr. Joe Kerwin turn up in one of my book-signing lines. I was astonished, though supremely pleased to see him there. I knew then I'd written a pretty good book.

The scientific and technological brilliance and love of adventure of all the Skylab astronauts were remarkable. This was also true of nearly all the people who worked on Skylab, such as Chuck Lewis, my former (and great, not to mention indulgent) boss at NASA, and Bob Schwinghamer who let me work in the NBS. Perhaps it was luck, or good fortune, but somehow the program got the people it needed and deserved. As a result, nearly every American-crewed mission since Skylab has been profoundly affected by the experiences gained by its nine crewmembers and the thousands of men and women who conceived, promoted, designed, constructed, rescued, and then made operational that magnificent facility. Just as the title of this book indicates, Skylab ultimately taught us how to make space our home. For a facility partially built from spare parts, I think that's prodigious!

Preface

If mankind is to travel from Earth to explore our universe, we will have to learn to live without the familiar experience of weight that is almost always with us on our home planet.

In the void between worlds, explorers will experience virtually total weightlessness. It's a strange environment without up or down, new to the body and with hidden threats, as big a step for us as was the classic emergence of life from the oceans onto dry land. They sputtered, we threw up, but apparently it won't take us as long to adapt. The point is that the process of *really* understanding "weightlessness" and *really* adapting to it was started by nine men in 1973. This is the story of that adventure.

Skylab was America's first step toward making space something other than a nice place to visit. Developed in the shadow of the Apollo moon missions and using hardware originally created for Apollo, the Skylab space station took the nation's astronauts from being space explorers to being space residents. The program proved that human beings can successfully live and work in space.

For many members of the public, Skylab is perhaps best known for two things—its beginning and its end. During the May 1973 launch of the Skylab workshop, an unanticipated problem damaged the station on its way to orbit. And of course, Skylab captured the world's attention with its fiery re-entry over the Indian Ocean and Australia in 1979.

But between those bookends lies a fantastic story of a pivotal period in spaceflight history. Skylab's three crews lived there for a total of six months, setting—and breaking—a series of spaceflight duration records. While previous U.S. spaceflights were focused on *going* places, Skylab was about *being* somewhere, not just passing through the phenomenal space environment, but mastering it. Everything that was to come afterward in U.S. spaceflight was made possible by this foundation—from scientific research in microgravity on the space shuttle to the on-orbit assembly of the International Space Station.

Even the unanticipated challenges that arose during the Skylab program turned into opportunities. The damage that crippled the spacecraft during launch became a rallying point for NASA and led to a repair effort that was unplanned and unprecedented—and perhaps still unparalleled.

This book is the story not only of the nine men who lived aboard Skylab but of all those who made the program a reality. And, like Skylab itself, this book depended on the contributions of a variety of people who shared their stories.

One of the pleasant surprises encountered in writing our story came in late 2005 when we showed Alan Bean (commander of the second manned mission) our draft of the second mission chapter. We had relied on the chronological account from Garriott's in-flight diary to tie together the events and to develop the story of that mission.

Much to our surprise, Alan said that he, too, had kept an in-flight diary and offered it to us for inclusion in this book! Naturally we took him up on that offer and were then absolutely amazed to find the extent of his handwritten account—more than one hundred pages of carefully written—albeit very difficult to decipher—print and script.

It covered not only events on board but also interpersonal relationships, his thinking and action to promote team spirit and optimum performance, his thoughts of home and family, and even more. We then incorporated as much of the "Bean Diary" in the story of the second mission as we thought appropriate and then added his full diary as an appendix to assure that all of Alan's thinking will be available to others.

Alan had kept the existence of the diary to himself for over three decades. Neither of his crewmates was aware that it had even been written. We are pleased and feel fortunate to include it here where others can better understand the thinking of arguably the most highly personally motivated crewman to fly in space.

Each of the eight living members of the Skylab crews has shared their stories with us, providing fresh perspectives of this unique experience. We deeply regret that the program's "Sky King"—first crew commander Pete Conrad—was not able to participate personally in this project. But his voice lives on in this book through previously recorded material.

You will also find portions of numerous interviews with Skylab engineers, scientists, managers, flight controllers, and other astronauts. We were struck by their unanimous view that Skylab was one of the most significant events in their professional careers—if not *the* most significant. Perhaps more to be expected, that is also true for all of the Skylab astronauts as well.

Yet, there has been very little written about the three missions themselves. Again almost all of our interviewees were most pleased to find that some of the crew were finally undertaking to report on these events from the perspective of those involved and, hopefully, that the contributions coming from all of the Skylab team would not be lost. Unfortunately we will certainly fall short of reaching the goal of recognizing even a modest part of their enormous contribution, but we do want to acknowledge their prime role in making the Skylab program the success we believe it came to be.

We hope that the dedication of this book reflects a little of that debt owed to the thousands of team members who really made it happen.

For all three of us, this book has been a true labor of love, and it is a story that we are very proud to be able to tell.

Acknowledgments

Just as it took a team of thousands working together to make the Skylab program, telling its tale would not have been possible without the generous contributions of many people. While the three of us struggled over the past several years to put everything in place and to make this story of Skylab both accurate and interesting for all readers, we have found that absolutely key elements required the personal contribution of additional members of the Skylab team.

Alan Bean's substantial contribution to this book, for which we are immensely grateful, was discussed in the preface.

And then there is Ed Gibson, the scientist pilot of mission three, who makes clear the major contributions made on the longest Skylab mission of all and who sets the record straight about some of the common misconceptions surrounding the mission. He is the principal author of most of chapter 10, "Sprinting a Marathon." He attacked the challenge passionately and went above and beyond our expectations.

Gibson's insight can also be found elsewhere in the book, particularly in his in-depth explanation of solar astronomy on Skylab. Gibson's knowledge of our sun, and observation thereof, is vast, and his expertise made for an invaluable addition to the book.

In addition, we would like to give particular thanks to the following people.

Vance Brand and Bo Bobko, who shared not only their personal experiences but also a wealth of resources they had saved over the years.

Chris Kraft, who provided us with unpublished Skylab material he had written for his memoir, *Flight: My Life in Mission Control*.

Lee Belew, Jerry Carr, Phil Chapman, Bob Crippen, George Hardy,

Charlie Harlan, Hans Kennel, Jack Kinzler, Don Lind, Gratia Lousma, Jack Lousma, Bob MacQueen, Joe McMann, George Mueller, Bill Pogue, Chuck Ross, Bob Schwinghamer, Phil Shaffer, Ed Smylie, Jim Splawn, J. R. Thompson, Bill Thornton, Stan Thornton, Jack Waite, and Paul Weitz, all of whom shared their experiences with us, either during in-person interviews or through written correspondence. (Some of these also extended and enhanced material from their interviews with the JSC Oral History Project for this book, particularly in chapter 10.)

Colin Burgess, our series editor, who got us started on this adventure and shepherded us along the way. Colin also contributed the story about Stan Thornton's experience finding a piece of Skylab; and he occasionally provided feedback on our manuscript when not too busy working on countless of his own.

The JSC Oral History Project, an incredible historical archive. Interviews from the project served as the foundation for the crew bios and the Skylab III chapter of this book and added additional insight to other areas.

Francis French, Gregg Maryniak, and Rob Pearlman, who looked through our in-progress manuscript and provided expert feedback.

Gary Dunham, who supported us graciously during this process.

Homer Hickam, who captured what we were trying to do in his excellent foreword.

Richard Allen of Space Center Houston, for letting us in at odd hours to review the Skylab trainer.

Genie Bopp; Sandra Brooks; Susanna Brooks; Eve Garriott; Bill and Leah Hitt; Lain Hughes; and Lee and Sharon Kerwin, who were kind enough to read through our developing book and point us in the right direction.

Many, many others who answered questions for us as they arose.

David Hitt would also like to thank his father, Bill Hitt, for setting his firstborn in front of the television on 12 April 1981 and fanning the flames ever since; Jim Abbott, for being the best mentor a young reporter could have hoped for; Nicole, for going along on an amazing experience; Jesse Holland; and last, but certainly not least, the good Drs. Garriott and Kerwin, for giving me the greatest adventure of my life by letting me share in one of the greatest of theirs, for being my patrons through Olympus, and, most of all, for their friendship.

Joe Kerwin would like to thank his wife, Lee; his daughters, Sharon, Joanna, and Kristina, for letting him be a part-time dad before the flight and for providing his main motive for coming back to Earth; and his grandsons, Christopher, Joel, Anthony, Brendan, and Joshua, for giving him a reason to help write this book—that they might be encouraged to go on adventures of their own.

Owen Garriott is most appreciative of the support provided by his family and children in his life both as a "flyer" and as a writer as he prepared this book. It is not an insignificant source of personal satisfaction to find that some of his enthusiasm for space adventure has carried over to his children.

Abbreviations

AAP	Apollo Applications Program
AM	Airlock Module
AOS	acquisition of signal
ASTP	Apollo-Soyuz Test Project
ATM	Apollo Telescope Mount
BET	beam erection tether
CDR	commander
CM	Command Module
CMG	Control Moment Gyroscope
CSM	Command and Service Module
EREP	Earth Resources Experiment Package
EVA	extravehicular activity
HOSC	Huntsville Operations Support Center
JOP	Joint Observing Program
JSC	Johnson Space Center
LBNP	lower body negative pressure
MCC	Mission Control Center
MD	Mission Day
MDA	Multiple Docking Adapter
MMT	Mission Management Team
MOL	Manned Orbiting Laboratory
MSC	Manned Spacecraft Center
MSFC	Marshall Space Flight Center
NASA	National Aeronautics and Space Administration
OWS	Orbital Workshop
PCU	Pressure Control Unit
PI	principal investigator
PLT	pilot

RC	Reaction Control System
ROTC	Reserve Officer Training Corps
SAL	Scientific Airlock
S-IVB	Saturn rocket stage used to construct Skylab
SM	Service Module
SMEAT	Skylab Medical Experiment and Altitude Test
SPT	science pilot
STS	Structural Transition Section (or, followed by a number, indicates a Space Shuttle mission designation, e.g. STS-9)
TACS	Thruster Attitude Control System
TAL	Trash Airlock
THC	translation hand controller
UCTA	urine collection and transfer assembly
USAF	United States Air Force
XUV	extreme ultraviolet

1. From the Ground Up

The task of turning a spent rocket stage into a livable space station was proving more difficult than anticipated. The man in the spacesuit was attempting to carry out the tasks that would convert the used, empty fuel tank into an orbital workshop. It was a daunting challenge. If the series of steps could be carried out, it would provide an expedient path to homesteading space. If not the station as designed would be worthless, an unusable husk. For the plan to work, when it came to these tasks, one of the agency's great truisms definitely applied—failure was not an option.

Almost immediately, he ran into problems.

Loosening the bolts before him was a simple enough task on the ground. Here though it was substantially more difficult. When he turned his wrench, instead of the bolts rotating, he did. The bolts were held in place, and since he was floating, there was nothing to keep him still. The gloves he had to wear only made things worse. Their bulkiness made it difficult to perform precise tasks. The fact that his suit was pressurized meant that it took effort to move the fingers of the glove. After a while, his hands would become sore from the effort. It was too much to ask, he realized. It couldn't be done. Reluctantly, he signaled to the safety divers to bring him to the surface.

That revelation was to be a turning point in the development of Skylab, America's first space station, and may well have saved the program. The man in the spacesuit was Dr. George Mueller, the National Aeronautics and Space Administration's (NASA's) associate administrator of Manned Space Flight, and the event took place in a water tank at NASA's Marshall Space Flight Center (MSFC) in Huntsville, Alabama. Mueller had been trying to find the best solution to the latest in a string of difficult decisions involving the orbital workshop. His quest for answers had led him to get hands-on experience himself with a simulated space station.

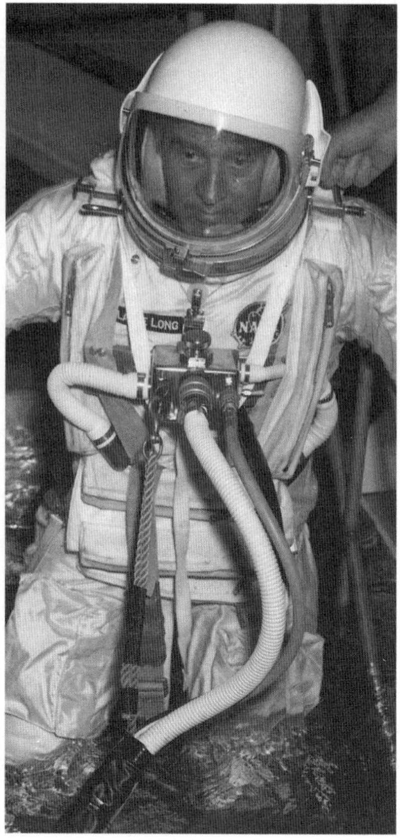

1. George Mueller (*left*) and Wernher von Braun prepare for dives.

The agency had already decided that a Saturn IVB rocket stage would be converted for use as the workshop. (Its name is a relic from early nomenclature for the Saturn rocket series.) Because launching more weight into space requires more fuel, every effort is made to reduce weight on a spacecraft. Dividing rockets into stages is one way that can be accomplished—when the fuel in one section is gone, that section separates, and the rest of the rocket continues. That way the rocket doesn't have to haul the weight of empty fuel tanks the entire trip.

Though there was agreement that the S-IVB stage should be used for the workshop, there were two schools of thought as to how that should be done. The initial idea was to launch the workshop as part of a Saturn IB rocket, the smaller of the two Saturn boosters. That rocket was not powerful enough to deliver a completed workshop to space, but it could place its S-IVB upper stage in orbit. Once the S-IVB was there, a crew of astronauts could convert the spent stage into a space station. Because this plan involved the station being launched full of fuel, it was known as the "wet workshop."

The other option was to use the larger and far more powerful Saturn V. That booster also used an S-IVB as its third stage. The workshop could be readied for use on the ground, and stacked on the Saturn V in place of the third stage. The first two stages would carry the heavy payload into orbit. This latter option was the "dry workshop."

As NASA's supply of Saturn V boosters was dedicated to the upcoming missions to carry men to the moon, the wet workshop option would allow the orbital workshop program to proceed simultaneously with the Apollo moon-landing program, using the more readily available Saturn IB rockets.

The plan, though, depended on the ability of astronauts to convert a fuel tank, which had just expended its supply of volatile liquid hydrogen and oxygen, into a home where they could safely live during the months to come. The crew would have to dock with the spent stage and then, working in bulky pressure suits, remove several bolts to gain access to its large liquid hydrogen tank. Then the astronauts would "passivate" the tank, making sure all of the propellant was gone and filling it with breathable gasses. The passivated stage would then have to be fitted with the equipment that would turn it into a laboratory and home. Those opposed to the wet workshop option argued that the required tasks would be too difficult for the astronauts to carry out while wearing spacesuits and working in a vacuum and in weightlessness.

Mueller, who initially supported the wet workshop, joined its detractors in 1969 following his visit to the Marshall Space Flight Center. He had been invited by the center's director, Dr. Wernher von Braun, who had been the leader of a team of German rocket scientists who were brought to the United States at the end of World War II. As happened more than once during their tenures as center directors, von Braun was in disagreement with Bob Gilruth, his counterpart at the Houston, Texas, Manned Spacecraft Center (MSC), later renamed Johnson Space Center (JSC). George Mueller recalled: "The resolution of the question of a wet workshop versus a dry workshop occurred when I was met at the Marshall Space Flight Center airport by Eberhard Rees. He said that Wernher had asked him to show me the newest facility at [Marshall]. He took me to an old hangar, which was most unremarkable.

And then he took me inside, and here was a gigantic tank.

"As we climbed up, he explained that they had decided that they needed a neutral-buoyancy facility to establish the feasibility of carrying out the refurbishment of the wet workshop. There were a number of technicians and several spacesuited divers working in the tank.

"Eberhard did not know what my reaction would be. This was an unauthorized capital expenditure and broke most of the rules for facility mods—typical Wernher. I guess to Eberhard's surprise, my first reaction was that I wanted to try out the tasks that the astronauts were being asked to do myself. So that's where I learned how to scuba dive. Once I tried even the simple task of closing the valves between the tanks, it convinced me that we couldn't rebuild and refurbish the tank in orbit, so that led me to the decision to go with the dry workshop."

After the dive Mueller began the process of making his decision a reality. "Bob Gilruth took a little more convincing, and Bob Thompson [the Skylab program director at MSC] was dead set against it. I really had to just say, we're going to do it, because I couldn't convince them," he said. "What they were trying to do, connecting all those things up, never would have happened."

The Prehistory of Skylab

The foundations that would eventually lead to the launch of Skylab in May 1973 had been laid much earlier. The idea of building a space station was nothing new. Not only did it predate Skylab, it predated manned spaceflight.

The first serious proposal for a manned space station was published in 1923 by German rocketry pioneer Hermann Oberth in his work, *The Rocket into Interplanetary Space.* In that work Oberth wrote, "Such a station could serve as a basis for Earth observations, as a weather forecasting satellite, as a communications satellite, and as a refueling station for extraterrestrial vehicles launched from orbit."

A few decades later the concept of the space station was familiar not only in the spaceflight community but also to the public at large. It hit the mainstream in a major way when it was explained by Wernher von Braun and others in a series of Walt Disney–produced television specials in the latter half of the 1950s. In a series of three *Tomorrowland* specials, the space station was presented not only as a place where humans would live and work in Earth orbit but also as a way station to other worlds. As a special-effects-laden enactment demonstrated, an orbital space station would be a key element in sending humans to the moon. The specials were based on concepts von Braun had presented at the First Symposium on Space Flight in October 1951, selected papers from which were published in *Collier's Magazine* under the title "Man Will Conquer Space Soon." Von Braun laid out what he saw as a logical progression for space exploration, beginning with simple orbital missions, moving on to the construction of a space station, which in turn would be used to support missions to the moon in the year 2000.

It was not to be. Two decisions sealed the fate of the idea of the space station as a steppingstone to the moon. Even before an American had been in orbit, the nation's space program was focused on a single goal. On 25 May 1961, less than three weeks after Alan Shepard became the first American in space, President John F. Kennedy issued the challenge to Congress that was to define the nascent human spaceflight programs of two nations for years to come: "I believe that this nation should commit itself to achieving the goal, before this decade is out, of landing a man on the moon and returning him safely to the Earth. No single space project in this period will be more impressive to mankind or more important for the long-range exploration of space; and none will be so difficult or expensive to accomplish." Billions of dollars, and more than eight years, were to go into making the dream a reality. Every spaceflight made during that time was dedicated to bringing the goal another step closer.

That decision effectively made any thoughts of a space station for its own merits a much lower priority and created a major stumbling block to the idea of using a space station to get to the moon. Von Braun's logical progression of infrastructure development required time that was a luxury that NASA could not afford in trying to meet Kennedy's deadline.

The second shoe dropped the next year with the decision to use the lunar-orbit rendezvous mission profile for the moon landings rather than the Earth-orbit rendezvous profile von Braun had outlined in the Disney specials. Lunar-orbit rendezvous involved sending two spacecraft to the moon instead of just one. While one descended to the lunar surface, the other remained in lunar orbit with the fuel that would be necessary to return to Earth. This technique made it possible for less mass to be sent to the moon. A lander that had to carry its Earth-return fuel to the surface would require even more fuel to lift that fuel back into space. Leaving the fuel for return to Earth on an orbiting spacecraft eliminated that need. As a result, both lunar-rendezvous spacecraft together were smaller than the one craft that would have been sent to the moon on an Earth-rendezvous mission. That meant they both could be launched on one Saturn V, unlike the larger craft that would have been launched on separate boosters and assembled in orbit. The lunar-rendezvous technique gave NASA a quick path to the moon but at the cost of a space station.

Kennedy's decision to pursue a bold fast-tracked lunar-landing program resulted in the most ambitious period in the history of space exploration and accelerated the achievement of the first human footsteps on another world. However, Mueller believes that in the long term human spaceflight would have been better served without that deadline. While it sped up the accomplishments of Apollo, he argues that haste was possible only by sacrificing the development of an infrastructure that would have supported continued exploration. "It's sort of unfortunate that the decision was made to go to the moon 'in this decade,' because that precluded the development of a real transportation system," Mueller said. "It never really got the sort of attention that it should have gotten, because it really, in my view, was the point in time when we had the opportunity to begin a true space civilization, a space evolution. In retrospect we would have been better off if we'd concentrated on a transportation system."

In an ideal situation, Mueller said, a combination of lunar orbit rendezvous and Earth orbit rendezvous would have been used for moon landing missions, creating a system in which the bulk of the spacecraft used to fly crews from the Earth to the moon and back would have remained traveling in a loop between the two, while smaller transfer vehicles carried astronauts to and from the surfaces of the two worlds. Such a system not only would have supported ongoing exploration of the moon, but it also would have helped develop techniques and infrastructure that later could have been used for human missions to Mars.

Though development of a space station had been placed on the back burner following the decisions of 1961 and 1962, it had not been abandoned, and several NASA centers continued to work on ideas for space stations. Langley Research Center in Hampton, Virginia, was working on the Manned Orbital Research Laboratory, which would support a crew of four to six astronauts for up to a year. The Manned Spacecraft Center in Houston was developing plans for Project Olympus, a large station that would remain in orbit for five years, where twelve to twenty-four crewmembers could live. Marshall, of course, was the home of von Braun, who had been outspoken about his own ideas for a space station. "There were a lot of concepts," Mueller said of the early space station discussions in NASA in the early-to mid-1960s. "Everybody was working on one. I'm afraid that a lot of them were just ideas."

The Birth of the Apollo Applications Program

Even as NASA strove to keep the fast-paced schedule necessary to meet Kennedy's deadline and beat the Soviet Union, there were those who realized that NASA needed to look past that mandate to the future. Since Apollo was a program with an established and concrete goal, Mueller and others believed that NASA needed to begin thinking about what the agency would do after the moon landing was achieved. In order to keep the space program going after the goal of Apollo was reached, NASA would need to begin planning years in advance to preserve continuity. Any new space program, any new missions, would need years of preparation. However, with around 5 percent of the national budget being funneled into NASA at the height of Apollo spending, they knew it would be difficult to win approval for another simultaneous space program. Basing that program around already-designed hardware would reduce the funding needed and would also make it easier to win support for the program. In July 1963 NASA headquarters commissioned a study by aircraft manufacturer North American Aviation of extended flight possibilities, including modifications to an Apollo Command and Service Module to support longer duration flights, creation of a small work-area module that could be used in connection with the Apollo capsule, and development of an independent laboratory module.

In 1964 President Lyndon B. Johnson asked NASA to report to him on its plans for the post-Apollo era. NASA administrator James Webb created a Future Programs Task Group to prepare a response to the president's request. The group's conclusion was that the agency should continue studies leading to flying extended Apollo missions by 1968 and continue long-range planning for space stations and human Mars missions in the 1970s.

While Webb was pleased by the report, others, including many in Congress, were not. It was criticized as insufficiently detailed and insufficiently ambitious. Mueller worked to address those concerns with the creation in August 1965 of the Saturn-Apollo Applications Program office at NASA headquarters. The office was to take the ideas generated during Extended Apollo research and use those as the foundation for a concrete program, including the development of a space station. The program name was shortened to the Apollo Applications Program (AAP), and the effort to find new uses for lunar-mission hardware began.

Mueller realized that he needed to act quickly to preserve the incredible team that was making it possible to reach the moon. In the latter half of the 1960s, while the Manned Spacecraft Center in Houston was still very much in the midst of the Apollo program, many of the engineers at NASA's Marshall Space Flight Center were completing their part of the Apollo program—the mighty Saturn V moon rocket. Unless something could be done to find new tasks for the team, NASA risked beginning to lose the expertise that had made the Saturn boosters possible. As other teams completed their Apollo duties as well, they would be in the same situation. But at Marshall the critical point was already rapidly approaching.

That difference in timetable was to be a key factor in the development of the orbital workshop, according to George Hardy, who was the chief of Program Engineering and Integration at Marshall and later the center's director of operations. "As time went on, and the idea of a follow-on to Apollo came up, I think von Braun and Mueller spent a lot of time together. [The Manned Spacecraft Center] was still busy with the lunar missions, and Marshall had pretty much delivered all their hardware, and certainly completed all the development work. And because of that, and I think also because von Braun's vision of the space station had been there for many, many, many years, Mueller and von Braun sort of collaborated on some of the early concepts of what could be done."

Mueller had one other goal in mind for Apollo Applications. For many within NASA, there was no question what the next major step should be after the flag had been planted on the moon. However, Mueller knew that before a flag could be planted in the red terrain of Mars, there were things that needed to be done. Apollo Applications would be the tool that would provide NASA with the knowledge and experience needed to forge onward to the Red Planet.

"The evolution really started with building something to go to the moon, and then, having built it, what do we do with it besides go to the moon," Mueller said. "We wanted to really lay the foundation for the future of space activities and really look at whether we can use that hardware to develop an understanding of what the next generation needs. So we looked at almost everything we can do in space. And one of the things that meant was the development of the space station."

The challenge of building a program around the Apollo equipment was made easier by the great potential that the hardware presented. Three major components made up the Apollo architecture. Of those the Lunar Module was the component most specifically designed for its Apollo task, landing men on the moon. Its thin, lightweight body was not designed for operations on Earth or for flight within the planet's atmosphere, and its propulsion system was engineered for the task of landing on and taking off from the surface of the moon. Despite the fact that its design was the most task-focused of the three major components, alternative uses for the Lunar Module were considered in the Apollo Applications Program.

Also developed specifically for Apollo but more easily lending itself to applications beyond carrying men to the moon was the Command and Service Module. Though relatively small compared to the American spacecraft that followed it, the Apollo capsule was downright roomy compared to its predecessors, the one-man Mercury vehicle and the two-seat Gemini. In addition to the three couches in which the crew members sat, the spacecraft featured a lower equipment bay that provided room for the astronauts to get out of their seats and move around. Attached to the rear of the Command Module was the cylindrical Service Module, which housed the spacecraft's primary propulsion system and four "quads" of maneuvering thrusters, each consisting of four thrusters at right angles to each other.

Rounding out these three components was the only one to predate, in concept at least, the Apollo program. More accurately, though, the final item was actually two—the Saturn IB and Saturn V launch vehicles. Work on the Saturn rockets had begun in the late 1950s, years before Kennedy had issued his challenge to Congress. So rather than designing them specifically to send men to the moon, von Braun had a broader goal in mind for the powerful launch vehicles: these would be the rockets that would open up the solar system for exploration. Mueller knew that the Saturns as well as the other hardware that had been developed for the Apollo program would lend themselves to a variety of applications beyond their original purpose.

Under Apollo Applications, the space station was given a new lease on life. The orbital workshop fit perfectly with several goals of the program. With its large volume and ability to remain in orbit for an extended period of time, such a workshop would be an ideal test bed for learning more about the effects of long-duration spaceflight and conducting microgravity science experiments. In addition, thanks to a concept that had been floating around NASA for several years, it was an excellent fit for another reason.

Well before Apollo Applications was created to find new uses for the hardware developed to go to the moon, that same idea had already occurred to Douglas Aircraft, the contractor responsible for the construction of the S-IVB, used as the upper stage of both the Saturn V and the smaller Saturn IB booster. The company suggested that the rocket stage they manufactured could be modified for use as a space station. While NASA would latch onto the Apollo Applications concept as an affordable means of developing new programs, Douglas of course proposed the idea in hopes of increasing business.

"That came from Douglas," Mueller said. "They were the ones that were pushing that. They had the S-IVB stage, and they were trying to figure out how to use it in the future." As early as 1959 von Braun had also advocated the spent-stage station concept. Many in the agency, including Mueller, supported the idea, which provided solutions to two major challenges in developing a space station. The proposal not only solved the issue of how to construct a facility with a large volume where astronauts could live and work, it also provided a convenient way for launching the structure by integrating it into the booster that would carry it.

The space station would also play an important part in pursuing Mueller's other goal for Apollo Applications, preparing the way to Mars. Mueller realized that an immediate post-Apollo manned flight to Mars would be impossible. There was yet too much to be learned before such a mission could be mounted since it would involve astronauts living away from Earth for months or even years. At the end of Apollo, NASA's longest spaceflight was the fourteen days astronauts Frank Borman and Jim Lovell had spent aboard Gemini 7 in December 1965, a record at the time but extremely brief when compared with the time needed to travel to a neighboring planet. NASA needed two things before it could even consider sending missions to the Red Planet. The agency would have to have long-duration spaceflight experience, and it would need a space transportation infrastructure. A space station would easily accomplish the former and would lay important groundwork for the latter.

"The decision back then was to get ready to go to Mars," Mueller said of NASA's response to a 1964 mandate by President Lyndon B. Johnson that the agency reveal its plans for the future. "All of this was really aimed at eventual Martian spaceflight. That was the real purpose of Skylab—to learn how to live in space for a year. That was really the drive behind Skylab, and really is the drive behind the Space Station. It's the only reason to build a space station, as a way-point, or to prove that you can live in space, or to find out how to live in space." While there were a few voices in the agency that wanted to press on to Mars immediately after Apollo, Mueller said that most realized that the agency would not be ready, that more experience would be needed before that goal would be attainable. "I don't know if there was any serious talk [of going straight to Mars] anywhere else, but there wasn't in Manned Space Flight," he said. "There was a great deal of concern in our scientific community dealing with human life, whether or not man could in fact operate and live and return safely from a trip to Mars. And that was really the incentive for the development of the Skylab program."

Working in the shadow of Apollo was to have advantages to go along with the disadvantages. The fact that the public and the powers that be were focused on AAP's older and sexier sibling often meant that support, both political and financial, could be hard to come by. But it also gave those involved in the program an opportunity rare in the world of government projects: the freedom to develop the program the way they felt it should be developed, largely free of bureaucratic involvement.

George Hardy, chief of Program Engineering and Integration at Marshall explained, "I always referred to Skylab initially as the little redheaded bastard out behind the barn because there were obvious political overtones to starting a new program. Skylab got started as a utilization of existing hardware as opposed to a congressionally endorsed program. That brought it a little more legitimacy, with people recognizing this makes a lot of sense.

"It was an evolutionary thing. It was something talked about, and concept studies were done. Compared to the way they do things today—with a lot of formal study, with various contractors competing, with different ideas, which is good, I'm not criticizing that—this was all done principally by the government, by Marshall and MSC. Studies were made, but it evolved. I can't tell you a single time, beyond [one], where there was some official direction. The program just evolved into doing it that way. It was different from programs before that or even programs after that. It was a program that was put together by people that were working on it as opposed to oral direction coming down from on high with a long set of objectives.

"This is somewhat of an exaggeration, but it was almost like, 'Look, we've got all this hardware and stuff here, we ought to figure out something to do with it.' And of course that's what they did, but they figured out something to do with it that was quite impressive.

"Once we got a program that was out in the open, so to speak, on Skylab, we were able to implement it because it was still not the primary focus of the space program for this country. The lunar landings were still occurring, and that obviously was the primary focus. The initial implementation of Skylab was what I felt was an absolute ideal situation. We had support of Mueller and others, the three manned space centers, but at the same time we had a lot of flexibility that other programs didn't have."

The Competition

NASA was not the only agency working on developing a space station. In the early 1960s, the U.S. Air Force had also begun work on its own space-station program, the Manned Orbiting Laboratory (MOL). The Air Force had already been involved as a partner in one successful space-related program, the X-15 rocket plane, which could carry pilots to the edge of space on suborbital flights and earned Air Force astronaut wings for several of them (awarded to pilots who reached an altitude of fifty miles). However, the Air Force was also interested in its own orbital spaceflight program. In the 1950s the Air Force was developing the Man in Space Soonest program, which lost out to Project Mercury to be America's first manned spaceflight program when President Dwight D. Eisenhower decided that he wanted a civilian agency, NASA, to be in charge of the first flights. In 1962 the Air Force began developing plans for Blue Gemini, which would involve military use of NASA's Gemini hardware. When Air Force officials realized in 1963 that their own next-generation rocket plane, the X-20 Dyna-Soar, could not be completed on a schedule competitive with NASA's Gemini spacecraft, the Air Force abandoned the X-20 in favor of the MOL, which would use Gemini technology as the basis for an orbital workshop program. Plans were made for a space laboratory to be launched in 1968.

Like the Apollo Applications Program, MOL was designed to make use of existing hardware. The launch would use the Air Force's proven Titan IIIC booster, and the crew would ride in a modified Gemini spacecraft.

On 25 August 1965 President Johnson gave his approval to the Manned Orbiting Laboratory program. In January of the following year, Congress strongly encouraged NASA to participate in the MOL program rather than pursuing its own Apollo-based space station program. NASA argued that the Air Force facility would be insufficient for supporting the scientific goals of the Apollo Applications Program and that modifying MOL to meet those requirements would generate costs and delays greater than moving ahead with NASA's Apollo Applications plans. The arguments were ultimately successful and bolstered support for the agency's program.

Karol "Bo" Bobko, an MOL astronaut who went on to join NASA's astronaut corps, said the two programs were very different: "Totally. But we'd have to shoot you before we told you," Bobko joked (probably). "The similarity was that it was a laboratory flying in space for a reasonable length of time. The dissimilarity was that missions were different." While the full details of the Air Force laboratory program have never been declassified, it would have involved conducting intelligence operations and establishing a military presence in space.

When the Air Force canceled the Manned Orbiting Laboratory program in June 1969 because of continuing delays and rising costs, NASA benefited in two ways. The cancellation of MOL meant an end to the political competition between the two agencies, allowing the Apollo Applications Program to be seen for its own merits. In addition some members of the Air Force astronaut team were accepted into NASA's corps. Some of these, most notably Bobko, Bob Crippen, and Dick Truly, went on to play important roles in the Skylab program.

Mueller said that his reaction toward the Air Force program was one of excitement. "Well, at least around me, we were all enthusiastic about the Air Force beginning to be interested in space with the Manned Orbital Laboratory. If that had [flown], I'm sure we would have had a much more vigorous space program." Competition between NASA and Air Force manned space programs, he said, would have forced each agency to be more aggressive in its efforts in an attempt to stay ahead of the other.

At the same time, a thought process similar to what NASA was going through in looking past Apollo was occurring on the other side of the world in the Soviet Union. The Soviet Union had started the space age staying one step ahead of the United States. The first Soviet satellite, Sputnik, was launched on 4 October 1957, ahead of the January 1958 launch of the first U.S. satellite, Explorer 1. The first Soviet cosmonaut, Yuri Gagarin, flew into space on 12 April 1961; the first American astronaut, Alan Shepard, followed less than a month later on a suborbital flight. It would be February of the following year before an American, John Glenn, would match Gagarin's feat of orbiting the Earth.

Knowing that NASA was developing a two-person spacecraft, the Soviet space program launched the first Voskhod capsule with three cosmonauts aboard on 12 October 1964. The Voskhod was essentially a modified version of the one-person Vostok capsule in which the ejection seat had been removed and three seats had been installed. Because of the cramped conditions in the spacecraft, the three cosmonauts flew without pressurized spacesuits. The gambit allowed the Soviet Union to beat the first multiperson U.S. spaceflight, Gemini 3, by more than five months, and the first U.S. three-person mission by almost exactly four years. Apollo 7, the first NASA flight to carry three astronauts, was launched on 11 October 1968.

On the next flight of a Voskhod spacecraft, this time with only two crewmembers aboard, cosmonaut Alexei Leonov became the first person to go outside a spacecraft for a spacewalk, on 18 March 1965. Less than three months later, astronaut Ed White made the first U.S. spacewalk on the Gemini 4 mission with a duration outside of over twenty minutes, besting Leonov's twelve.

Following Voskhod 2, however, the momentum shifted. It would be two years before the Soviet Union launched another manned spaceflight, and during that time, NASA's Gemini program established some firsts of its own, including the first orbital rendezvous and dockings. Gemini flights also set new records for altitude and duration.

The year after the Gemini program ended, 1967, was a tragic one for both nations' space programs. In the United States, the crew of the first Apollo mission was lost in a fire during launch-pad tests only weeks before they were to launch. Less than three months later, the Soviets suffered a disaster of their own. Cosmonaut Vladimir Komarov launched on 23 April 1967 in the first flight of the U.S.S.R.'s new Soyuz spacecraft. The Soyuz 1 mission was all but complete, and Komarov was almost home when the parachute system for the spacecraft failed, killing the cosmonaut.

When the two nations resumed manned spaceflight in 1968, the momentum in the race to the moon had definitely shifted. Before that year was out, NASA had reached the moon, successfully placing the crew of Apollo 8 in lunar orbit on Christmas Eve. Five months later NASA returned to lunar orbit, this time to test the Lunar Module that would be used to land on the moon. Earlier that year, the Soviets had made the first test launch of the booster with which they hoped to send cosmonauts to the moon. The first of the Soviet Union's powerful N1 moon rockets was launched on 21 February 1969 and exploded around sixty-nine seconds after launch. A second test was conducted in July 1969, and this time the first stage engine shut down prematurely immediately after liftoff. Seventeen days later, Neil Armstrong made humanity's first footsteps on another world. Even after the United States had won the race to the moon, two more tests were made of the N1 booster, but like the first two, these were also unsuccessful.

Though their lunar objective was slipping away, the Soviet space program was still going strong. In January 1969 two Soyuz spacecraft docked in orbit, and for the first time, members of the crew of one spacecraft transferred to another spacecraft. Soyuz 4 launched with only one cosmonaut aboard but returned to Earth with three. In October of that year, the Soviets achieved another first when Soyuz 6, 7, and 8 were launched within two days of each other. Though there was no docking involved, it was the first joint mission involving three spacecraft. In June 1970 the two cosmonauts aboard Soyuz 9 set a new spaceflight endurance record of eighteen days, besting the fourteen days set four and a half years earlier on Gemini 7.

These missions involving multiple spacecraft and longer durations were paving the way for a new era in spaceflight. The United States had won the race to the moon, so the Soviet Union had set a new goal for itself. In early 1970 Soviet general secretary Leonid Brezhnev himself ordered that a civilian space station program be fast-tracked, using technology under development for a military orbital facility so that it could beat Skylab into space.

On 19 April 1971 the Soviet space program took a major step toward that goal with the unmanned launch of Salyut 1. About forty-three feet long and with a diameter of over thirteen feet at its widest point, the twenty-ton spacecraft was launched on a Proton booster. The spacecraft could support three cosmonauts and carried a complement of military and scientific equipment. It was designed to be used by multiple crews on successive missions. However, though the Soviet space program succeeded in placing a workshop in orbit more than two years before NASA did, it failed to man a station with multiple crews before the United States.

After the successful launch of the Salyut 1 facility, the program hit a series of problems. Three days after the launch, the crew that was to be the first to man it was launched on Soyuz 10. Upon reaching the facility, the crew found that they were unable to dock with Salyut and returned to Earth.

The problems that prevented the docking were worked out, and on 6 June 1971 a second crew was launched to Salyut aboard Soyuz 11. This flight was able to successfully dock with the station, and the crew lived on Salyut, spending a total of twenty-three days in space. Then tragedy struck at the end of their mission. Their capsule returned to Earth successfully, but when the hatch was opened, its crew was found dead inside.

Skylab II astronaut Jack Lousma explained why the death of the Russian cosmonauts was a cause for concern: "We were already selected for the Skylab missions and were in serious training. A serious part of that was medical experiments. It was a time when not a lot was known about the effects of weightlessness [on the human body]; that's why we were there. Then the cosmonauts came back after about twenty-three days, and when the capsule was opened, they were found to have died. And this was the longest flight to that date.

"They launched three people up for whatever reason, and they couldn't all fit in their Soyuz, or the descent module, with spacesuits on, so they didn't take them. So something caused them to die. Apparently they hit the ground with a nominal landing. The question was what caused it. One option was that they had in that period of time developed some sort of health problem or space malady that was a result of being in weightlessness for twenty-two or twenty-three days, and the other was that they had an accident of some sort.

"Chuck Berry was our doctor at that time, and so he kind of explained all this to us. We talked about it, and the question was what [had] happened to the crew. There was a lot of disinformation flying back and forth during that time because this was still the Iron Curtain days, so we didn't know if we could get an answer from the Soviets or not.

"We were pleased when they did respond. And they came back and said that they'd had a space accident. A valve had stuck open when they separated their modules just before reentry, and had depressurized the spacecraft, and they had died of being unable to breathe in vacuum. That then was disheartening news for the space community at large, but as far as we were concerned, it gave us the go ahead to continue onward.

"We felt badly for the Russians. I think the sense was, as long as the Russians were successful, we'd be successful too. We really cheered them on. Because we knew whatever success they might have would be superseded by ours. But we were relieved that was how that turned out."

Following the loss of the Soyuz 11 crew, further Soviet spaceflights were canceled for the immediate future, and no more flights could be made to Salyut 1 before it deorbited in October 1971. In 1972, still before the launch of Skylab, the Soviets were ready to resume the space station program. However, on 29 July of that year when a second facility was launched, it failed to reach orbit because of a problem with its Proton launch vehicle.

In the month and a half before the Skylab workshop was launched, the Soviets made two more efforts to place a space station in orbit, one military and one civilian, but both were also unsuccessful. It would not be until July 1974, after the Skylab program was complete, that a Soyuz crew would again successfully dock with a Salyut station. In that month, the crew of Soyuz 14 docked with Salyut 3, staying in space for almost sixteen days.

Skylab Takes Shape

Well before the key decision was made to launch a dry workshop space station with a Saturn V, NASA went through a series of other decisions that shaped what was to eventually become Skylab. Though there had been many different ideas within NASA in the first half of the 1960s as to just what the space station should be like, one thing that many of those ideas had in common was the idea that a proper space station would involve the use of artificial gravity to be generated by the creation of a rotating station. By causing the facility to spin around a central axis, the centrifugal force generated would pull crewmembers toward the outside of the station, creating a sensation of some fraction of Earth gravity.

"There was quite an argument in the early stages, do you have artificial gravity or not," said Mueller, who found himself a rare dissenter from the conventional wisdom. To make his point, Mueller decided to give others in the agency an idea of what life would be like for astronauts on such a facility. He had them join him at the Slow Rotation Room at the Naval Aerospace Medical Institute in Pensacola, Florida. The room spins around, generating a centrifugal force pulling its occupants toward its perimeter. The experience is tolerable at first, even novel as the room seems to shift as the centrifugal force causes the direction of gravity to seem to change. However, after time the spinning can become increasingly uncomfortable. "And I had Bob Gilruth, and Wernher, and Sam [Phillips, head of the Apollo Program Office], and I riding in one of these rooms that spin, and after about half an hour, the great desire for artificial gravity dissipated," he said. "I was having trouble convincing Wernher and Bob of [the disadvantages], but I inherently knew it." Mueller said that his conviction regarding artificial gravity applied to missions to Mars as well as space stations—that it would be better to use techniques to mitigate the atrophying effect of weightlessness than to subject a Mars-bound crew to the rotation necessary to generate artificial gravity.

Because the Marshall-managed S-IVB stage was to be used as the basis for the workshop, Marshall was given the responsibility for hardware development for the workshop. Houston's Manned Spacecraft Center, which had previously had the responsibility for spacecraft development, was tasked with overseeing the crew operations for the space station. "That really started the Marshall Space Flight Center into the space station business," Mueller said. This arrangement had the additional benefit of allowing Mueller to keep many of the engineers on the Saturn team in NASA's workforce. "Marshall was running out of work," he said. It also, Mueller said, took the most advantage of the centers' management resources. "Wernher was very enthusiastic about space stations, and Gilruth was sort of not very enthusiastic."

Leland Belew, an engineer who had been involved in Saturn propulsion development at Marshall, was tapped to serve as the center's Skylab program director. Belew said that he was talked into taking the job during an offsite discussion in 1966 during which he was relieved of concerns about how the program would be developed, including how much discretion Marshall would be given in a program centered around crew operations, then the exclusive domain of the Manned Spacecraft Center. "I got involved in Skylab basically with a meeting with von Braun and George Mueller down on Guntersville Lake [about a half-hour from Huntsville], and they talked me into taking the job after some arm twisting," Belew said. "We stayed up all night in that activity. And, let me tell you, they absolutely stood behind everything that they said. No questions asked. They lived up to everything they said they'd do."

Belew said that the relationship between Marshall and MSC evolved over the course of the Skylab program, with the Houston center gradually sharing its traditional crew operations duties with its Huntsville counterpart. "It changed with time. I know that from time to time they would pose a question of, 'We can't do that because . . .' and we'd take it over. That sort of thing. We took them over one by one."

However, he noted that in addition to the Apollo spacecraft, there were other areas where MSC led the effort. "On the biomedical, we did everything that they wanted, no questions asked," he said. "We absolutely did everything possible in all the biomedical stuff. That was their main thrust. They wanted to baseline that. We knew that was priority number one. And on the solar stuff, we did everything that they wanted."

As the program developed, the two centers began working more and more closely together. The cooperation was facilitated by an aircraft that NASA scheduled to make daily runs between Huntsville and Houston, allowing Skylab team members at either center to work face-to-face with their counterparts at the other. "We used that airplane a lot," Belew said, "because we had a close relationship with them."

For a center that had been focused almost solely on developing propulsion hardware prior to that point, the transition to working with crew operations proved not to be a difficult one, Belew said. "No, it sort of flowed pretty naturally. Astronauts were always here, in the neutral buoyancy, and that sort of thing. And they enjoyed the heck out of it."

On the subject of the origin of that neutral-buoyancy tank, George Hardy's recollection echoed Mueller's: "There is a story, too, about how that tank came into being. I don't know all the details of that. In fact, I don't want to know them; I'm not sure it's safe to know them. That's about as much as I know. There were some facilities that were approved and so forth, but somehow or another it turned out to be a water tank. And it was a very, very, extremely useful tool."

Hardy also recalled the relationship between the two centers as one that evolved over time. He said, "There was some reluctance on having Marshall, who knew about big boosters with fire coming out the end of them but nothing about astronauts and what it takes to keep astronauts alive and working. There were occasionally some pretty heated debates on how things were going to be done. I don't know that there was a single time that we tried to contradict JSC on anything having to do with flight crew.

"But we started working closer together. Initially JSC had a contract with McDonnell Douglas for the Airlock Module, but that was transferred to Marshall. So Marshall ended up with basically all the workshop: the Airlock Module, the Multiple Docking Adapter, and the ATM [Apollo Telescope Mount], which was basically the cluster, short of the Command Module.

"We worked very closely with JSC, and astronauts were very much involved in it. I can remember when we first started working with flight crew on the control panel on the ATM, it seemed like we changed the configuration of the switches and the location of the switches by the week. Crew would come up, and they'd want it this way, and we'd fix it that way. Next week, they'd want it that way, we'd fix it that way."

The decision to use an S-IVB stage as the basis for the workshop established its basic parameters, but the rough design for the station was formalized on 19 August 1966 in a meeting at Marshall Space Flight Center's headquarters, Building 4200. Debate over the design of the station had been going on for some time. "That was the culmination of a series of meetings that we had," Mueller said. "But we were not closing in." During the meeting, Mueller did a quick sketch of what the space station was going to look like. The crude felt pen drawing on a flip chart showed the large cylindrical workshop with an Apollo spacecraft connected to it via a smaller docking cylinder. Connected to the docking cylinder by a tether was an Apollo Telescope Mount solar observatory. Mueller had to leave the meeting early, but before the meeting adjourned, his deputy, Maj. Gen. David Jones, initialed it for him, and the sketch became law—NASA had a design for its space station.

George Hardy described the meeting in which Mueller introduced the cluster concept: "It was here at Marshall, again at a management council meeting, where again the primary subject was a lunar mission, and Skylab got tacked on to the end of the day for a little discussion. Mueller was great at that; he could take one meeting and put an add-on to it for something that he wanted to spend a lot of time on. That's when he first introduced us to the so-called cluster concept. Because even though a lot of missions had been brought together and integrated, we still had a so-called orbital workshop mission, and we still had an ATM mission.

"So that's when he got up from his chair, and he went up to a flipchart with a magic marker, and he actually drew the sketch. And General Jones, who was his deputy, came up and signed it. We [had] all kind of kidded about it, talked about it, and said 'Is that our new specification?' And he said, 'Yeah, that's your new specification,' and Jones got up and signed it. That's what we got as our direction. That was the original direction. That really solidified the program."

In 1965 NASA was still pursuing the wet workshop option, which was then seen as the best way to develop an orbital workshop program as quickly as possible. Mueller had been working to get a wet workshop–based space station into orbit as quickly as possible with an original target date set for early 1968, which would have established an orbital workshop during the early phase of the Apollo flight program. Everything changed though on 27 January 1967 with the Apollo 1 pad fire. During a routine rehearsal for their upcoming mission, astronauts Gus Grissom, Ed White, and Roger Chaffee were killed when a fire started in their spacecraft and spread rapidly in its pure oxygen atmosphere. "It obviously had a real effect. We were scrambling to get AAP pulled together," Mueller said. "That abruptly disappeared from the agenda."

The delay allowed time to further think through the debate involving the wet and dry workshops. One of the arguments for the wet workshop program had been its perceived benefits in fast-tracking the workshop program, allowing it to begin concurrent with the early Apollo missions. After the fire, however, with the efforts to get Apollo back on track, AAP became a lower priority in the agency. With the loss of that supposed advantage, the pros and cons of the debate received a closer look. Each side had its proponents. "The dry workshop was really pushed forward by a scientist, Homer Newell," Belew said, adding that Newell was to heavily influence another major decision in the program as well. "Homer Newell insisted that we have two workshops, fully equipped. He insisted that we make them identical in every way." Newell was to get his way in that also; two workshops were built.

The astronaut office also came out in support of switching to the dry workshop, largely at the recommendation of Apollo 7 astronaut Walt Cunningham, who had been the corps' representative on Apollo Applications. "I give him credit for supporting it from our office," said Skylab II commander Alan

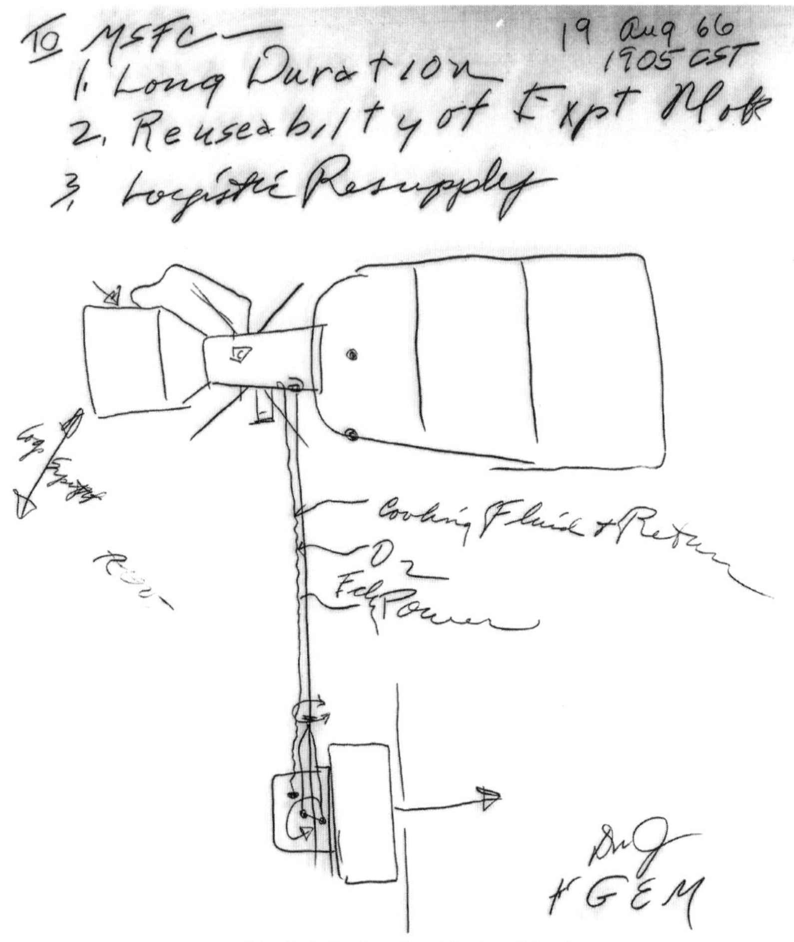

2. Mueller's flipchart "specifications" drawing.

Bean, who had been involved in Apollo Applications himself prior to being moved to the Apollo program, leading up to his assignment to Apollo 12. "He was the guy that was pushing to have it on a Saturn V and everything, which I thought was real great. And also I give him credit for going against the general office hierarchy position and convincing them that would be a better way. I think that's good, too, if you can struggle against Deke [Slayton] and Al [Shepard] and others and turn out to be right."

In addition to Gilruth another outspoken advocate of the wet workshop option was NASA spacecraft designer Max Faget, who had been instrumental in the development of the Mercury, Gemini, and Apollo vehicles. "We sort of pinned him down in one of our management council missions, and he backed off a little bit," Belew said.

The debate culminated with Mueller's hands-on experience in the neutral-buoyancy tank at Marshall. From that moment the future was set. Mueller made the recommendation to the agency's administrator that the Apollo Applications Program space station be a dry workshop and be launched on a Saturn v. The announcement that his recommendation had been accepted was made on 22 July 1969, two days after the Apollo 11 moon landing. "Maybe my credibility went up enough [after the landing]," Mueller said. "After the fire, it took a while to get my credibility back."

The decision was to prove one of the most vital, if not the single most important, turning points in the development of Skylab. It not only made the program possible by avoiding tasks that may have proved impossible, but it also gave the workshop new purpose. From that moment all the various parts of what had been the diverse Apollo Applications Program missions began becoming a part of one unified program—the space station. (It was in February 1970 that the consolidated space station program was given the name "Skylab." An Air Force employee working with NASA, Donald L. Steelman, had submitted the idea when the agency solicited suggestions. While other suggestions included such things as continuations of the mythological nomenclature used for Mercury, Gemini, and Apollo, Steelman's straightforward suggestion was based on the fact the facility would be a laboratory in the sky.)

What had been planned as an entire series of Apollo Applications spaceflights could all be manifested aboard a single launch of a Saturn v. Missions included in the bailiwick of Apollo Applications included microgravity science, long-duration spaceflight, solar astronomy, and Earth observations. A number of the flights originally discussed were lunar missions, including continued surface exploration after the initial landings. These flights were later transferred to the Apollo program. Originally, each of those fields was to be conducted independently, each with its own flight program. The dry workshop decision allowed the series of dozens of flights to be consolidated into just four—the launch of the Skylab space station and the three crew launches. Experiments that would have required numerous Saturn IB boosters to carry them into space were all launched on just one Saturn v.

While the consolidation of the various Apollo Applications plans into one facility was of huge benefit to the space station program, in retrospect Mueller believes it had a long-lasting detrimental effect on American spaceflight. Before the decision was made to use the Saturn v, he envisioned Apollo Applications as an ongoing series of scientific missions. With the Saturn v, however, that ongoing series was transformed into one complete package. As a result the focus shifted toward what would be included in that package and away from continuing research afterwards. "Unfortunately, that was when we quit," he said. "It's a great mistake to get an endpoint without working out what's going to happen then. One thing I learned is you ought to have the next two generations in planning."

Nonetheless, there is no question that the Saturn v and the dry workshop enabled a Skylab scientific program that would not have been possible otherwise. An example of the changes the use of the Saturn v enabled is the Apollo Telescope Mount, a battery of eight astronomical observation tools that was attached to the outside of the Skylab facility. "The ATM started out as one of the AAP ideas, which was a major telescope in space," Mueller said.

The Apollo Telescope Mount, he said, had its roots in a conversation he had with an official from the Mount Wilson Observatory, during which Mueller noted that "with the Saturn v, we could put the Mount Wilson Observatory up in space, and really get some real good views. And that led to a number of different looks at space observation. The ATM came from that kind of set of thoughts." Mueller said the subsequent research and development served as the foundation for the work that later led to other space-based observatories including the Hubble Space Telescope.

Though that conversation and the ATM played an important role in the history of space telescopes, the idea of the space telescope predated them. In 1946 eleven years before Sputnik became the first object placed in Earth orbit, astronomer Lyman Spitzer Jr. wrote a paper in which he proposed that a telescope placed in orbit, beyond the interference of Earth's atmosphere, would be able to perform observations superior to a terrestrial facility.

Originally, the Apollo Telescope Mount missions were not going to be a part of the space station program. Instead an orbital observatory was to be used in a series of independent Apollo missions. In keeping with the philosophy of the Apollo Applications Program, the ATM was intended to make use of existing Apollo hardware and was originally designed to incorporate the telescopes into a free-flying spacecraft. There was a debate as to which

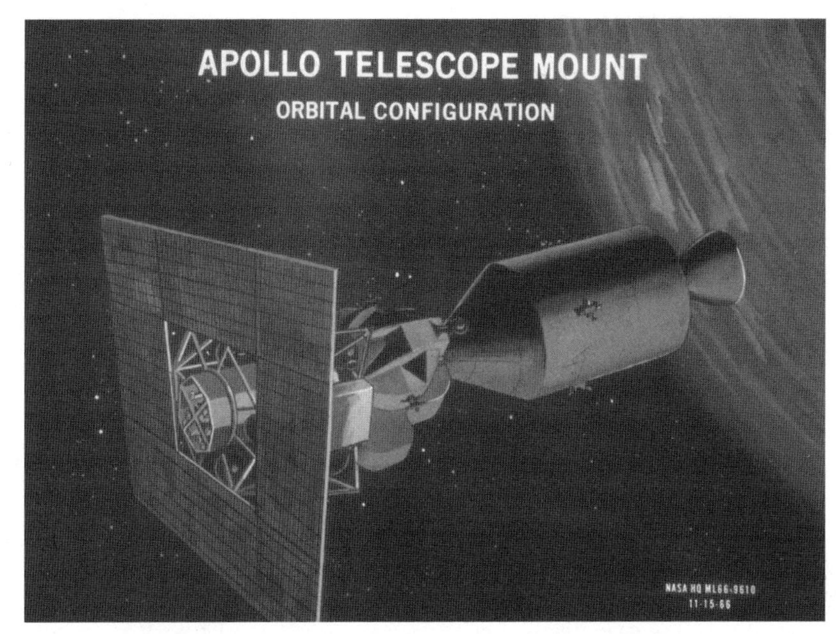

3. An early concept for a nonstation-based Apollo Telescope Mount.

existing Apollo spacecraft should be used to carry the telescopes: the Service Module or the Lunar Module. Some argued that it would be cheaper and easier to integrate the ATM into the Service Module. Mueller though was dedicated to the idea of eventually using the then-still-in-development Lunar Module as a multipurpose space-based laboratory and saw the ATM as a step toward that goal. In addition the Lunar Module provided another advantage. While the Service Module reentered the Earth's atmosphere and burned up when the Command Module landed, the Lunar Module could remain in orbit. This meant that while a Service Module–based ATM would have a maximum life span of a fourteen-day Apollo mission, a Lunar Module–based facility could be reused. However, concerns were raised about the idea of a free-flying ATM, arguing that it would be unsafe for the astronauts involved. Since the Lunar Module was designed for descent to the surface of the airless moon, it was incapable of operating within Earth's atmosphere and would not withstand reentry. If a problem developed during a free-flying Lunar Module–based ATM mission, the crew would be trapped, unable to return to Earth.

In order to avoid this dangerous situation, the proposal was made to incorporate the Apollo Telescope Mount into the orbital workshop program.

When plans still centered on a wet workshop, this would have required separate launches. The Saturn IB that would carry the S-IVB stage into orbit would not be able to carry the telescope mount as well. Instead, the ATM would still be configured as a separate free-flying spacecraft, which would be launched later and would then rendezvous with the workshop. With the decision to go with the dry workshop and the introduction of the Saturn V to the equation, however, this became unnecessary. The larger booster was capable of carrying the ATM into orbit along with the station. As a result, the ATM no longer needed to be designed as an independent spacecraft but could be incorporated into Skylab.

The evolution of the Apollo Telescope Mount contributed to the development of another of Skylab's modules. Though not needed for this purpose in the final configuration of the station, Skylab's Multiple Docking Adapter was originally intended to support the ATM. When the decision was initially made to incorporate the telescope into the space station program, mission planners were unsure exactly how to do this, since they weren't sure how to attach the ATM craft to the station. The proposal was made to tether the telescope mount spacecraft to the workshop, essentially floating freely but "tied" to the station. However, this would also have suffered from some of the disadvantages of the independent Apollo Telescope Mount missions and was thus still considered too dangerous.

To avoid that situation, the docking adapter was developed—essentially an empty metal cylinder with ports for multiple spacecraft to dock with the station. This would provide a place for not only the Apollo spacecraft crew taxi to dock with the facility but the ATM spacecraft as well, allowing astronauts to transfer easily between the workshop and the observatory. With the use of the Saturn V, however, this too became unnecessary. Since the Apollo Telescope Mount was to be attached to Skylab at launch, there was no longer any need for a docking port for it.

However, mission planners realized that the docking adapter could still be beneficial to the space station. Since it would allow two Apollo spacecraft to be docked with Skylab at the same time, it could be used if there were ever a need for a rescue mission. As it turned out, the second docking port was never needed, and the module ended up being used primarily for the additional enclosed volume it provided.

Another decision that Mueller said had a major impact on Skylab—and a lasting impact on human spaceflight—was the involvement of the Raymond Loewy/William Snaith, an industrial design firm, which looked at human factors on the space station. Loewy, who was seventy-four years old when approached to work on the space station, was a legend in the field of industrial design, having had a hand in everything from Coke bottles to office buildings.

Mueller described Loewy's effect on the program, "One of the things that probably people don't appreciate is, early on, I took a look at what Marshall was planning to do with Skylab, and I said, 'I don't think people are going to want to live there.' I said we needed to get a human factors genius. So I got them to bring Raymond Loewy to really take a look at how to make it habitable. And I must say I think he had a positive effect on the design and probably had a lasting effect on how to go about designing for space living. His contributions to livability were one of the keys to the success of the Skylab." Loewy's contributions were "mostly man-machine interface-type things," everything from a warmer color scheme for the interior of the station to a more agreeable design for its toilet system. Skylab's wardroom—the area where the crew could eat, relax, and take care of routine tasks—was another of Loewy's suggestions. Mueller said that Loewy later told him that his work on Skylab was one of the accomplishments of which he was most proud.

Not all of Loewy's recommendations were readily received, however. "He wanted a window," Mueller said, but he was unable to get engineers to agree that it was worth carving a hole in the side of the S-IVB stage for it. "He tried convincing them."

According to George Hardy, however, concerns about the crew's happiness on the station eventually won out: "I know of one meeting that was quite amusing. This was when we were in the process of going from the wet workshop to the dry workshop, and it was a meeting at Kennedy [Space Center in Florida], a management council meeting. It was a week or so before one of the latter lunar missions. [The workshop] got put on the agenda; it was just an add-on late at the end of the day. It wasn't the major focus; not too many people interested in it. But von Braun and Mueller wanted to talk about it, so they did. Bob Gilruth acknowledged the subject, but he didn't have a lot to say about it. He still had some lunar missions to work. So the discussion was we're going to keep this simple. We're not going to do a lot of things to modify this hardware; we're going to do minimum modifications on it and so forth. And we had some artist's concepts of an S-IVB stage that could be inhabited in some way.

"And von Braun got up in the discussion; this was late in the evening, maybe 6:30, 7:00 or so. And anyway, von Braun got a little carried away with it, 'cause he was so excited about it. He went up to one of the sketches that was on the board and said, 'We can put a porthole right here where the crew can see out.' And Gilruth said, 'Yeah, there we go, now we're going to modify the whole thing.'

"So Mueller kind of called the meeting to a halt, and he said, 'Well, it's getting about cocktail time.' He wanted to keep those two apart, so he worked on both of them, I think, that evening. Then came back the next day, and everybody was back together again. And after all that, they got a window."

The process of making the decisions regarding the project that became Skylab proved to be more complicated than implementing those decisions. "I can't think of one thing that I'd say was a real hard task," Hardy said. The biggest challenge came not in the process of the engineering work for any one component or system but in integrating all the components together.

And in order for the systems to work together, the people responsible for them had to work together as well. "I used to tell my people that the technical integration of the hardware was about 30 percent of the job, and the integration of the people was about 70 percent of the job," Hardy said. "To me, that was one of the most fun things about Skylab, the integration part of it. You had the integration of two centers that had worked together, but worked together in a much different way on this program."

Working to integrate those two centers, Hardy said, is one of the things he's most proud of in the role he played in Skylab. "This was going to be the first manned spacecraft that was going to be operated in space with MSC astronauts, of course, and MSC flight controllers and that was designed and built under another center. The tradition and the history was that the flight controllers started getting involved, not necessarily at the very beginning of the design but in the design process, such that flight operations requirements flowed into the design and development in a very real way in much the same way as flight crew requirements." About a year before the launch of Skylab, the Marshall engineering team realized that while there had been some collaboration with the Houston operations team, it had not occurred on the scale that it would have on a program handled entirely in-house at the Houston center. Hardy got in touch with Gene Kranz, chief of Flight Operations at MSC, and planned a massive review process, which lasted around three or four months. "For each major system area, like crew systems, instrumentation, electrical systems, propulsion systems, we took the designers of that system, whether they be Marshall guys or [contractors] or whatever, and the operators, the MSC guys who were going to handle primary flight operations, and we put those people together with some structure to the review but a lot of opportunities for interaction and questions and discussions, and conducted these reviews."

The review process provided two major benefits. First, it enabled the two teams to provide feedback to identify and correct problems so that they wouldn't arise after the station had already been launched. But in addition the process provided a second benefit. When the mission controllers had problems or questions during the operational phase of Skylab, they were able to ask them of engineers whom they had met face-to-face and worked alongside. Those relationships had developed naturally on spacecraft programs managed entirely in-house at MSC but had to be forged deliberately on such an unprecedented cross-center program.

"That was very, very effective," Hardy said. "I think because of that, we developed a relationship with flight operations people that was invaluable during the course of the mission, and I think it paid off from Day 1, when we lost the heat shield. And then of course the continued operations of Skylab were again a testament to the centers working together."

Those continued operations required that Marshall, which had already been providing engineering support to mission controllers for the boosters it managed, greatly increase its operations support efforts. Hardy began discussing with Kranz what would be needed for this project. The two quickly reached the opinion that the level of support that would be required was much greater than either center had realized.

Hardy set about gearing up the Huntsville Operations Support Center (HOSC) at Marshall to provide twenty-four-hour-a-day, seven-days-a-week engineering support for the operations team at MSC, setting up a three-shift-a-day rotation schedule, with a fourth shift set up to give the other three time off.

In addition to the integration of the two groups from the two centers, the teamwork between NASA civil servants and contractors proved to be vital to the success of the engineering support efforts. Hardy cited as an example contractors from IBM, which worked on the ATM and the instrument unit. While NASA had three senior civil servants (as it worked out, one for each shift) with expertise in the gyroscopes used to control Skylab's orientation, most of the know-how for the gyros on the operations support team resided with the IBM contractors, and that knowledge proved extremely useful as the program progressed.

"We got down to losing rate gyros and losing control moment gyros," Hardy said. "And those guys, along with the Marshall guys, would come in with different control schemes that just a week before we said would be impossible. We'd say, 'If we lose this rate gyro, then we're done.' And we would lose it, and they'd come in with a control scheme that would work. The greater the demand got for more ingenuity in that area, the more they produced."

None of that is to say, though, that the relationship between the two centers in the area of operations was always seamless. While Marshall and MSC would eventually establish good working guidelines for what responsibilities fell into which bailiwick, there were a few snags that had to be ironed out in the process of reaching that point. MSC director Chris Kraft recalls his involvement in making sure that the program made the best use of each center's strengths: "As we in Flight Operations began to come to grips with Skylab and the fact that we were going to be responsible for taking care of it and the astronauts who would be involved, several sticky issues came to light. In typical MSFC fashion, they had set up what they intended to be their own flight operations group with the intent of flying the machine themselves. There is no doubt in my mind that they saw this as their opportunity to get directly involved in manned spaceflight.

"They were building what they perceived as their own Mission Control Center [MCC]. This became obvious as their budget submissions were made and in their relationships with us. When this became apparent, I challenged this at both the management council meetings and directly with the management at MSFC. I don't think headquarters was willing to support this approach, but I wasn't about to sit still for what I saw as an infringement on the center responsibility (roles and missions). The lead manager for MSFC in their flight control work was Fridtjof A. Speer. Fred, as we addressed him, was a typical German from the old school. That is he was very competent, knew his business, but was equally stubborn in his approach to the issue. He had obviously been given his marching orders and was adamant about taking over this aspect of the program.

"After several meetings with him and his engineers, it was obvious we would have to confront him head on, and we did. He insisted that it was MSFC's responsibility to do the flight control on all of the hardware they were responsible for, and he intended to carry out these functions. I and my people were equally adamant that MSC was responsible for all NASA manned spaceflight control and that we intended to control Skylab from the MCC in Houston. It took a number of months to bring this to a point where headquarters recognized the duplication of effort going on here, and I'm sure the press of the budget for operations activities finally forced their hand to step in and adjudicate the issue. Speer did not give up easily, but I think Eberhard Rees recognized they would lose this argument and directed Speer to cooperate.

"From the beginning we had made it clear that we had no objections to MSFC having their people familiar with the various Skylab systems perform these flight control functions in the Houston MCC. Speer, after a direct meeting with me on this subject, finally agreed to this concept. Even so, Huntsville Operations Center spent a lot of NASA's money on an elaborate set up and manned it throughout the program. As might be expected, as the time for detail flight control training became imminent, Speer backed away from having his people in Houston full-time and had the MSFC rep in Houston fulfill most of these responsibilities. We ran the program, but MSFC had their HOSC operational practically full-time. In all fairness, Fred Speer supported the MCC very well and spent a lot of time there personally throughout the program. He was also very helpful in the conduct of the missions."

In one of the more interesting aspects of the relationship between the two centers, the materials laboratory at Marshall Space Flight Center essentially took on the role of primary contractor to the Houston center on some of the biomedical experiments. The Manned Spacecraft Center had sought a company interested in developing and building the flight hardware for the experiments but found none willing to take on the work. "They couldn't get anybody to take that research on to build the bicycle ergometer and the metabolic analyzer, because industry viewed it as a one-shot," said Bob Schwinghamer, who was the head of the Marshall materials lab. "They didn't know what the potential was for it, and hell, they didn't want to do all that stuff. So von Braun goes down there and tells Gilruth, 'Yeah, we can do that for you.' Those guys in Houston just couldn't get anybody interested. But von Braun could do anything. 'Yeah, we can do that.' He was something else.

"So he calls me up there in the office one day, and he's got the project manager from here [Lee Belew] sitting in there with him. And he said, 'Bob, we want you to build some biomedical experiments. I want you to deal directly with Houston on this. Go to [MSC's Skylab program manager] Kenny Kleinknecht and whoever the doctor is down there.' He looks at Lee, and said, 'Lee, I want to do it this way 'cause I want to cut all the horseshit paperwork out of this in between.' And we were off and running the next day."

The MSFC materials lab was tasked with developing and building the lower body negative pressure device, the bicycle ergometer, and the biomedical experiment support system. While he sent his Marshall superiors courtesy copies of his paperwork, Schwinghamer said things worked well with his lab reporting directly to officials at MSC on the work. "It got to where they were demanding a report once a month," he said. "And that was pretty frequently when you're trying to build hardware. So one time, I told John Massey, 'Go right in the middle, and leave thirty-two pages out, and ship it.' And so he shipped it. Then I waited a couple of days, and I called down there and said 'What did you think of that last report?' They said, 'Everything's all right, just keep it coming.' I said, 'You didn't worry about the thirty-two pages missing in the middle?' We both had a big laugh. After that, things got a lot easier."

Schwinghamer noted that private business missed out on a great opportunity by not developing some of the equipment. The metabolic analyzer that his team built proved to have Earth-bound applications for doctors, who used it for tests on patients. "That spin-off turned out to be a pretty good thing," he said. "But they just weren't interested."

It was during work on a urine sample volume measuring system that Schwinghamer first developed a relationship with a man who was to provide him with some rather unique—but ultimately invaluable—assistance. "One day, a guy called up, Bill Thornton," Schwinghamer said. "He says on the phone, very innocently, 'I'd like to help you if I could.' I said, 'I'm looking for any help I can get.' What can I say?"

Thornton was a member of the sixth group of astronauts selected, the second group of scientist astronauts. Even before joining the corps though, Thornton already had a background in spaceflight and specifically in microgravity biomedical research, having performed research for both NASA and the Air Force's space program. A brilliant researcher and a physically powerful man, Thornton proved quite capable of both building equipment during the development phase and breaking it during the testing phase. His tall, muscular build contributed to his distinctive ability for finding potential bugs by pushing equipment to—and past—the breaking point, which would play an important role in the Skylab program.

The materials lab had been working to try to figure out how much volume the urine system should be able to handle. "I had some really good people," Schwinghamer said. "I even had some biologists in my lab. And I thought, 'Okay, how much volume should it handle?' And nobody seemed to know."

The male members of the materials lab workforce were called upon to help determine the answer. Cups were placed in the men's room, and the staff was asked to use those cups whenever they needed to urinate. Each worker initialed his cups so that his daily output could be tracked. They were then asked to estimate how their output at home in the evening compared to the amount being measured at work.

"We did the means and standard deviation of the cup and added a third or something," Schwinghamer said. "Over a sample of fifty-sixty people, I came up with a mean and a standard deviation, and I took the mean, and added three standard deviations, and it came out to be about 900 to 920 cc [cubic centimeters], just under a liter. And Thornton comes up, and guess what, he pees 1,200. He peed 1,200, and I had to redesign the damn thing."

It would not be the last time Thornton would send one of the Schwinghamer's lab's projects back to the drawing board. "Then he said, 'I'd like to look at this bicycle ergometer you're doing.' And I thought, 'Oh man, I'm in trouble,'" Schwinghamer said. "So I took him back to the lab, and Dr. Ray Gauss was working on it back there, doing a good job. He had that feedback circuit so he could maintain a constant heart rate and all that stuff and adjust the torque on the pedals automatically. Thornton gets on that, and what does he do, he breaks the shaft holding the pedals." The team went back and increased the thickness of the pedal shaft from a half-inch to five-eighths of an inch. "And that was enough, he couldn't break that one. He tried like hell, but he couldn't break that one." (Thornton would have a continuing relationship with the bicycle ergometer later in the program, however, during which he picked up some unfinished business with it.)

"He was something else," Schwinghamer said. "But you know what, I'm glad he was there. That stuff functioned flawlessly. I kept an eye on how the biomeds were doing [during flight], and they did very well. Nobody reported any deficiencies or malfunctions even. And I always thought, 'Thornton, you did me a big favor, buddy.' He was a ninety-ninth percentile man. If he couldn't break it, nobody could break it. The guys out in the lab, they'd say, 'Hey Pop, when's King Kong coming back?'—they called him King Kong.

"After it was all said and done and over and the missions were accomplished, I thought, I don't know if everything would have functioned that flawlessly without having him really run it through its paces. And he did the job superbly. He was a good one."

In theory, countdowns are pretty straightforward things—time goes by, and the launch gets closer. For quite a while, however, that didn't seem to be the case with Skylab. Time would pass, but the launch would get no closer, something that became known as "the T minus one year phenomena," Flight director Phil Shaffer recalled, "Lots of us, as Apollo started stretching out, would finish an Apollo thing, and then we'd go work on Skylab for awhile. And when we came back to it, it was still T minus one year. Part of it was this craziness with the funding, but it was like they couldn't get their act together to decide what they were going to go fly. It really got to be a joke. 'Maybe we'll get to work on the Shuttle, and it will still be T minus one year.' We weren't making any progress. We'd go away for two or three months and work on the next Apollo flight, and we'd come back, and the schedule would have been reworked, and we were at T minus one year again. It seemed like it just went on forever. It was never going to be anything other than T minus one year."

Finally, though, the countdown did tick away. The Marshall materials lab was to perform one of the last steps in the development and construction of Skylab. "Three days before liftoff, somebody noticed that we didn't have an American flag on Skylab," Schwinghamer said. "So somebody said, 'Oh hell, that's no problem. We'll just make a nylon flag and stick it on there with Velcro.' I said, 'Nylon? You gotta be kidding me. You ever have your nylon-covered chair sitting out in the sun? That ain't gonna work.'"

Among the equipment in the lab was a solar irradiator, which could not only simulate the light and radiation produced by the sun but could be turned up to levels several times greater than actual exposure levels experienced at Earth. Schwinghamer had some nylon tested in the machine at a ten-sun level. "In about two hours, the nylon crapped out," he said. "So I knew that wasn't going to work.

"I had this boat that I used in the Gulf of Mexico to fish all the time, about a twenty-footer, white," he said. "And I had found this paint, and you could buy it in the paint store in Huntsville. So I gave the guy some money. I said, 'Go down and get red, white and blue. I know that stuff can stand the ultraviolet.' So he did that, and we started building the flag."

The team used the paint to create a flag on a thin sheet of aluminum. The completed flag was then tested in the solar irradiator at ten suns, but did not change color.

"Two days before launch, they flew it down there and stuck it on the side with Velcro," he said. "It didn't come off in flight, either. That was another thing everybody got excited about, but that Velcro hung in there like gangbusters."

Schwinghamer recalls informing the owner of the store of the unusual application of his paint. "I went back after it was in orbit, and I said, 'Hey, you know what, your paint's in orbit on the Skylab.' 'Oh, is that right?' He didn't think a thing of it. I'll be. Can you imagine: paint from a little old paint store in Huntsville, Alabama, is in orbit, and he doesn't think a thing about it."

2. The Homesteaders

The nine astronauts selected to serve on the Skylab flight crews represented three different demographics. Only two, Alan Bean and Pete Conrad, had flown in space before. Three of them, Owen Garriott, Ed Gibson, and Joe Kerwin, were members of the first group of scientist astronauts NASA had selected. The remaining four, Jerry Carr, Jack Lousma, Bill Pogue, and Paul Weitz, were unflown pilot astronauts.

The Moonwalkers

Not only were Bean and Conrad the only two flown astronauts on the Skylab flight crews, they had flown their last mission together; on it, the two had walked on the moon.

With three previous spaceflights under his belt, Pete Conrad was far and away the senior member of the three Skylab crews. Born in June 1930 in Philadelphia, Conrad at an early age developed a love of flying. After earning a bachelor's degree in aeronautical engineering from Princeton, Conrad pursued that love as a naval aviator. He went on to earn a place at "Pax River," the Navy Test Pilot School at Patuxent River, Maryland, where he served as a test pilot, flight instructor, and performance engineer.

It was there that Conrad first applied to become an astronaut in the initial selection process that brought in the original Mercury Seven. Though he was not selected in that round, the experiences of friends who were chosen inspired him to try again, and in 1962 Conrad was named as part of NASA's second class of astronauts, a group of nine men that also included Neil Armstrong, Frank Borman, Jim McDivitt, Jim Lovell, Elliot See, Tom Stafford, Ed White, and John Young.

His first spaceflight came three years later when he served as pilot of the third manned Gemini mission in August 1965, commanded by Mercury astronaut Gordon Cooper. Gemini 5 was to have a mission length of eight days, the first of two times in his life that Conrad would set a new spaceflight duration record.

4. Members of the first Skylab crew: (*from left*) Joe Kerwin, Pete Conrad, and Paul Weitz.

Just over a year later, Conrad moved up to a command of his own, flying the Gemini 11 mission with pilot Dick Gordon in September 1966. The second-to-last Gemini mission, flown just months before manned Apollo flights were then scheduled to begin, Gemini 11 was intended to gain more experience with rendezvous and extravehicular activity (EVA), two areas which would be vital for Apollo.

When Conrad flew again three years later, the success of Apollo was a *fait accompli*. Four months earlier NASA had fulfilled Kennedy's mandate "of landing a man on the moon and returning him safely to the Earth" before the decade was out. Neil Armstrong and Buzz Aldrin had become the first and second men on the moon on 20 July 1969, and next it was Conrad's turn. The Apollo 12 mission reunited Commander Conrad with Command Module pilot Gordon, and Bean joined the two as Lunar Module pilot. On 19 November Conrad and Bean left Gordon in lunar orbit and touched down on the surface. As he became the third man to walk on the moon, Conrad referenced Armstrong's famous "That's one small step for man; one giant leap for mankind" line in his own first words on another world: "Whoopee! Man, that may have been a small one for Neil, but that's a long one for me."

Alan Bean was born in 1932 in Wheeler, Texas. Like Conrad, Bean earned his bachelor's degree (in aeronautical engineering at the University of Texas) and followed that with service in the Navy, having been in Reserve Officer Training Corps (ROTC) while in college. After a four-year tour of duty, Bean also attended Navy Test Pilot School and then flew as a test pilot of naval aircraft.

He was selected as an astronaut in NASA's third group in October 1962—a class almost as large as both of its predecessors combined—along with Buzz Aldrin, Bill Anders, Charles Bassett, Gene Cernan, Roger Chaffee, Michael Collins, Walt Cunningham, Donn Eisele, Ted Freeman, Dick Gordon, Rusty Schweickart, David Scott, and C. C. Williams.

Bean's first crew assignment was as backup for Gemini 10, along with C. C. Williams. While his crewmate preceded him in getting an Apollo assignment, as backup for Apollo 9, that slot went to Bean after Williams's death in a crash of one of the T-38 jets used by the astronaut corps. From that assignment, Bean rotated up to the prime crew (the flight crew, as opposed to the backup crew) of Apollo 12.

It was his participation in the Apollo 12 mission with Pete Conrad that led to their joint involvement in Skylab. Bean had previously worked on Apollo Applications, supporting the program as a ground assignment while waiting to be placed on a crew. He served as the astronaut head of AAP until he became a member of Conrad's backup crew for Apollo 9. After transferring from AAP to Apollo, Bean maintained his interest in the program and kept up with its development (noting with approval, for example, the change from the wet workshop to the dry).

Alan Bean recalled the decision to pursue a Skylab mission: "We were starting to talk about what we wanted to do; this was on the flight home from the moon. Dick wanted to stay in Apollo because we knew we were cycling threes, so he could be commander of Apollo 18. [Under the regular rotation, an astronaut, after a mission, would skip two missions, be on the backup crew for the third, skip two more, and then be on the "prime" crew for the next mission.] First, we decided we'd divvy up every flight, and we'd swap around. This was Pete: Dick would be the commander of the next one, and the three of us would run the space program. But then as we got to talking about it, Pete wanted to do Skylab. And we both felt that we didn't want it to get crowded, other people deserved chances too. So we thought, well, we'll try to be part of Skylab.

"So Pete says, 'That looks like it'd be a good thing to do, looks like it'd be fun.' I don't think Dick was interested. A lot of the astronauts weren't interested in flying for twenty-eight days or fifty-six days. We were; we thought it'd be good adventure.

"I never did go and see Deke. I should have done it, but I never did it. But Pete went over and talked with him. It seems to me the announcement in the meeting of me and Owen and Jack [as a crew for Skylab] was a surprise to me, or maybe Deke phoned me and said this is what is going to be announced. But he didn't consult me about Owen and Jack. It turned out great. We ended up with the best crew, no doubt about it."

After Apollo 12 the three members of its crew were sent by NASA on a goodwill tour of the world, and upon returning Bean and Conrad transferred from Apollo to Skylab. In addition to their common background as moonwalking spaceflight veterans, the first two Skylab commanders shared another trait as well. Each has been described by members of their Skylab crew as being one of the most motivated men in the astronaut corps. In Conrad's case a lifelong drive to succeed had been increased by his rejection from the Mercury astronaut selection. "Pete was rejected, and the basis for his rejection was a psychiatric evaluation that he was psychologically unsuited for long-duration space missions," Kerwin recalled in an oral history interview for Johnson Space Center in 2000. "So here's Conrad; he's gone to the moon, he's up here in Skylab with us on the first-ever long-duration space station mission, and he's saying, 'I'll show that son of a gun who's psychologically unsuited for what!'

"So he was very motivated to do a great job on Skylab. Just the kind of commander you want. He exercised more than we did and kept us all up to a very high level, even coming home. He said, 'Guys, we're going to walk out of this spacecraft. There's going to be none of this carrying us out on stretchers stuff. . . . When that hatch opens, I'm outta here, and I want you guys to follow me.'" Bean's drive was an extremely important factor in the direction the second Skylab mission took. Owen Garriott said that a major reason for the incredibly high productivity of his crew was that "We had one guy that was better motivated than anybody in the astronaut office."

Despite having accomplished things as a Navy test pilot and astronaut that many other people only aspire to, Bean continually pushed himself further. Even during his days in the astronaut corps, Bean was a devotee of motivational tapes. Three decades after the time of Skylab, Bean continues to listen to the tapes, still working to motivate himself to accomplish all he can, to be the best he can. When his spaceflight days were behind him, Bean channeled that drive into his devotion to capture in his paintings the emotional aspects of his unique experiences. "I've always had a point of view that you don't have to be the smartest person, or the healthiest, or the brightest person to really do good work," Bean said. "I've never felt like I was that, but I always felt like I could do good work. Like these paintings, I never was the best artist in class, but I can do better art than anybody that was ever in any of my classes because I just keep doing it."

Dick Gordon, who flew with both men on the Apollo 12 mission, declined to speculate as to which was the more motivated, saying only that each was very motivated in his own way and that each had his own distinctive leadership style. He added, "If the space program doesn't motivate you, you're in the wrong place."

The Scientist Astronauts

Each of the three Skylab crews also included one of the members of NASA's first group of scientist astronauts, selected in June 1965. Six men had been selected in the group of scientists: Owen Garriott, Ed Gibson, Duane Graveline, Joe Kerwin, Curtis Michel, and Harrison Schmitt.

By 1962 the recommendation had been made to NASA that it should add scientists to the crews it would be sending to the moon. It was argued they would be able to more effectively conduct research there than the pilots that then made up the corps. The idea that scientists should be included in the first lunar landing crew was soundly rejected by management who argued that spaceflight to another world was a challenging prospect, requiring the skills of expert pilots. Including a scientist on the crew to conduct research on the lunar surface would be of no use if they were unable to reach the surface safely to begin with.

However, the agency conceded that there would be benefits to recruiting scientists into the astronaut corps for future missions and in 1964 partnered with the National Academy of Sciences to open its first scientist astronaut application process.

To be eligible to apply, candidates had to have been born no earlier than 1 August 1930 and be no more than six feet tall. Applicants had to be U.S. citizens and most importantly for this round had to hold a doctor of philosophy

5. Members of the second Skylab crew:
(*from left*) Owen Garriott, Jack Lousma, and Alan Bean.

degree (PhD) or equivalent in natural sciences, medicine, or engineering. While no flight experience was required, it would count in an applicant's favor.

Within two and a half months of announcing the selection process, NASA had received 1,351 applications. The agency screened those applications and submitted 400 of them to the Academy of Sciences for review. Hoping to bring roughly ten to twenty new candidates into astronaut training at the end of the process (to ensure enough made it through the training), NASA asked the academy to select fifty finalists from which it could pick its candidates.

After its review though, the National Academy of Sciences only felt that sixteen of the applicants were sufficiently qualified to recommend to NASA. The agency then put those finalists through its selection process of medical and psychological testing and interviews and ended up with only six men it found worthy of bringing in as astronauts. "For nine months NASA and the National Academy of Sciences screened over thirteen hundred applicants and, as I joked at the time, in all of the U.S., NASA could find only six healthy scientists," recalled Ed Gibson.

One of the six, Duane Graveline, left the corps very shortly after reporting for duty because of concerns over publicity concerning his wife's decision to file for divorce. Kerwin and Michel were already jet qualified, but the other three began their astronaut careers by going through flight training at Williams Air Force Base in Arizona. "Two of our group had pilot wings from the military," Gibson said. "NASA sent the remaining four of us off to flight school to get Air Force wings. We all did reasonably well. I was second in my class of forty-two; I would have been first but I screwed up an aerodynamics exam. It was very embarrassing for a guy with a PhD that included a lot of theoretical aerodynamics. Since then I acquired 2,200 hours of flight time in the T-38 and additional hours in other aircraft including helicopters. I felt that in a flash my lab stool had been ripped out from under me and replaced by a T-38 ejection seat."

Much had been made of the role of the scientist astronauts within the astronaut corps. Certainly the members of Group 4 were treated differently by management than their pilot counterparts, but with reason: they were different. Some of the scientist astronauts, particularly in the next group selected, chafed at a treatment they saw as relegating them to second-class-citizen status within the corps. Others believed that it made sense that the two types of astronauts would perform different functions and did not mind the role they'd been assigned. Yet others fell somewhere in the middle.

Joe Kerwin recalled: "There was a pilots' meeting in the office conference room every Monday morning at eight o'clock. At my first one I sat in the back of the room while Al Shepard told the group that we were here. Then he said, 'Headquarters has agreed that we can select another group to report next year.' Dick Gordon asked, 'Are they gonna be pilots?' Al said, 'I certainly hope so.'

"A couple of weeks later Shepard said, 'We'll be putting together crews for the last three Gemini flights soon. Any volunteers? (a pause) Put your hand down, Kerwin.' We both smiled. It was clear that these were not the flights they had in mind for us. Nor was I ready for a flight."

Whatever their relationship to the powers that be, the scientist astronauts' personal relationships with their fellow astronauts was generally positive. "In my case, one of the latest [Group 5 astronauts], Joe Engle, was my neighbor on the right, while another, Al Worden, was my neighbor to the left at our homes in Nassau Bay," Garriott said. "My relationship with them and others in the office has always been excellent."

Kerwin explained that while their classmates were in flight training, he and Michel were in a sort of limbo status while awaiting the return of the others and the selection of Group 5 so their official training could begin. "I was given a nice, big office and shared a secretary with about three other astronauts," he said. "It was explained that training for the two of us would have to wait until the arrival of the next group to be selected, the 'Original Nineteen' as they would call themselves, in the spring of 1966. So I was left pretty free to roam the center, learning what I could on my own. The other astronauts were always friendly, but they didn't pay much attention to us (and Curt spent a lot of time back at Rice University). Only two, Charlie Bassett and Neil Armstrong, made it a point to drop by my office, welcome me aboard, and offer to answer any questions I had. But two was enough. That was a great morale booster.

"I thought about spending some time in the clinic, keeping my medical skills fresh, and asked Captain Shepard for his concurrence," Kerwin said. "Al thought about it for a minute then said, 'I don't think that's a good idea. We'll have a lot of other things for you to do.' I accepted that as a dual message. One, my first priority had to be to learn, contribute, and prove myself as astronaut material. Two, maybe it wasn't a great idea to spend too much time with the doctors. And there was some sense to that; I might put myself into a conflict of interest situation treating fellow astronauts or their dependents.

"It wasn't long before Jim Lovell, who'd been in my squadron at Cecil Field, Florida, before he came to Houston, dropped by and asked me to help design him a primitive exercise program. He was training to fly with Frank Borman on the longest spaceflight planned to date—Gemini 7, which would orbit the Earth for fourteen days. The cockpit was about the size of the front seat of a Volkswagen Beetle, so Frank and Jim would get pretty well acquainted during the flight, and they had very little room for exercise gear. They'd selected an Exergenie—a compact device consisting of ropes passed through a core where the pull friction could be set. You looped two ropes over your feet and pulled on wooden handles at the other ends with your hands against the resistance. I sat down with Rita Rapp, a NASA physiologist and a wonderful worker, and together we designed a routine for Frank and Jim to use to stretch those unused back and leg muscles.

"At that time and for a long time thereafter, the astronauts considered exercise in flight to be their prerogative—an operational activity, not a medical one. So supplying their own hardware and protocol was business as usual to them. But Dr. Chuck Berry, the chief flight surgeon at MSC, thought otherwise. He considered the fourteen-day Gemini flight to be NASA's one opportunity to certify humans for the upcoming flights to the moon and wanted control of and data from exercise. I was called to Chuck's office on the eighth floor of the main building at MSC (it was Building 2 then), and he told me that meddling in medical business without his concurrence could adversely affect my career. I said 'Yes, sir,' and walked down to the other end of the hall where Deke Slayton, Al Shepard's boss, was located. Deke listened to my story thoughtfully and responded with five words: 'Keep doing what you're doing.' I did. And from then on, I got a lot of assignments to go to meetings and participate in teams where medicine and operations met and sometimes clashed. It was a lot of fun, and most of the time we all got along famously. I was accepted as a loyal member of the astronaut corps, and I had an opportunity to learn a lot about life-support systems, spacesuits, bends, and exercise that was valuable later on."

Alan Bean recalled that he and the others already in the corps were uncertain what to make of the new arrivals when they were brought in. "I guess it would have to be said that we were kind of wait and see," he said. "You tend to not want any other people to come in because you want to take all the flights. So any time some new group of anybody shows up, even though you know you have to have younger people, you still haven't had your fill.

"And of course, scientists. We're all test pilots; we're saying I don't know if those guys can cut it. But they don't show up; they go off to flight training. By the time they come, we're aware that they've gone through military flight training. We also know their grades and stuff, sort of. So we're then changing our attitude a little. They got through flight training, and some of those guys were better than we were, and that's good. And, of course, then we started to fly with them, and our attitude began to change even more."

The use of the term "scientist astronaut" surely affected the corps' initial perception of its newest members. "I still think the word scientist wasn't a good word," Bean said, explaining that it likely prompted a "knee-jerk reaction" among the pilot astronauts. "Over time, though, that distinction lessened as their flying proficiency was recognized and some even qualified as 'instructor-pilots' in a T-38 jet. Then too their contributions to their assigned crews in geology, medical, or solar science training became very positive points in their relationships to other pilots. Although members of Group 4 may have come in as 'scientists' rather than 'pilots,' well before flight their complementary talents earned them both acceptance and respect from their peers.

"And so by the time we worked together, and they were assigned, I thought of Owen as a scientist when we did science, but as far as flying airplanes, we thought of him as just as good as we were. So it was more like, there was never any flying thing that I would have said 'I'd better do that, or Jack should do it, but not Owen.'"

By the time of Skylab, there remained only three unflown members of Group 4 as it rather nicely worked out, one for each of the three missions. Michel, realizing that an assignment on one of the Apollo flights was unlikely and unsure when another mission would be available, had decided about two months after the Apollo 11 mission to leave the corps and return to teaching and research.

Schmitt, considered the best fit of Group 4 for a lunar mission by merit of his background as a geologist, was assigned to Apollo 17 as Lunar Module pilot and walked on the moon in December 1972. That left Garriott, Gibson, and Kerwin to fill the role of science pilot for the three Skylab crews.

Owen Garriott said, "Occasionally I'm asked if I was disappointed in not having a chance to go to the moon—only into orbit around the Earth (even though [the flight was] many times longer than a lunar flight). In fact, the answer is 'no,' and if given the choice of only one or the other, I would pick two months on Skylab. Why?

"There are several reasons. First, that is where my background training (electrical engineering, physical science research on the Earth's ionosphere) can be of most use. In fact, all scientist astronauts have found that regardless of their backgrounds, what the scientist astronaut job most requires is the skills of a scientist-generalist, someone who thinks like a researcher and has broad enough knowledge and experience to interpret what he sees. I would like to think that I fit the role of the generalist placed in a position to work with world authorities in several disciplines in the conduct of their research.

"Secondly, all of us in the astronaut office had a marvelous opportunity to travel the globe with world-class geologists studying (principally) volcanic regions thought to resemble conditions on the moon. We all greatly enjoyed these 'geology field trips.' I also soon realized that the pilot astronauts with whom we traveled were excellent observers and keenly interested in the research objectives of our instructors. For the three nongeologist scientist astronauts, I believe we would have been hard pressed to do any better job than the pilots while on the moon's surface, whereas we might have had (arguably, I must admit) a modest advantage in Earth orbit with many disciplines to represent.

"And finally, there is the issue of personal satisfaction. World-record durations, working in several fascinating disciplines more suited to my background, more time for reflection, and camaraderie all make a Skylab mission the first choice for me."

Owen Garriott's path to the astronaut corps began at the dawn of the space age. Garriott was born in Enid, Oklahoma, in 1930 and received a bachelor of science (BS) degree in electrical engineering from the University of Oklahoma in 1953. He had earned the degree on a Naval ROTC scholarship, and so he served from 1953 until 1956 as an electronics officer in the U.S. Navy. After completing his obligation, Garriott continued his education, earning a master of science (MS) degree and a PhD from Stanford University in electrical engineering in 1957 and 1960, respectively.

After completing his master's degree in 1957, Garriott was working on choosing a research topic for his PhD. Inspiration came in the form of the "beep-beep" heard 'round the world. After Sputnik was launched on 4 October, almost all of the graduate students and professors in the Radio Propagation Laboratory went out to the equipment set up at the field site and listened to the signal sent back by the Soviet satellite as it orbited the Earth. Garriott selected his topic: propagation of signals from orbiting satellites through the planet's ionosphere.

After earning his PhD, Garriott stayed on at Stanford, teaching and conducting research, eventually becoming an associate professor. He continued to follow the space program, and his interest grew when, after Alan Shepard became the first American in space, he realized that there might be a need for astronauts with research backgrounds in the future. Looking ahead to what might make a candidate more appealing if that were to come about, Garriott acted on a long-held ambition to earn a pilot's license.

When NASA decided to seek applications for scientist astronauts, Garriott was ready and waiting. "In May of 1965, I was waiting hopefully for a decision from NASA as to whether my life (and my family's) might undergo a major reorientation," Garriott recalls. "I was teaching a class at Stanford University and coming up on the end of the quarter when a call arrived from NASA wanting to verify that I would be available for a telephone call later that day. 'Yes, of course!'

"But I also had a lecture scheduled later in the afternoon. So I asked the secretary to whom the call should come to be alert for a call from NASA and to be sure and let me know about it. But if I was giving a lecture, just to come to the door and signal hand to ear that a call had arrived. Naturally, the call came in the middle of the lecture, Sally signaled as planned, and I decided to complete all (or most) of the lecture and call them back. Not knowing who for sure was calling and not knowing what the decision might be was more than the usual distraction!

"But I returned the call in fifteen minutes or so and apologized profusely for being unable to come to the phone immediately. Al Shepard did not seem concerned and provided the hoped-for question—'Would you like to come to work for NASA as a scientist astronaut?' Again 'Yes, of course,' started the brief exchange. A quick telephone call home alerted the wife, and we waited for an official announcement because I never felt certain of selection until NASA had made some public commitment."

"I started out being president of my first grade class two years in a row," joked Ed Gibson, in a NASA oral history interview, of his inauspicious academic beginnings. Self-described as "not a good student" in elementary school, Gibson said the only subjects that really captured his interest were science and astronomy. He recalls, as a young child, drawing pictures of the solar system. Though Gibson, born in 1936 in Buffalo, New York, improved his academic performance in high school, the interest in science remained. After high school he earned his bachelor's degree in engineering at the University of Rochester. The choice was inspired by his father, who wanted his son to work at his marking-devices company and thought engineering skills would be a valuable addition to the business.

A desire to fly for the Air Force was shot down by a bone condition that was then a disqualification for being a pilot. Unable to fly planes, he decided to pursue building them. Rather than joining his father's business after earning his bachelor's degree, Gibson went on to earn a master's and then a doctorate in engineering from the California Institute of Technology.

His childhood interest in astronomy and space never went away, and while in graduate school, he followed the Mercury and Gemini programs with great fascination, "never thinking [he'd] have a chance to be involved in them." After completing graduate school, he took a job as a senior research scientist with the Applied Research Laboratories of Philco Corporation at Newport Beach, California. It was while he was working there that his wife, Julie, read him an article at breakfast one morning saying that NASA was looking for scientists who wanted to fly in space. "I thought long and hard about it, and 8 o'clock that morning, applied," Gibson joked. "I had no qualms, whatsoever."

Of the four scientists astronauts who ended up flying, Joe Kerwin's path had the most in common with that of the first groups selected—it involved many hours in the cockpit of a military jet.

Born in 1932, Kerwin is a native of Oak Park, Illinois. After earning a bachelor's degree in philosophy from Holy Cross, followed by his doctor of medicine degree from Northwestern University Medical School in Chicago in 1957, Kerwin completed an internship at the District of Columbia General Hospital. At that point, under the Berry Plan, which allowed medical students to be exempted from the draft while completing their school or internships, Kerwin was called up for service. Among the options he was offered was the last seat in flight surgeon training at the U.S. Navy School of Aviation Medicine in Pensacola, Florida. Though it would mean an additional six months of service, Kerwin was intrigued by the prospect of getting some flying time and signed up. After flight surgeon training, he was assigned to the Marine Corps Air Station at Cherry Point, North Carolina.

During his tour, the Marines with whom he was assigned would allow him to start their fighters and taxi them around. "The bug really bit me," Kerwin said, and he applied for a Navy program in which a select number of flight surgeons were trained to become naval aviators with the idea that it would provide them a better background for performing their duties. He was accepted to the program and transferred from the naval reserve to the regular Navy. While assigned to an air wing at Cecil Field, Florida, a couple of friends he had made among the aviators asked him for a favor—help filling out the medical portion of their applications to become astronauts. Those two pilots were Alan Bean and Jim Lovell.

When the scientist astronaut program was announced in 1964, his wife asked him whether he wanted to try for it. He was skeptical of his chances but finally submitted his application, and the combination of a physician with two thousand hours of jet flight time proved too good to pass up. Garriott recalled, "At our first meeting for the ten-day physical examinations at the School for Aerospace Medicine leading up to scientist astronaut selection, I had a 'funny' in one electroencephalogram test. The physicians required that I stay up all night—as an extra stressor—for a repeat test the next morning. New acquaintance and probable competitor Joe Kerwin graciously offered to stay awake about half the night with me just to help me avoid falling asleep. He ended up staying until 5:30 in the morning. It worked, and we were both selected. But those gestures are never forgotten!"

The Group 5 Rookies

Of all of the groups of astronauts, the fifth—jokingly self-dubbed "The Original Nineteen" in a nod to the Mercury "Original Seven"—had the most diverse fates when it came to their eventual spaceflight assignments. Nine of them would make their first flights during the Apollo program (Charles Duke, Ronald Evans, Fred Haise, James Irwin, T. K. Mattingly, Edgar Mitchell, Stuart Roosa, John Swigert, and Alfred Worden), with three of them (Duke, Irwin, and Mitchell) walking on the moon. Four would make their rookie flights during the Skylab program (Jerry Carr, Jack Lousma, Bill Pogue, and Paul Weitz); one during the Apollo-Soyuz Test Project (ASTP) (Vance Brand); and three would not fly into orbit until the Space Shuttle began operations (Joe Engle, Don Lind, and Bruce McCandless; Engle had flown to the edge of space on the experimental X-15 plane before joining the astronaut corps). One of the members, Ed Givens, died before flying a mission, and another, John Bull, developed a medical condition that disqualified him from flight status.

Group 5 member Paul Weitz, who made his first flight as part of the first Skylab crew, said that he does not remember any indication being made when his class joined the corps what their role would be, which he said he could live with: "I can't speak for anyone else, but I just wanted to get an opportunity to fly in space."

Even if no formal promise regarding their role had been made, some members of the group had some expectations. "When we were brought onboard, there was no end at Apollo 20," Jack Lousma said. "We were going to land on the moon before the end of the decade, and we're going to explore the moon. And there was no end to the number of Saturn missions there was going to

6. Members of the third Skylab crew: (*from left*) Jerry Carr, Ed Gibson, and Bill Pogue.

be, the number of Saturn flights, Saturn vs, or the number of landings on the moon. It was going to start with the landing, and then it was going to be increased duration, staying on the moon up to a month. And they were going to orbit the moon for two months; I guess that was more of a reconnaissance type of thing."

The members of the Original Nineteen were competing for missions both with one other and with their predecessors still in the corps, a competition made even tighter by the cancellation of Apollo missions 18 though 20. "I was very much disappointed that the last three missions to the moon were canceled—I thought I had a chance of making one of those missions," Weitz said.

While there may not have been much overt competition among the Group 5 astronauts for the remaining Apollo berths, Weitz said, "It became obvious that some of the folks were doing their best to position themselves to punch the right cards to be considered for early flights. But everyone wanted to fly as soon as possible, and I think that no one was consciously considering giving up on Apollo-era flights so that they might get an early Shuttle flight."

Though he had been disappointed with the cancellation of the last three Apollo missions, Weitz said that he was pleased with his eventual rookie assignment. "I really enjoyed flying Skylab with Pete and Joe, and I thought we did our part to further the benefits of human space research."

Jack Lousma said, "I was a little naive. I wasn't a politician; I'd never been in a political organization. I just figured, the harder I work, the better I do. That wasn't good for this particular system, but I wasn't smart enough to know that. Moreover, I didn't know anybody when I came. A lot of the other guys had worked with each other on different projects in the Air Force and things, and I was just kind of a lone ranger. And I was the youngest guy selected, most junior, least experienced. So it seemed as though the people that had more experience, were a little bit older, or had done different things than I had were selected before me. Like Fred Haise was selected for the Lunar Module. Ed Mitchell worked on that. They were senior guys. And Fred was such a competent guy in aviation. So I was just going to do my best and work as hard as I could and see what happened, and if it came to a point where I had to be more overt about this, I would have felt that I'd earned the right to speak up.

"There was some amount of politics to the system of selection, but for the most part I felt that Deke Slayton was a fair guy, and I thought he selected the right people for the right job. I felt that everybody there was qualified to handle any mission that would come their way, but I felt that the selections as Deke made them were fair. He would come in and assign maybe two crews for a couple of Apollo flights, assign a backup crew, make a few guys happy and a lot of guys mad. But he would always say, 'And if you don't like that, I'll be glad to change places with you.' And nobody could refute that. 'Cause poor old Deke, he still hadn't flown at all."

Lousma was one of the members of Group 5 who reached a point where he was confident the moon was within his grasp, only to have it snatched away. Upon joining the corps he had originally been put on a different track as one of the first members of his class to be assigned to Apollo Applications rather than to Apollo. Even before he had completed his initial astronaut training, Lousma was tapped to work on an instrument for one of the lunar-orbit AAP flights that was planned at the time. Lousma had come to the astronaut corps from a background in military reconnaissance, and so was a perfect fit for the project, which involved using a classified Air Force high-resolution reconnaissance camera attached to an orbiting Apollo capsule to study the surface of the moon.

After Lousma had spent a year working on that project, however, the planned mission was canceled, and he was back to square one. It seemed, though, that fortune was smiling on him. Fred Haise, who had been the corps' lead for Lunar Module testing and checkout, had been assigned to the backup crew for Apollo 11, which meant he would then be rotating up to prime a few flights later. (Under the standard rotation, Haise would have been Lunar Module pilot for Apollo 14, but Alan Shepard's return to active flight status bumped the original 14 crew up to Apollo 13.) The corps needed someone to take the Lunar Module assignment that Haise had vacated, and at the same time Lousma needed a new ground assignment. The fit was perfect.

While serving as Lunar Module support crew for Apollo 9 and 10, Lousma was also scrounging time in the lander simulator whenever he could. "By the time all of that was said and done, I had seven hundred hours in the Lunar Module simulator, plus all the knowledge that went with the systems of how the lunar module worked," Lousma said. "Al Bean, when he flew, wanted the Lunar Module malfunction procedures revised, so I did that for him. I knew how this thing worked. So I figured I was probably destined for a Lunar Module mission.

"Before they canceled the last three missions, as I recall there was a cadre of fifteen guys that all worked together to somehow populate the last three missions; and some of them were guys that had flown already, and some of them weren't. Once it got to this cadre of fifteen guys, I don't remember there being any politics there. I wasn't involved if there was. I was just really focused on doing the best I could and qualifying on my own. I felt I was ready to do it. I was not a political person. I don't think Deke was much of a political person at all. He was somewhat predictable in that the guys who had been there the longest were going to fly first. And in that group of people, probably the guys that were going to fly first were a little bit more senior, militarily speaking, and I was a junior guy; I was twenty-nine years old.

"That's the way Deke worked. I think Jerry Carr would have flown to the moon before me. Jerry was senior to me in the Marines, so I'm not going to fly before him. The way it worked out on Skylab was I ended up flying before Jerry. And to offset that, Deke assigned Jerry to be the commander, not just the ride-along guy on the third Skylab mission. That's the way his mind worked. So to some extent, he was kind of predictable for people who were coming up through the ranks. He was kind of unpredictable for guys who had already flown. So there was probably more politics between those guys than there was between me and my friends. Deke never wanted much politics, I don't think.

"I don't know if I'd have been going to the moon or not, but there were three flights there for which I was eligible, and I was a Lunar Module–trained guy, so I thought I was definitely going on one of them."

But, of course, it was not to be. The final three planned Apollo missions were canceled and with them went the hopes of Lousma and the others that they might reach the moon. With those missions canceled, Skylab became the next possible ticket to space for the unflown members of Group 5 (and their Group 4 predecessors). But even after the nine Skylab astronauts were told they had been assigned to the flights, some still had a sense of uncertainty, particularly in the wake of the cancellation of the last three Apollo flights.

"The Skylab missions were always threatened to not fly for a long long time," Lousma said. "It was never sure until the last six months or a year that the Skylab was going to get to fly. There were those that said, 'Let's save the money and put it in the Shuttle.' So I always felt like I could lose that Skylab mission, even when I was training for it.

"Somehow I found out that there was probably going to be a flight with the Russians and probably Tom Stafford was going to command that. Maybe it was common knowledge, maybe it wasn't. But I decided that if Skylab doesn't fly, I'm going to be ready for the next thing. So I went while I was training for Skylab and took a one-semester course in the Russian language, took the exams and documented it and all that sort of thing, turned it in with my records, and said, 'Here you go, Deke, in case you're looking for a guy to go on the Russian flight, well, I've got a little head start on the language thing.' It turns out that before we went on the Skylab mission I knew I was going to be on the backup crew for the Russian flight. It didn't matter then, because Skylab was going, so backup was okay. But I remember being concerned about Skylab not flying."

Lousma said he is frequently asked if he felt that Skylab was a poor consolation prize for the lunar flight he missed out on. "'Going on a space station mission instead of going to the moon, did you feel bad about it?' Heck no, I didn't," he said. "I thought any ride was a good ride. And I felt that there weren't that many rides to go around, so this was going to be all right. But moreover, we were doing things that hadn't been done before. This is what I think the lure was for most of our guys, to do things that hadn't been done. All the things that Apollo did, we became operationally competent in.

"We knew we could fly in space. The question was, could we survive in space for long periods of time in weightlessness, and moreover could we do useful work. [Skylab] proved that this could be done, and it also demonstrated that EVAs in zero-gravity were doable if you were properly trained and had the right equipment and had been properly prepared. And we didn't really know that either."

The term "zero-G" is frequently used to indicate when things appear to be "weightless." When flying in space, crewmembers do feel weightless, and they really are, but not because there isn't any gravity at their altitude, about 435 km (270 miles) above the Earth for Skylab. In fact, the force of Earth's gravity is about 87 percent as strong there as it is on the ground. The essential difference is that on Skylab the crews were in what physicists call "free-fall," meaning that there was nothing to support their weight, like the ground or a chair. Instead, they were falling freely toward the center of the Earth, our "center of gravity," just like a diver or a gymnast does until they hit water or the ground. So to be more precise, the free-fall toward the Earth's center is the source of the apparent weightlessness in space. Space flyers do not hit the Earth because they are traveling so fast that their orbit just matches the curvature of the Earth and the Earth's surface drops away from beneath them just as quickly as they fall toward it. When "zero-G" is used to mean weightlessness, this is what the correct explanation should be.

While in-space EVAs had been carried out successfully in the final Gemini flights and then in Apollo, Lousma noted that they were nothing like the spacewalks performed during Skylab. The Skylab spacewalks were much longer and, unlike the earlier carefully planned and prepared spacewalks, involved responding to situations as they occurred.

"We were the first real test of whether you could have a useful space laboratory, and you could do scientific experiments of all sorts that we never did in Apollo, and get useful data back, and investigate things that had not been investigated before," Lousma said. "I didn't get to go to the moon, but I got to do something which was one-of-a-kind, first of a kind, to demonstrate all of those things that we were wanting to know and had to learn to go to the next step, and that's the one that we're going to take in the next couple of decades. From that point of view, I think Skylab is NASA's best-kept secret. We learned so many things that we didn't know, and we did so many things for the first time."

Like most of the members of the early groups of astronauts, Jerry Carr's path to becoming an astronaut began with a love of aeronautics developed in his youth. For Carr, who was born in 1932 and raised in Santa Ana, California, that interest was first spurred during World War II, when he would spot airplanes, including experimental aircraft flying overhead through the southern California sky. He and a friend would ride bicycles fourteen miles to the airport on Saturdays, where they would spend the entire morning washing airplanes. In compensation for his efforts, he would be paid with a twenty-minute flight. During his senior year in high school, he became involved in the Naval Reserve. He was assigned to a fighter jet and was given the responsibility of keeping it clean and checking the fuel and fluid levels.

After high school Carr attended the University of Southern California through the Naval ROTC and earned a bachelor's degree in mechanical engineering in 1954. Following his graduation, Carr became a Marine aviator. After five years of service, he was selected for the Naval Postgraduate School, where he earned a second bachelor's degree in aeronautical engineering in 1961. Carr was then sent to Princeton University and earned a master's degree in the same field a year later.

Three years later, he saw the announcement that NASA was seeking candidates for its fifth group of astronauts. He made the decision to apply on a whim. He was friends with C. C. Williams, who had been selected in the third group of astronauts in 1963. Carr figured if Williams could make it, he was curious to see how far through the process he could go. On April Fool's Day in 1966, Carr learned just how far he had gone when he received a call from Capt. Alan Shepard informing him that he had been selected to the astronaut corps.

Jack Lousma wasn't always going to be a pilot. Sure, Lousma, who was born in 1936 in Grand Rapids, Michigan, loved airplanes as a kid, building models and going out to watch them land and take off. And he had two cousins who were pilots and still remembers the time one of them flew a jet over his farm so low "I could almost see his eyeballs," as he recalled in a NASA oral history interview. But Lousma wasn't always going to be a pilot. Through high school and into college, he planned to be a businessman. However, during his sophomore year, he says, he decided he just couldn't figure the business classes out and decided to get into engineering. And as long as he was going into engineering, he did love airplanes, and the University of Michigan did have a great aeronautical engineering program.

While completing his bachelor's degree in aeronautical engineering, though, he saw a lot of movie footage of fast-flying jet airplanes. And so he decided the best thing for an engineer who planned to design planes to do would be to learn to fly them. After being turned down by the Air Force and Navy because he was married, he found out that the Marines not only had airplanes, they had a program that would take married people. After completing his training and deciding he wanted to make a career of the Marines, he went on to attend the Naval Postgraduate School, earning a master's degree in aeronautical engineering.

Lousma had reached the point where he was starting to look for new challenges when he heard that NASA was selecting its fifth group of astronauts, and he decided that fit the bill perfectly. His application process was nearly cut short, however, by a requirement that candidates not be over six feet tall. According to his last flight physical, Lousma was 6 feet 1. The Marines Corps, which screened his astronaut application, agreed to give him a "special measurement" by his flight surgeon. This one came out to 5 feet, 11 7/8 inches. "I was really 5 foot, 13 inches, but I didn't tell anybody," he joked.

Garriott notes that at the time of their Skylab flight, Lousma had only had nine birthdays, having been born on 29 February. "But 'the Marine' acts in a much more mature manner, and if there ever was a true 'All American Boy' in a quite positive sense, this is your man," he said.

Bill Pogue also was fascinated by airplanes in his youth but like Lousma had plans for his life that did not include being a pilot. Fate, however, had other plans, and Pogue ended up thrust into following his childhood fascination. Born in 1930 in Okemah, Oklahoma, Pogue planned during high school and college to be a high-school math or physics teacher, following in his father's schoolteacher footsteps. Pogue earned a bachelor's degree in education at Oklahoma Baptist University. After the Korean War broke out, though, Pogue decided it looked like he was going to be drafted and enlisted in the Air Force. He was sent to flight-training school and then to Korea, where he was a fighter bomber pilot. In the six weeks before the armistice, he flew a total of forty-three missions, bombing trains and providing air support for troops.

An assignment to leave Korea to serve as a gunnery instructor at Luke Air Force Base in Phoenix, Arizona, proved to be fortuitous when after two years there he was asked to join a recently formed group based at Luke—the Thunderbirds air demonstration unit.

When he left the Thunderbirds, Pogue was given his choice of assignment, and he asked to be allowed to earn his master's degree. He was reassigned to Oklahoma State University, where he earned his master's in mathematics. About halfway through a five-year tour of duty as a math instructor at the Air Force Academy, he successfully petitioned to be allowed to work toward a goal of becoming an astronaut. He attended the Empire Test Pilot School in England. After completing that, he spent a couple more years there and then transferred to Edwards Air Force Base. Before he even moved there, however, he learned that NASA was selecting a new group of astronauts, and from his first day at Edwards, he was already working to leave Edwards to join NASA.

Unlike Lousma and Pogue, Paul Weitz did always want to be a pilot. Specifically, Weitz, born in 1932 in Erie, Pennsylvania, wanted to be a pilot for the Navy. His father was a chief petty officer in the Navy and during World War II was in the Battle of Midway and the Battle of Coral Sea. That made a deep impression on Weitz, and by the time he was around eleven, he had decided he was going to be a naval aviator.

Toward that end he attended Penn State on a Navy ROTC scholarship, completing his time there with a bachelor's degree in aeronautical engineering and a commission as an ensign. An instructor there advised Weitz that if he wanted to make the Navy a career, he should begin by going to sea, so he spent a year and a half on a destroyer before going to flight training. From there he spent four years with a squadron in Jacksonville, Florida, where he met Alan Bean.

The next few years of Weitz's life were not necessarily as he would have planned them. After initially being turned down for test-pilot school, he was accepted for the next class but not until he had already been ordered across the country for an air development squadron. Since the Navy had just moved him from one coast to the other, they refused to move him back for test-pilot school. At last after two years he received unsolicited orders to attend the Naval Postgraduate School, where he was in the same group as Jack Lousma and Ron Evans, who went on to be the Command Module pilot for the Apollo 17 mission. Further complicating the situation, Weitz found himself allowed only two years at a school where a master's degree was a three-year program. With the aid of sympathetic professors, he was able to earn his master's in aeronautical engineering in the two years he had.

The next year, he made a combat tour in Vietnam. While in the western Pacific, he got a message from the Bureau of Naval Personnel asking if he would like to apply to be an astronaut. Though Weitz had never given the matter any thought before, he decided that he would, indeed, like to be an astronaut.

Those nine men would make up the crews of the three Skylab missions. Pete Conrad and Al Bean, the two veteran astronauts, became the commanders of the first two Skylab crews. Conrad was joined by pilot Paul Weitz and science pilot Joe Kerwin. Bean's crew consisted of himself, pilot Jack Lousma, and science pilot Owen Garriott. Rookie Jerry Carr was assigned as the commander for the third crew, joined by pilot Bill Pogue and science pilot Ed Gibson.

One veteran and five rookies made up the backup crews for the three Skylab missions. Rusty Schweickart, the Lunar Module pilot for the Earth-orbit Apollo 9 mission, was the commander of the backup crew for the first mission, joined by pilot Bruce McCandless and science pilot Story Musgrave. Commander Vance Brand, pilot Don Lind and science pilot Bill Lenoir (Lenoir and Musgrave were members of the second group of scientist astronauts NASA selected) served as the backup crew for both the second and third Skylab missions.

The first crew chose for their mission patch an image depicting the Earth eclipsing the sun, with a "top-down" view of a silhouetted Skylab in the foreground. The patch was designed by science-fiction artist Kelly Freas.

The second crew's patch, with a red, white, and blue color scheme, featured Leonardo da Vinci's famous Vitruvian Man drawing of the human form in front of a circle, half of which showed the western hemisphere of the Earth, and the other half depicted the sun, complete with solar flares. The patch reflected the three main goals of their mission—biomedical research, Earth observation, and solar astronomy.

7. (*Clockwise from top left*) The Skylab I mission patch, the Skylab II patch, the Skylab II "wives' patch," and the Skylab III patch.

The third crew's patch featured a prominent digit "3" with a rainbow semicircle joining it to enclose three round areas. In those three areas were depictions of a human silhouette, a tree, and a hydrogen atom. The imagery on the patch symbolized man's role in the balance of technology and nature.

There was one other Skylab "mission patch," a companion to the second crew's patch. The wives of the three Skylab II astronauts had been involved in the creation of that mission's patch, and decided they wanted to do something a little extra. Working with local artist Ardis Settle, who had contributed to the official patch, and French space correspondent Jacques Tiziou, they created their own Skylab II "wives' patch." The central male nude figure drawn by Leonardo had been replaced with a similar but much more attractive female nude and the crew names altered to Sue, Helen Mary, and Gratia.

"One of our first tasks when reaching orbit was to unpack our 'flight data file,' carried up in our CSM [Command Service Module]," Garriott said. "What we did not expect to see when we unpacked our individual 'small change sheets' and 'check lists' was a new crew patch with a much more memorable female nude in the center!"

"A very pleasant 'gotcha'!"

It is important to note that there are two different systems of nomenclature for the Skylab flights. During the planning phase for the program, there was debate as to whether the unmanned launch of the Skylab workshop should be numbered as one of the flights or whether only the three crewed missions should be counted. Ultimately, it was decided that the launch of the station would be numbered; it would be Skylab 1, and the three manned flights would be 2, 3, and 4.

That decision, however, had not yet been made at the time that the crew patches were designed and ordered, so the flight suits were produced bearing patches marking the three manned missions as Skylab I, II, and III. As a result, both numbering systems were used in different places. Frequently, the former system is written using Arabic numerals, and sometimes with the two-letter mission abbreviation used for Skylab, thus SL-1 through SL-4. The latter system is generally written with roman numerals and almost exclusively with Skylab written out, thus Skylab I through Skylab III.

For the purposes of this book, we abide by the conventions of using Arabic numerals, and generally the "SL" abbreviation, when using the former system and of using roman numerals for the latter system. However, to the greatest extent possible, we have avoided using either, referring to the missions with less-ambiguous terminology ("the first crew's mission," for example).

Pogue explained the numbering system for their mission: "When the Skylab crews were announced in 1971, the prime crews set about designing their mission insignia or 'patch' as it was usually called. The missions were officially designated as Skylab 1, for the unmanned launch of Skylab on the Saturn V, and Skylabs 2, 3, and 4 for the three manned visits, which were launched on Saturn IBs.

"That seemed simple enough, but mischief was not long in coming. We began receiving flight training and procedures documents labeled SLM-1, SLM-2, and SLM-3 for the three Skylab manned missions. Other documents were labeled SL-2, SL-3, and SL-4, which conformed to the *official* mission designations. We began receiving mail and documents clearly meant for one of the other crews and the astronaut office mailroom became as bewildered, confused, and uncertain as the rest of us.

"In the meantime we had designed our mission patches incorporating the *official* numeric designations of Skylab 2, 3, and 4. During a visit by the NASA headquarters director of the Skylab Program, Pete Conrad asked him, "Are we 1, 2, and 3 or are we 2, 3, and 4"? He said, "You are 1, 2, and 3". All of us went back to designing new patches to incorporate the numerals 1, 2, and 3. Skylab 1 and 2 used Roman numerals and Jerry, Ed, and I used the Arabic numeral 3. The designs were rendered by artists and sent to NASA headquarters for approval. The whole process took several months, and the artwork didn't arrive at NASA headquarters until about six months before the scheduled launch of the Skylab.

"The associate administrator for Manned Space Flight took one look at the artwork and disapproved the design because he said the *official* flight designations, '2, 3, and 4' were to be used. Thus informed, we dug out our original designs (2, 3, and 4) and were in the process of getting the artwork done when we were informed by headquarters "not to bother." We could use the designs for 1, 2, and 3. Then we found out why the change of heart.

"The people who had manufactured the Skylab flight clothing (to be worn onboard) had already completed their work several weeks earlier in order to get the clothes packaged and shipped to the Cape to meet their deadline for stowage onboard Skylab, which was already in prelaunch processing. Furthermore, they had already used the designs submitted earlier for the mission patches. They didn't have time in their schedule to wait for *official* approval. The designs using the numeric designation 1, 2, and 3 became approved by default because items with these patches were already manufactured and stowed in Skylab lockers at the Cape. Removing them for patch change-out was considered much too expensive and disruptive during launch preparations.

"So, although officially designated as Skylab 2, 3, and 4, the mission insignias bear the numeric designations as follows: Skylab 2 (Roman numeral I), Skylab 3 (Roman numeral II), and Skylab 4 (Arabic numeral 3). When traveling in Afghanistan in 1975, I presented some Afghan VIPs with our Skylab

4 mission patch. One lady looked thoroughly confused and asked about the numeral 3 on the Skylab 4 patch. I gave her this long-winded explanation, and by the time I finished, the Afghans were roaring with laughter.

"Today it is especially confusing to autograph collectors who still scratch their heads trying to sort out their trophies."

3. Getting Ready to Fly

Joe Kerwin recalled: "Here's the story about my first brush with Skylab: One day in January 1966, Al Shepard said, 'Kerwin and Michel, I want you to go out to the Douglas plant in California. Marshall's working on an idea of using the inside of an S-IVB fuel tank as an experimental space station.' So we called out to Ellington for a T-38 jet and flew to Huntington Beach. At the plant they made us put on bunny suits and slippers, then showed us to the end hatch of a freshly manufactured S-IVB lying on its side. The hatch had been removed, leaving an opening about forty inches in diameter into the fuel tank.

"We noted that the hatch was secured with seventy-two large bolts. 'How will the astronauts remove it in flight?' we asked. 'We'll give you a wrench,' they replied. We climbed into the tank. It was big enough, all right—about thirty feet long and twenty feet in diameter. It was empty except for a long metal tube along one side—the 'propellant utilization probe'—and a couple of basketball-sized helium tanks. There was a faint chemical smell coming from the fiberglass, which covered the interior. It felt like standing in the bare shell of what was going to be a home someday after the builders had finished with it.

"'What would we do in here,' we asked. 'You can fly around in your suits.' Perhaps you'll test a rocket backpack. (That was prophetic.) And Marshall was even considering a plan to pressurize the tank with oxygen, so we could remove our spacesuits. That was a start!

"Curt had a conversation with the project rep about what experiments could and would be performed. After our return to Houston, he wrote Al a memo which likened the experiment selection process to 'filtering sand through chicken wire.' We were both inexperienced, glad to have something to do, and skeptical. I did not dream that seven years later I'd spend a month inside that tank, in space."

8. Joe Kerwin tests the vestibular-function experiment during Skylab preparations.

From a crew perspective, the development of the Skylab space station and the training of the astronauts who would live there are in many ways the same story. Usability is a primary concern in developing new space hardware. To ensure usability engineers would turn to the people who would be using that hardware. Throughout the development of Skylab, crewmembers would be brought in to give input on hardware as it was being designed and tested. So in many cases, they learned to use the equipment by helping its designers make it usable. Crew involvement began early in the development with the first Apollo Applications Program assignments being made in the astronaut office years before the first moon landing.

"Of course, those were early days for Skylab, and we'd looked at a tiny sample of 'bottom-up' planning, while the 'top-down' planning was taking place elsewhere and would answer a lot of our questions," Kerwin said. "'Elsewhere' was largely at the Marshall Space Flight Center. Not long after our trip to Huntington Beach, I was invited to observe a meeting between a visiting delegation from Marshall and MSC managers. The Marshall people gave a briefing on their plans for the 'Apollo Applications Program,' as it was then called. They sketched several missions on an ambitious schedule and asked for operations and training participation. The MSC managers basically said, 'That's great, but we're busy going to the moon.' So the team from Marshall left, saying over their shoulders, 'This is going to happen!' And so it did. It was still seven years from launch, but activity got started, and astronauts began to participate. We all had various assignments then, supporting Gemini, Apollo, and Skylab, and they changed fairly often, but Skylab began to take more and more of my time and attention."

Kerwin recalls standing around with a group of colleagues one evening in 1967 in the mockup building at MSFC. Someone had drawn with chalk a big circle on the floor, twenty feet in diameter, representing a cross section of the S-IVB tank. In the circle the astronauts worked with Marshall engineers on deciding how best to arrange the sleeping, eating, bathroom, and experiment quarters. "Al Bean was our leader at that time, and Paul Weitz, Owen Garriott, Ed Gibson, and a few other astronauts were there too, with several engineers," Kerwin said. "We had a great time and began to develop a friendly relationship with that S-IVB fuel tank."

In the earliest days of the Apollo Applications Program, the astronauts working with the program were a loosely defined group, with members rotating in and out as they began and completed projects for other programs. While the official flight crew rosters were not announced to the public until 18 January 1972, the group from which the assignments were made had been assembled about two years earlier.

"Pete Conrad had just come off his Apollo 12 flight, which was November '69, so this had to be around January or February of 1970 when Slayton came into a pilots' meeting on a Monday morning," Kerwin said in a NASA oral history interview. "He had a sheet of paper in his hand. He said, 'The following people are now formally assigned to crew training and mission development for the Skylab program.' He read the names of fifteen people. He didn't say who was prime, who was backup, who was what mission or anything else. All he said was that Conrad was going to be 'Sky King'; he was in charge, and he would tell us all what he wanted us to do."

The list included not only the nine astronauts that would make up the Skylab prime crews—Conrad, Kerwin, Weitz, Bean, Garriott, Lousma, Carr, Gibson, and Pogue—but also the six astronauts who would form the backup crews. "We had no idea what that list meant," Kerwin said. "There was a lot of speculation going on about who was going to be on what mission. There were fifteen of us, which meant that there were three prime crews, but only two backup crews. So somebody was going to have double duty as a backup crew it looked like unless the first prime was going to be the last backup. Deke didn't say. Deke was not a man of many words. He didn't say more than he thought was necessary at the time. It turned out, again in retrospect, that the way he had read that list was first prime, first backup, second prime, second backup, third prime, exactly in order."

In April 1971, "Sky King" Pete Conrad sent a memo to all of his "Skytroops" specifying who would be responsible for what. He made the assignments based on experience and on equalizing both the training and the in-flight workload.

The commander (CDR) would have overall responsibility for the flight plan and training; he'd also be responsible for the Apollo spacecraft systems and spacewalks. Estimated training hours: 1,411.

The science pilot (SPT) would be responsible for medical and ATM hardware and experiments and would be the second spacewalk crewman (in the end all three crewmen trained to make spacewalks). Estimated training hours: 1,500.

The pilot (PLT) would be responsible for airlock, MDA (Multiple Docking Adapter), and workshop systems and for the Earth Resources Experiment Package (EREP) hardware and experiments. Estimated training hours: 1,420.

Each of the fifteen men on the prime, backup, and support crews was also assigned specific experiments and hardware. This was as much for the benefit of the rest of the training, engineering, and flight operations world as for the astronauts themselves; it meant other organizations knew which astronaut to call to get an office position on a procedure or a hardware change. To keep those calls from becoming too much of a burden, training managers were assigned to the crews to help organize their schedules. "Bob Kohler was our crew training manager, an energetic but calm man able to steer us through the months of competition for our precious time," Kerwin recalls. "I think we burned him out; he left NASA after Skylab and became an optometrist."

The activity planning guide Kohler put together for the first crew for April and May of 1973 was typically busy. "We'd already done our multiple-day on-orbit simulations and were now concentrating on launch, rendezvous, and entry integrated sims ('integrated' meant the simulations included full Mission Control participation)," Kerwin said. "Saturdays were full, but we had most Sundays for family, unless we were traveling. There were more and more medical entries: exams, blood drawing, and final preflight data runs of the various experiments. Saturday, April 24 was listed as 'Crew Portrait Day—flight gear?—check with Conrad.' It was all a blur. Sometimes things happened on schedule, but often not. I have a handwritten sheet of paper from March of 1972 that says the following:

3/6/72: Joe—m133 Interface Test has slipped to Saturday, per Dick Truly. Bob Kohler.

Joe—it slipped back to Friday—keep checking! Richard.

Friday it is—as of 3/7/72. Kohler.

Would you believe Monday the 13th—Kohler—3/8.

3/10: cancelled until further notice."

After the first crew launched, Kohler put together the SL-2 Crew Training Summary, showing exactly how many hours each of the three astronauts had actually spent in trainers and simulators during the two years of "official" crew training. Conrad had the least, at 2,151 hours, but he'd been on three spaceflights already. Kerwin was next with 2,437 hours, and Weitz had the most at 2,506 hours. Those times don't count the many hours they spent flying, in meetings, reviewing the checklists, and trying to memorize all the switch locations and functions—the "homework" that had to be done to prepare for the simulator work. ("This would explain why none of your children recognized you after the flight," joked Kerwin's daughter, Sharon.)

Another of the activities on the busy astronauts' schedule was spacecraft checkout. "In early June of 1972, we strapped into our T-38s and hustled to St. Louis, to the McDonnell Aircraft plant, where the flight Docking Adapter had been mated to the flight Airlock Module and was waiting for final checkout [McDonnell had merged with Douglas Aircraft in April of 1967]," Kerwin said. "The next morning, June 6, we briefed, put on our bunny suits and slippers, and entered the flight unit. Outside was a large team of McDonnell engineers led by the test director. Every switch throw was in the test plan, and its effects would be watched and measured.

"The test was scheduled for twelve hours, but we accomplished it in half that time, flying from panel to panel and reporting over the intercom, 'Roger . . . in work . . . complete.' The spacecraft was clean, beautiful, and completely functional. We felt that industry had finally learned how to build them and test them, and we partied that night at the motel with our contractor teammates."

There seemed to be no limit to the tasks requiring the crews' attention during the period of the station's development and their training, everything from the overseeing the functional requirements for the triangle shoes to fighting with the Public Affairs Office over television shows on Skylab. (The astronauts weren't opposed to doing them, but they'd had no training and there was no time in the flight plan for them.) And of course an astronaut wouldn't want to find himself heading out for a spacewalk if, while on the ground, he hadn't customized the fit and comfort of his UCTA—the urine collection and transfer assembly worn under the spacesuits. One could change the location of the Velcro, add a snap, wear a suitably perforated athletic supporter, and wear the UCTA over or under the liquid cooling garment. Then there was the task of designing, and redesigning, the crew clothing to be worn in-flight.

"Testing and modifying the clothing was fun, although it dragged out a bit because clothing was a matter of both requirements and personal tastes," Kerwin said. The following excerpts from a series of internal memos exemplify this:

To: CB/All Skylab Astronauts
From: CB/Alan Bean
Subject: Skylab Clothing

a) Would it not be better to remove the knitted cuffs completely from our Skylab flight suits, since it looks like the temperature will be warmer most of the time than we would desire? [That was a prescient guess by Al!]

b) There seems to be a difference in philosophy as to what constitutes proper uniform for the "cool Beta Angle" and the "warm Beta Angle" on the Skylab mission. [Beta Angle was essentially the angle between Skylab's orbit and the sun; it varied with the season and determined how much of each orbit was spent in sunlight.] For the warm case our only option is to take off some of the cool weather garments. Taking off the jacket is all right because we end up with a cool polo shirt. However, if we wanted to take off our pants, we end up standing around in our underwear. I don't personally have anything against running around in my underwear, I do it all the time at home; but it would be better to at least have something more military in appearance planned for the warm case. . . .

To: CB/Skylab Astronauts
From: CB/Joe Kerwin
Subject: Al Bean's Clothing Memo

a) The knit cuffs are there to retain the sleeves and trouser legs under zero-g. They can be snipped off by a crewman at his option. Recommend they be retained, as a better military appearance will result.

b) The "warm weather uniform" question was a good one. . . . Unfortunately, all the clothing will be up there before we know the answer. We looked, briefly, at bermuda shorts last fall, and nobody thought they were needed. . . . Alternatively, we can ask Crew Systems Division to engineer the longies for easy cutting off. Pete, you decide. (Incidentally, Admiral Zumwalt says we can wear frayed pants in the wardroom now.)

c) Lip buttons will be provided for complainers.

To: CB/Skylab Astronauts
From: Gerald P. Carr
Subject: Skylab Clothing (Another shot across Medinaut's bow) (that's Kerwin)

a) Agree that the cuffs make the suit a bit too warm, but Joe's answer is fine. We can snip them out if they get too warm.

b) . . . I have no objection to making my own Bermuda shorts out of a "cold case" set of clothing.

c) Disagree with Joe's proposal for lip buttons. Zippers or Velcro are much more appropriate in the space biz.

Eventually, the Skylab astronauts all agreed on a clothing set. It contained cotton T-shirts for warm-weather wear and provisioned a change of underwear every two days and of outerwear once a week. The outerwear was made of a fireproof cloth, polybenzemidazole (called PBI; "We couldn't pronounce it either," quipped Kerwin) that only came in a golden brown. But it was comfortable. Rejected were the proposed small-bore fiberglass (called "beta cloth") items, which itched.

On the lighter side, the crewmembers all got to pick the music for tape cassettes they would carry with them on the mission. Each would have a small tape player, with Velcro on it to attach to a handy wall so that they could accompany their various experiment chores with music. For example, on the first crew, Conrad was a huge fan of country; his cassettes featured the Statler Brothers, Lynn Anderson, and other favorites. Kerwin liked classical; some of his favorites were Rachmaninoff's *Rhapsody on a Theme of Paganini* and Ravel's *Piano Concerto for the Left Hand*. He also snuck in a few folk songs recorded by his brother, Ed. Weitz's selections proved popular with his entire crew—Richard Rodgers's *Victory at Sea*, the Mills Brothers, Glen Campbell, Andy Williams, and the Ink Spots. Selecting the music was one of those last-minute chores like completing the guest list for our launch," Kerwin said. "It felt good; we were getting close."

Of course, not all Skylab training took place in the relatively comfortable confines of NASA centers and contractor locations. For example, as with Apollo, the Skylab crews went through training to prepare them for the contingency of an "off-nominal" reentry that could return them to Earth far from where they were supposed to land. "Although they never had to be used, the water egress, and desert and jungle training were lots of fun," second crew science pilot Owen Garriott said.

The jungle training took place in Panama under the guidance of local Choco Indians. "They were expert trackers and, of course, knew the jungle as their own backyard," Garriott said. "We were given an hour or so head start and told to evade capture and meet some twenty-four to forty-eight hours later on the beach some distance away.

"We all took off in groups of three—I was with Tony England and Karl Henize—at a fast trot, trying to get as far away as possible before darkness descended. The Chocos would set out after us and try to 'grab our hats,' equivalent to a capture.

"We succeeded almost too well," Garriott said. "We didn't get 'captured,' but we ran for so long that it got dark before we had properly made camp. We hurriedly gathered sticks to try to make a lean-to to be covered with a nylon sheet and to make a fire from small pieces of wood, but the every-day rains made a fire impossible. But darkness and more showers arrived before we had anything like a dry shelter. That night has been long remembered as the most uncomfortable, mosquito-plagued night of my life.

"Of course, we had to have a graduation celebration (after we were all finally recovered) on the banks of the Panama Canal," he continued. "Scientist astronaut Story Musgrave, always the adventuresome explorer, thought it would be fun to swim across the canal—in pitch darkness. So he stripped down and paddled off into the night, with numerous warnings about avoiding the alligators. In an hour or so, back he came, none the worse for any animal encounters."

Ed Gibson also had a memorable experience during his survival training. Despite all the challenges of living in the wild, Gibson decided the biggest threat to his own survival was one of his own teammates. "People ask me what is the most dangerous thing I've ever done in the space program," Gibson said. "Well, we went on a jungle survival trip, and I was out in the forest with Jack [Lousma] and Vance Brand. And after a couple of days or so, Jack was getting pretty hungry, and he kind of came up and started feeling my flesh. And I realized my objective for that whole time was to find enough food to feed him so I wouldn't get eaten."

Marshall's Neutral Buoyancy Simulator

We kidded about, we may have a dry workshop on orbit, but you're going to go through a wet workshop in training, that being underwater.

Jim Splawn

Joe Kerwin recalled: "From, I'd guess, 1968 onward, we traveled ever more frequently to Huntsville—for engineering tests and design reviews, but more and more to do EVA training in the new, bigger, and better water tank. I remember going there with Paul Weitz. We'd fly up together in a T-38. You'd take off from Ellington, point the nose to a heading of just a little north of east, climb to 17,500 feet, and go direct. We could make it in an hour if all went well. When we landed at Redstone Arsenal [the Army base in Huntsville on which Marshall is located], there'd be a rental car waiting, and we'd hustle off to the Tourway Motel; $7.50 with black and white TV, $8.50 with color.

"Bright and early the next morning we'd go to the neutral-buoyancy tank. That was always a professionally run organization and always a pleasant experience. We'd suit up in the dressing room, brief the test, and make our way up to 'poolside' and into quite a crowd—with divers, suit technicians, mockup engineers, and test personnel. Hook up the suit to communications, air, and cooling water. Down the steps into the water. Then float passively while the divers 'weighted us out.' They did this by placing lead weights into various pockets to counteract the buoyancy of the air-filled spacesuit, until we were neither floating to the surface nor sinking to the bottom. I recall gazing idly up through the bubble-filled water to the bright lights above and imagining that I was a medieval knight, being hoisted on to my charger before the tournament.

"Then the two of us, each accompanied by a safety diver (ready to assist us instantly in case we lost air or developed a leak) would move over to the Skylab mockup, laid out full size in the forty-foot-deep water and practice film retrieval from the ATM. We'd evaluate handrails and footholds, opening mechanisms and locks, how to manage the umbilicals, which trailed out behind us as we worked. After two or three hours we'd quit, return to the locker room, and debrief. It was wonderful training. By the time we launched, each of us could don and zip his own suit unassisted and move around in it with the same familiarity as a football player in his helmet and pads."

The idea of neutral-buoyancy simulation of the microgravity environment had arisen at the Manned Spacecraft Center in Houston before it was developed at Marshall, though neither center would implement the concept until the mid-1960s. Mercury astronaut Scott Carpenter had proposed using a water tank for astronaut training early in the space program, but management did not pursue the idea at the time.

A water tank was constructed for astronaut training at MSC, but not initially for neutral-buoyancy work. Rather it was used to prepare astronauts for the end of their missions. Since Mercury, Gemini, and Apollo flights all concluded with water landings, the MSC tank was used to rehearse the procedures that would be performed in recovery of the astronaut and spacecraft.

When Ed White made the first U.S. spacewalk in 1965 on the Gemini 4 mission, his experience seemed to belie the need for intense training; for White, the worst part of the spacewalk was that it had to end. When Gene Cernan made the second American spacewalk the following year, however, his experience was quite different. He found it difficult to maneuver, his faceplate

9. Astronauts practice for spacewalks in the neutral-buoyancy tank.

fogged up, his pulse rate soared, and he got overheated. It was obvious that changes were needed in spacewalking technology and procedures, and that included training. The idea of neutral-buoyancy training was revisited and implemented in time to prepare Buzz Aldrin for his Gemini 12 spacewalk, five months after Cernan's. With the changes that had been made and the intervening experience, things went far more smoothly for Aldrin's attempt on the final Gemini flight. Underwater training continued during the Apollo program; spacesuits weighted past the point of neutral buoyancy allowed astronauts to simulate the one-sixth gravity of the lunar surface.

At Marshall neutral buoyancy development came about from a grassroots initiative, at first as almost a hobby among some of the center's young engineers in the mid-1960s. "Some of us young guys got to talking about, we really are going to be in space, and if you're in space, you're going to need to do work," said Jim Splawn, who was the manager of space simulation at the Process Engineering Laboratory at Marshall. "And if you do work, how do you keep up with your tools? How do you train? So that started the discussion about how are you going to practice. How are you going to simulate the weightlessness of space? And we talked and talked for weeks, I guess, about that.

"And so one guy said, 'Hey, have you ever watched your wife in the swimming pool?' And we all giggled and said, 'Yeah, you bet, we watch our wives and other wives too.'

"But he said, 'No, no I'm serious. Have you ever looked at her hair while she's underwater, how it floats?' And that started a whole 'nother discussion, and so we said, 'Well, why does it do that? It's sort of neutrally buoyant—it doesn't sink; it doesn't necessarily float to the surface.' So then we started talking about how we could do that. We started coming up with the idea then of going underwater. That was the first concept that we had, the first discussion about going underwater."

The group thought the idea had potential and decided to use some of their free time to pursue it, and Marshall's first neutral-buoyancy simulator was born. Of course official facilities and equipment require funding, so the first phases of their research relied on using whatever they had available. The first exercises were done in an abandoned explosive-forming pit. The pit had been used to create the rounded ends of Saturn I fuel tanks and was about six feet in diameter and about six feet deep. Initial dives were done in swimsuits until the group felt like they needed more duration underwater, at which point they began using scuba gear.

Their experiment was showing promise, and they were ready to graduate out of the six-foot-diameter tank. Once again, though, their almost nonexistent budget forced them to make use of what was on hand, which was once again leftover Saturn hardware. The tank was based around an interstage for a Saturn rocket, the short, hollow cylinder that connects two booster stages together. "It was like a ring, probably twelve-feet vertical dimension," Splawn said. "So we had a backhoe, and dug a hole in the ground, and positioned the interstage and backfilled the dirt around it. And, guess what, we had a swimming pool now made out of excess Saturn hardware to become our next simulator for underwater work."

The extra volume meant that they could take the next step in their underwater evaluation. Just as they had moved from swim trunks to scuba gear in the first tank, the second allowed them to move on to pressure suits, simulating the gear that astronauts would be wearing in orbital spacewalks.

"We had to go to Houston to try and get pressure suits," Jim Splawn said. "Pressure suits in the mid- to late-60s were few in number and of great demand and expensive and were very, very well protected by the Houston suit techs. So we took an alternate route; we contacted the Navy, and a couple of us went to San Diego one Friday, worked with the Navy on Saturday, and they put us in high-altitude flying suits, and then they had huge oversize suitcases that they put these high-altitude pressure suits in, complete with gloves, helmet, everything, there. They trained us in a large swimming pool that they had; in fact, we had to jump off of diving boards into the water, and we took the helmets off, and we had to learn how to take a hookah [breathing apparatus] for underwater diving, so they taught us how to get the helmet off and take the hookah and still survive. So anyway, they taught us how to do that, so then we flew home on Sunday afternoon; we brought back four pressure suits, just on commercial air. So that's where we got our first pressure suits."

The "hookah" is a rubber full-head covering that is used underwater, similar to scuba. Instead of coming from a tank, air is pumped down from the surface by a hose to maintain a certain airflow into the rubber "helmet," regardless of the depth of the diver. It is particularly useful in tanks like Marshall's neutral-buoyancy trainer because it allows voice communications to the surface. However, one must be careful to not turn upside down, as air goes out and water comes in.

Up until this point, Splawn said, Marshall and MSC had not had any discussions about the work each was doing on neutral buoyancy. "We had absolutely no interaction at all," Splawn said. "We knew nothing at all about Houston and the type of simulations or training or anything else that they were doing. I really don't know the timing between what Houston did and what we did. I just don't have any data point there at all. Once it became known what we had and what we had done, there was competition, and some pretty heated discussions between Houston and us. But we ended up doing the crew training for Skylab."

In fact the first astronauts came to check out the work when the team was still using the second tank. Alan Bean, at the time an unflown rookie, was one of the first astronauts to perform a pressure-suited dive in the interstage tank. It was also during the experimentation with the second tank that the team decided they could let the Marshall powers that be in on their work. Von Braun himself made a dive in a pressure suit to evaluate the potential of neutral-buoyancy simulation.

Bob Schwinghamer, who was the head of the Marshall materials lab, recalled a nerve-wracking incident that occurred during one of Bean's early visits. "I was safety diving, and I was floating around in front of him. He was in there unscrewing those bolts off of that hatch cover. And all at once, it said, 'poof,' and a big bubble came out from under his right arm, a stream of bubbles. I thought, 'Oh my god, I'm going to drown this astronaut.'" Schwinghamer said he attempted to cover the hole in Bean's suit, but he could see the suit collapsing—first near Bean's feet, then up to his knees, then his thighs. Since he didn't have a communications system at that time, Schwinghamer left Bean and surfaced, and told the operators to give him more air.

"He never lost his cool," Schwinghamer recalled. "By then, he wasn't neutrally buoyant anymore; he was about sixty pounds too heavy. So he walked across [the tank], and he just climbed up the ladder and got out. That's all there was. And I said, 'Oh my goodness, what if we had drowned an astronaut?' But he was just cool."

Working with pressure suits complicated the situation. The pressure suits, representing spacesuits, were basically balloons containing divers. That meant the air caused the suits to tend to float. In order to make the suits neutrally buoyant, weight had to be added to balance out the effect of the air. This had to be done very carefully. Putting too much weight in one area would cause that area to sink more than the rest of the body, invalidating the simulation of weightlessness.

"After many, many stop-and-go kind of activities, we settled in on a low-profile harness of small pockets of lead strips, so that we could move the lead about depending on the mass of the human body that's inside the suit and consequently what kind of volume of air you had inside that suit," Splawn said. "We could move the lead weights around until we could put the test subject or flight crewman into any position underwater and turn him loose, and he would stay there.

"We started off just in a room, so in order to get some data points, we put the guy in the pressure suit and then lay him flat on the floor and tried to get him to lift his arms—Is the weight distributed?—and lift his legs—Is the weight sort of distributed correctly?" Splawn said. "And so we said, 'OK, get up,' and he couldn't get up, he had so much weight on him. That was in the very early days." Typically, he said, about seventy to eighty pounds of lead weights were needed to achieve neutral buoyancy. To make sure the weighted, pressured-suited divers didn't encounter any problems, each one was accompanied by two safety divers who could help out in an emergency.

Once the team had enough experience in the interstage tank, they were confident that neutral buoyancy could be used for weightlessness simulation. They were ready to move on to the next step. "From that we graduated to what we called the big tank," Splawn said. "The big tank is seventy-five feet diameter; it's forty feet deep; 1.3 million gallons of water as best I remember. It was complete with underwater lighting, underwater audio system, umbilicals that would be very much like the flight crew would use to do an EVA on orbit."

This tank, Marshall's Neutral Buoyancy Simulator, was designed to take the work to the next level. Unlike previous facilities, which were experiments designed to evaluate the efficacy of neutral buoyancy as a microgravity analog, the Neutral Buoyancy Simulator was a working facility. The theory had been proven and now was being put into practice. The facility was designed to be large enough to submerge mock-ups of spacecraft in order to test how easily they could be operated in a weightless environment.

"We sort of had the vision of building a facility large enough to accommodate some pretty large mock-ups of hardware, and it really proved out to be very, very beneficial," Splawn said. "Because once we had the difficulty at the launch of the Skylab itself headed towards orbit, it really proved its worth because of all the hardware we had to assemble underwater."

The origin of the "big tank" was rather unconventional. In order to hasten the process of building the tank, Marshall leadership found a way to circumvent the bureaucratic requirements of creating a new facility. "The facility was not a 'C of F,' or construction of facilities type project," Splawn explained. "There is a small tool tag that is on the side of the tank, and it has a number stamped on that tag, and so that designates the seventy-five-foot diameter tank as a portable tool. There were a lot of eyebrows raised at that." While the tank was not technically secured into place, saying it was portable was somewhat of a stretch.

"I don't really remember how that happened," Splawn said. "I know there was great interest in having a facility, and we thought we had the right idea of how to simulate weightlessness and how to train. We needed a facility, and the schedule of when we needed it just could not be supported through the official construction of facilities kind of red tape that you had to go through to get a facility approved, and then all of that kind of business that occurs to acquire a facility. So that's why we went this alternate route."

As a result of the way the Neutral Buoyancy Simulator was built, many people elsewhere in the agency did not know what Marshall was doing until it had been done (as was the case with associate administrator George Mueller, who was not aware of the tank until his "wet workshop" dive).

"The tank was built in-house," Splawn said. "We used the construction crew out of a test lab because they were equipped and they were accustomed to doing construction work. So the steel segments of the tank, of course, were rolled steel. They were shipped in, and then the government employees welded the tanks together, and we installed all the systems, electronic, mechanical, filtration, all of that was worked internally."

The tank attracted some unusual visitors, Splawn recalled: "It was very interesting to have some of the caliber people come through our area that came through. Of course, starting with von Braun—back when we had just first started the thinking and the dream of going underwater to do evaluations in a weightless environment, we found out that von Braun was a scuba diver. So once we had been through the early stages and thought we could sort of reveal our thoughts a little bit, we contacted his secretary, Bonnie, and told her we'd like to have Dr. von Braun come and see what we were doing.

"I guess the first time he ever knew anything about it, we were on the twenty-foot tank. He didn't know about it up until about then. Because us bunch of young guys, what we would do is work our regular kind of work through the day, and then we would go out in the late evenings and play, and I say 'play' in quotes. But we would try to figure out just exactly what we were trying to do. We didn't know if we had a cat in the bag or not. But we finally revealed the cat to Dr. von Braun and got him to come down, and he thought it was wonderful. He said, 'Ja, ja, keep going, keep going.'

"I remember one day that von Braun had been to the Cape for a launch, and we got a call from his secretary again. Bonnie said, 'Dr. von Braun has just called me from the Cape, and he is bringing a guest on the NASA aircraft back with him from the Cape to Huntsville, and they want to go to the neutral-buoyancy facility this afternoon, and this guy's name is Jacques Cousteau, and can you accommodate him?' And I said, 'Yes, ma'am, we sure can.'

"So von Braun dressed out in swim trunks, and Jacques Cousteau dressed out in swim trunks, and they went for a dive in scuba gear, and von Braun showed Jacques Cousteau some of the things that we were doing underwater, put him through a few paces with some of the hardware that we had mounted in the tank at that point in time. So it was sort of interesting."

As an additional safety precaution, the Marshall facility also included a decompression chamber, which could be used if a diver surfaced too quickly. The medical term is "dysbarism"—Greek for "pressure sickness"—but to divers it's simply the "bends." Bends has affected divers since humans began to dive for pearls centuries ago. It doesn't just happen underwater; workers building the foundations of the Brooklyn Bridge a hundred feet beneath the surface of the East River developed the strange pains and disorientation of "caisson disease." The doctor hired by the company to look into the problem noted with interest that the pains often went away when the men went back down to the diggings. But it was another twenty years before other doctors figured out what was happening.

When a diver descends in water, the water's weight increases the pressure against the body; at thirty-three feet it's double the pressure at the surface. In order to breathe, the pressure of the air the diver breathes also must increase. And that pressure drives nitrogen into the lungs, blood, and tissues. That's not normally a problem; nitrogen is inert except at very high pressures, when it exerts a narcotic effect.

But if a diver ascends rapidly to the surface, the pressure suddenly diminishes. Then the absorbed nitrogen reverses course and comes out of the tissues. The diver is able to breathe some of it out, but if the pressure was high, some of it forms bubbles in the blood and tissues, and these can have dangerous effects—bubbles compressing nerves in the joints cause bends, bubbles blocking capillaries in the lungs cause chokes, bubbles in the blood vessels of the brain can mimic a stroke. To prevent these things, it's essential to reduce the pressure on the body slowly enough to allow for "breathing out" the nitrogen without letting bubbles form.

The dives in Marshall's tank never caused the astronauts to have any problems. However, the recompression chamber was used once, Splawn said, for a Tennessee Valley Authority utility diver in the area who had been doing work underwater and surfaced too quickly and was rushed to Marshall. Splawn said that, while it was too late to prevent lasting harm, the chamber may have saved his life.

Concerns over rapid decompression did affect the crews training in the tank in one way, though. "In our dives, we never went deep enough for long enough that we couldn't safely return to the surface of the tank in a hurry," Kerwin said. "But climbing into the cockpit of a T-38 and flying home at reduced cabin pressure was another story. Flying after diving sets pilots up for bends. So we did a study, and came up with rules for how long a diver had to loiter on the surface before launching for home. It varied from a few hours after one dive to an overnight stay after two days' work underwater."

Blood, Toil, Sweat, and Teeth: Memories of Skylab Medical Training

Until Skylab, crewmen had worn biomedical sensors pretty much all the time during flight. On the early Mercury and Gemini flights, when ground stations in the Manned Spaceflight Network (known by the time of Skylab as the Spacecraft Tracking and Data Network) were scattered around the world, the flight surgeon attached to each station crew would study those heartbeat and respiration traces intently as the spacecraft passed overhead, looking for signs of stress. Heart rates during spacewalks were useful as they were a pretty good indication of crew workload and oxygen consumption.

As the NASA doctors looked at the heart rates of astronauts under the stresses of launch acceleration, weightlessness, spacewalks, and just hanging around, they inevitably witnessed the occasional irregularity—usually a premature beat or a run of two or three of them. They came to accept these as within the limits of normal. But the arrhythmias they saw in the Apollo 15 crew on the way back from the moon were more marked and a cause of considerable anxiety on the ground. Future Apollo flights carried medications for such arrhythmias.

With this background and the greatly increased duration of the planned Skylab flights, a medical desire for as much data as possible remained, as exemplified by the following excerpts from NASA memos:

To: EA/Manager, Apollo Applications Program October 3, 1968
From: CA/Director of Flight Crew Operations [Deke Slayton]
Subject: Bioinstrumentation for Apollo Applications Program (AAP) Missions

The long duration, large volume and required crew mobility of AAP core missions will require different guidelines for the transmission of biomedical data. Continuous-wear instrumentation will not be feasible. Numerous medical experiments will be performed which require instrumentation, and which will give medical monitors the information needed to assess crew status.

Therefore, the following guidelines are recommended: Bioinstrumentation will be worn for launch, entry, EVA and medical experiments. It will not be worn at other times unless required for diagnostic purposes. . . .

To: CA/Director of Flight Crew Operations Oct 16, 1968
From: DA/Deputy Director of Medical Research and Operations
Subject: Bioinstrumentation Requirements in the
Apollo Applications Program

. . . I feel it is inappropriate for you to propose guidelines for the acquisition of biomedical data without full coordination of these guidelines with our Directorate. The following comments regarding your memorandum are offered in a constructive vein in the hope that you may be persuaded to address future recommendations to this Directorate. . . .

It is our present hope that the principles enunciated in your two proposed guidelines can be fully satisfied but we do not have sufficient technical or operational information to accept these guidelines as program constraints at the present time.

The doctors had a point; it was pretty early in the program. Deke withdrew the memo, and the problems were worked out amicably. Not without a glitch or two along the way, however.

To: CB/Pete Conrad
From: CB/Joe Kerwin
Subject: Medical Operations Requirements

DA memo of 5-15-70 (on file) presents instrumentation requirements and guidelines for Skylab. . . . Wearing of bio-harness during sleep is a new requirement, is not feasible or useful, and should be discouraged!

At about this time, the question of dental treatment on Skylab surfaced. The astronauts' dentist, Dr. Bill Frome, recommended putting a dental kit onboard and training two men on each crew to use it, in light of his experience with astronaut patients. He argued that palliative treatment, even up to extracting an abscessed and painful tooth, was preferable to terminating a mission. Deke asked Kerwin to review it.

To: CA/Donald K. Slayton
From: CB/Joseph P. Kerwin
Subject: Pulling Teeth

A one percent chance of a serious dental problem on a 28-day mission is not surprising. That's (28 x 3 =) 84 man-days, which is one percent of 8,400 man-days or 23 man-years. If we have 46 astronauts, one of them will need emergency dental care every six months—which matches Dr. Frome's experience.

I have asked Dr. Frome to set up his proposed 1.5-day training program and run me through it as a guinea pig. . . .

I believe that the right thing to do is to let them put the hardware on board, agree to train one of three crewmen (which cuts the risk but does not eliminate it) and reevaluate after the first mission.

"Management decided to go ahead and train two members of each crew, and we had a ball," Kerwin said. "We traveled with Dr. Frome to San Antonio, to the U.S. Air Force Dental Clinic at Brooks AFB. Bill and the dental staff had recruited a number of volunteers who needed to have a tooth extracted. (One of the first lessons was that you didn't pull teeth, you extracted them.) So there we were, six of us, wielding syringes filled with xylocaine and wicked-looking dental forceps (and much more nervous than the patients were), getting those jaws numb and those molars out under the watchful eye of our dentist instructors.

"Paul Weitz drew a retired Air Force general. My patient's molar broke in two during the procedure and had to come out in pieces. We were *very* glad when it was over. But I believe we could have done the deed in flight had we needed to. (We didn't, and no dental emergencies arose during any mission.)"

The dental kit became part of a medical kit for taking care of illness and injury aboard the Skylab space station. It was called the In-Flight Medical Support System. In retrospect, it looks like supplies for a pretty modest doctor's office, but at the time it was quite a leap forward. It contained minor surgical instruments, a laryngoscope and tracheostomy kit, intravenous fluids, and lots of medications including injectables. Diagnostic equipment included equipment to make and examine blood smears and do cultures and antibiotic sensitivity tests on various body fluids. Kerwin, the doctor of the group who was quite familiar with the tools, was very much in favor of carrying the equipment to Skylab. Some of the others, familiar with medical equipment primarily from being on the receiving end, were less so.

To: CA*/Donald K. Slayton*
From: CB*/Joseph P. Kerwin*
*Subject: In-Flight Medical Support System (*IMSS*)*

It's clear from glancing through the list that this is mostly a doctor's bag, not a first-aid kit. The document doesn't say that, and it even proposes to train pilots to use all the equipment, which I find unrealistic. (Med school was easy, but not that easy!) It's also apparent that to justify the more elaborate equipment operationally—from the standpoint of mission success—is darn near impossible. Major medical catastrophes just aren't that much more likely to happen in eight weeks than they were in two. Minor illnesses are, but not heart attacks, etc. . . .

But that's not the only point of view. Let me give, from my point of view, some reasons for carrying a doctor's bag:

1. *Up to now, the medical program has been unbalanced in the direction of pure research instead of treating illness and injury in space. This is a capability we don't need today, but we certainly will need it in space station times—for economic reasons at the least. It seems prudent to start using Skylab to develop equipment and procedures to meet this need, just as we used Gemini to develop a rendezvous capability.*
2. *It's true that a doctor isn't mandatory on any Skylab flight. But if you do happen to have one along, you ought to allow him to do a little good for the program in his spare time by providing him with some of the tools of his trade. He could do an occasional physical exam on his buddies, and try out the simple laboratory tests on himself, by way of proving that they work. It would sure beat looking out the window.*

In retrospect Kerwin found that last statement to be really dumb—nothing in Skylab beat looking out the window. But the In-Flight Medical Support System was approved, and the same two crewmen who wielded the dental forceps were taught to use an otoscope and an ophthalmoscope, palpate and percuss, and report their findings to a doctor in Mission Control. "It was a wild experience for the pilots and a valuable refresher for me," Kerwin said. "We were even taken to the trauma unit at Ben Taub Hospital in

Houston on a Friday night, where under the skilled tutelage of Dr. Pedro Rubio, the chief resident, we watched one of the best emergency medicine teams in America deal with life-threatening trauma and illness."

Trauma training at Ben Taub Hospital proved a memorable experience for the astronauts. It was always scheduled on a Friday or Saturday evening when the probability of gunshot or knife wounds was apparently the highest. Sure enough the crew saw their share but usually kept their distance from the emergency team engaged in what was a life-or-death procedure for some incoming patients. More relevant to their Skylab situation, they also had personal discussion and training with the experts in ear, nose, and throat; gastrointestinal tract; and eye and other specialties about how to handle in-flight emergencies. Even in these early days, they could expect to have experts in prompt voice contact and even with TV downlink to provide images to the ground. So they ended up with reasonable confidence that most emergencies could be handled if they should arise.

The astronauts were also introduced to a fine team of consultants from the Houston medical community—specialists who would be on call during all the Skylab missions to advise the NASA flight surgeons should trouble arise in flight. Drs. Page Nelson, Hiram Warshaw, Everett Price, Kamal Sheena, and Jules Borger gave freely of their time and talent. Knowing they were there provided the crew with a feeling of security.

One of the best things to come out of the In-Flight Medical Support System, Kerwin said, was the checklist. Stimulated by the need to explain medical equipment and procedures to a bunch of pilots, the medical team linked up with the training team to produce a fine, very graphic, and explicit manual showing with simple line drawings what everything looked like and what to do.

"We had one more treat in store," Kerwin said. "Drunk with enthusiasm by the opportunity to experiment in space, the medical research team pushed for one final capability—to take and return blood samples. Not a big deal, you say; but it was, first because it had never been done before and second because it posed some hardware challenges in weightlessness."

It was done. The crews agreed to give blood weekly; one member of each crew was trained to be the "vampire"; and an assortment of air-evacuated tubes, a centrifuge to separate cells from plasma, and arrangements to freeze and return the samples were designed and flown. It all worked quite well. "I drew my own blood, not wanting to put Pete or Paul to the trouble of learning (and perhaps forgetting) how," Kerwin said. "Pete hated being stuck and on the ground tended to become light-headed. But the blood couldn't rush from your head in zero-G, so Pete was fine. He just looked away from the needle."

The first crew, by benefit of being first and of having the physician of the group among its number, bore much of the hard work in planning for crew participation in the medical experiments (with a lot of help from Bill Thornton, also a medical doctor and a Skylab guinea pig himself during simulations). Therefore the training activity for the second and third crews followed much the same protocol as developed for the first flight team.

"Of course there were always some personal differences in practice," Garriott said. "Whereas the first mission would have a doctor on board who knew the medical objectives and protocol in detail, as he had helped devise them, plus the fact that some of his other crewmembers were apparently not too enthusiastic about some of the procedures (e.g., blood draws), the second flight team all started substantially at the same level in terms of medical experience."

Garriott described his crew with respect to the medical procedures as being all novices but with a keen interest in the protocol and personal results. No deference was provided to the scientist astronaut in this area, he said; everyone wanted to know about and participate in all that they could. They were all trained to draw blood and planned to do it in flight. They started with practice puncturing the skins of oranges or grapefruit with a hypodermic needle to simulate that of a human arm. Next came human volunteers, usually from life-science workers in the MSC laboratories. As it turns out, there were more female than male volunteers ("Perhaps tougher constitution, or more highly motivated?" Garriott remarked), and this often made the task more difficult—perhaps having less visible and accessible veins to attack. But all three of the crewmen successfully accomplished the blood draws a number of times, finally even drawing their fellow crewmen's blood at least once. "It was good practice and we actually enjoyed the training," Garriott said.

During flight all three crewmembers put their training into practice. Garriott routinely drew the blood of Bean and Lousma, while one or the other would draw his blood on the desired schedule, every week or so. On one occasion in the middle of the crew's two-month stay, the ground asked to have a video of the actual procedure. Lousma was scheduled to draw Garriott's blood.

"We got all the cameras placed properly and the video recorder running for later dump to the ground," Garriott said. "With all the paraphernalia in place, I bared my left arm, got the tourniquet tight, Jack made an excellent 'stick,' and the blood flowed freely just as desired. When finished, we withdrew the needle and blood promptly squirted all over the place! I had forgotten to remove the tourniquet first and all the blood pressure trapped in the lower part of the arm took the path of least resistance into space. So we cleaned up the mess I had made, rewound the tape recorder and did it all over again using my right arm. The physicians on the ground seemed happy with the demonstration."

Mission Control and Training

The astronauts assigned to the flight crews were not the only ones having to train for the mission. In February 1972, over a year before the launch of the Skylab station, the Mission Control Center team began running their first simulations for the missions.

The long-duration aspect of the Skylab program presented new challenges for the MCC team that would require advance preparation. On the ground every moment that the crews were in space, a team of people would be supporting them around the clock in Mission Control. In fact the control team would be operating Skylab even when the astronauts were not aboard it. And for the Mission Control team as much as for the astronauts, Skylab was a new spacecraft, completely unlike anything flown before, with its own unique parameters and requirements. In addition, the work the crews would be doing on Skylab would be unlike anything done in space before, so new procedures would have to be learned in order to support them.

According to Phil Shaffer, the lead flight director, operations control for Skylab was a mixture of old and new for the flight directors, with some elements being very similar to those in Apollo, and others being different from anything flown before. "The part that is similar to prior programs is that there was a trajectory function and there were the systems functions," Shaffer said. "There was an electrical guy, a communications guy, there was an environmental guy, you know, each with their support staff and in that sense was all very similar. The manning level or the expertise requirement was the same as if we were doing a lunar mission.

"The teams, if you stood away a little ways, looked like Apollo teams or Gemini teams in the way they were structured because there was a flight director who literally was responsible for everything, there was a capsule communicator for air-to-ground voice, there was a surgeon, and there was a networks guy," Shaffer said. "And all of those positions, you know some of them had slightly different names. Like GNC [guidance, navigation, and control] for the CSM was called GNS [guidance navigation system] for the Skylab to distinguish different positions. Different names were required when both the CSM and Skylab were up and active at the same time. There was a limited on-orbit team for when the CSM was powered down. There were five on-orbit teams that did planning, preparation, and support execution for the experiments, EVAs, maintenance and repair, or whatever else was going on. These teams were led by [Phil] Shaffer, Don Puddy, Neil Hutchinson, Chuck Lewis, and Milt Windler. There was also a trajectory team led by Shaffer that was decidedly different from the on-orbit teams. It supported launch and rendezvous, and deorbit and entry, and maintaining orbital lifetime by raising the vehicle orbital altitude. They did all those calculations. So, there were six teams: five on-orbit teams and one trajectory team, basically, for the year of the program."

Differences began with the launch. The crews flew into space on one spacecraft that was essentially a taxi carrying them to another spacecraft where they would spend the bulk of their mission. "Another thing that was different was having two very dissimilar vehicles, with some of the time both being active, so that you had two com guys and two environmental guys and two electrical guys on occasion," Shaffer said. "Certainly until you got the Skylab powered-down for leaving or the Command Service Module powered-down for the habitation period. The situation on Apollo was similar during the lunar-landing sequence with the Lunar Module and CSM being involved. It was a bit of a zoo keeping all of that business straight."

The attitude control systems for the massive Skylab space station were also very different from both a conceptual and an operational standpoint than any of their predecessors. "The new for Skylab was not new in name but new in type and that was an attitude control system with Control Moment Gyros [CMGs]," he said. "That was a whole new business in place of small rockets, reaction control thrusters, to control the attitude. You had these giant

CMGs that were wonderful. The CMG system was assisted by a cold gas system called TACS [Thruster Attitude Control System]."

Attitude control—which basically amounts to which way the spacecraft is pointing—on Apollo was pretty straightforward, a basic application of Newton's law that states for every action there is an equal and opposite reaction. That law is what allows rockets to travel through space, even though there is nothing there to push against. A rocket engine burns fuel to generate thrust, and the action of the engine spewing flame backwards leads to the opposite reaction of the rocket moving forward. The same principle that pushes a large rocket through space also, on a much smaller scale, allowed the Apollo spacecraft to control its attitude. Rocket engines burned fuel, and the spacecraft turned in the opposite direction. The Skylab Thruster Attitude Control System took that simple concept and applied it in an even simpler way. Rather than burning fuel, the TACS simply vented cold gas into space. The action of the gas being vented produced the opposite reaction needed to control attitude.

The CMGs worked on a more arcane principle of physics—angular momentum. Tilting the spinning rotor of a Control Moment Gyroscope resulted in a torque that would rotate the entire station. Attitude control via CMG had the additional benefit for a long-duration mission of requiring no fuel, relying instead on the power produced by Skylab's solar panels.

In addition to the new attitude-control techniques, Shaffer said, new Mission Control responsibilities were added to provide support for the science operations on Skylab. "And then there were the experiments," he said. "We had a control function for Earth sensing. We had a control function for the celestial viewing. One looked up, the other one looked down. We had a control function—a control position—for all the biomedical activity, a control function for materials science."

While Mission Control had been involved in science support before, notably during the lunar research during Apollo, Shaffer said that the support needed to coordinate the Skylab research was substantially more complex. For example, both Skylab and Apollo missions included making surface observations from orbit. Skylab had its Earth resources observation package and Apollo carried equipment in the Service Module's SIM [Scientific Instrument Module] Bay that imaged the lunar surface. Although there was a general similarity in function, they were very different in operation. "The

Earth resources guy [in Mission Control], for instance, had a huge coordination activity he did with the aircraft overflight, and with the ground truth people, and with the weather service going on with his planning. This was dramatically different from the equivalent function on Apollo. The guy in the Command Service Module was not running the SIM Bay."

Another change for Skylab that was worked out before flight was the real-time mission planning that would have to take place while the crews were in orbit. On prior missions extremely detailed plans were laid out ahead of time. On Skylab more activities were scheduled on a day-to-day basis during the mission. Every day the flight control teams would plan out what the crew would do the next day. "The evening shift did the detail preparation for the next workday's activities," Shaffer said. "The midnight shift did the overall plan for two days hence. And in part I think that was done to provide shelf life for both the support data that was going to go to the crew for the upcoming day and to give negotiation and preparation time for the structure of the plan two days hence."

That's not to say no planning was done further ahead. Rough outlines of activities were put together for a week in advance, structured around such things as astronomical or Earth resources observations that were to be made. Since those had to take place at a very precise particular time, they were placed on the schedule first, and other activities that were more flexible were filled in around them.

"All of that was all done by the time we entered the upcoming twenty-four-hour thing; then the remaining pieces were put in," he said. "The surgeons would have to get their requirements in. Life sciences was a really big deal, so significant effort was needed to get all of their activities in within their constraints. Vehicle maintenance had to be done, including servicing the ATM and the associated EVA activity. All of that got dropped into the plan. All of that happened on the evening shift. And that was new. The nearest thing to it may have been the lunar excursion planning activity while crews were on the lunar surface for two or three days. It evolved, and we all got really comfortable with it."

There was some concern about why there had to be so many levels of advanced planning, but the system proved effective. Among its strengths was that getting a good bit of the planning done early freed up more time to react to any unexpected situations or to finish any previous scheduling that needed adjustment. "If we needed more time to get the detail flight plan support stuff ready for the crew, you had it," Shaffer said. "There was basically another whole shift available to finish up that work. And if something was wrong with your big plan for the day, then you had time to renegotiate whatever problems that created."

Of course, no matter how much planning was done in advance, there were always times the plan had to be changed as new circumstances arose. "The classic case, to me, happened on one of my watches," said Shaffer, "and it comes up under my title of 'surgeon's rigidity and the bologna sandwich.' A volcano in Central America decided serendipitously to start a major eruption while we were on orbit with all of our wonderful EREP equipment. Of course the geologists and geophysicists were going nuts because it was an opportunity to use much of the EREP sensor equipment we had to really get new and significant information about an erupting volcano that they had never had the opportunity to get before. It would be like looking 'down the gun barrel' right through clouds. They really wanted to do this.

"The conflict was that the orbit track that was going to go over the volcano happened during an already scheduled meal. The surgeon, because of his dietary scheduling requirements rigor declared that they were critical, and he couldn't change the mealtime. That might change the digestive processes results, and there was no compromise for it. And I had a lot of sympathy for both parties, but here was a one-time event and we were going to be up there for many, many meals.

"Finally after much debate, I resurrected mission rule one dash whatever it is that says the flight director is in charge in real time. It means he can do whatever he needs to. So I decided to do it, and I told the surgeon on the loop that we are going to do the data take over the volcano, that his dietary concerns are not equal in terms of return. Plus, everybody knows ya'll have the wrong diet. Everybody knows the best diet for in-flight work is a bologna sandwich.

"The surgeon kind of imploded. I think he thought I had impugned him, and so he stopped objecting. We did the data take, and it was wonderful. Lunch was about a half-hour late. It was no big deal. I believed that. I believed it didn't make any difference. We got all of that done.

"A curious thing happened the next day. When I came on shift there on my console was a bologna sandwich, which honest to goodness was a foot and a half long and six inches wide and had at least an inch of bologna in it. Nobody ever 'fessed up to where it came from. So I don't know whether the surgeons did it or somebody who had heard the conversations. I always hoped that the surgeon did it. But it changed the dynamic. We got along better after that. Not a lot, but . . ."

During flight, this issue was greatly alleviated by the addition of another level of coordination within the science community. The initial structure in which the various disciplines each advocated their own concerns to Mission Control was putting substantial strain on the flight directors, who had to weigh and balance those concerns. "So what we did was invent a tsar—a 'science tsar,'" Shaffer explained. The first science tsar was Robert Parker, a member of the second group of scientist astronauts. "At that point we refused to listen to all those people any more; we only listened to Robert. He brought the finished product into the planning shift, which we then implemented. That all worked well in the planning cycle, though it didn't help a lot if you ran into something happening in a real-time conflict, because Robert wasn't always available to us."

At one point during Skylab mission preparations, Shaffer said, the ultimate authority of the flight director for dealing with real-time situations as they occurred was challenged by a visitor from NASA headquarters. "This is another one of those stories people don't know anything about," he said. "During the Skylab 2 sims [simulations], this guy showed up, badged and everything, and walked into the control center. Because I was launch flight director, I was running the sims.

"And he said, 'Where's my console?'" Shaffer said. "And I said, 'Who are you?' He said 'I'm the mission management representative from Washington.' I said, 'What do you do?' And he says, 'I am from NASA headquarters, and I have the final say in all of the decisions we'll make in this program.' And I said, 'Well, I find that pretty interesting. I've never heard of you before, and there's really no place in my flight control team for you to do that, particularly during a dynamic phase. Frankly, you'll be a lot more trouble than you're worth no matter how good you are.' And he says, 'Be that as it may, I am here to stay.' And I said, 'Very well.'"

Shaffer said that he considered calling director for flight operations—and NASA's first flight director—Chris Kraft to come deal with the situation, confident that the original "Flight" would back him up. However, he decided to try and handle the problem himself before resorting to calling for help. "I went back to my console and got on one of my secondary voice loops to the simulations supervisor, and said 'I want you to give me the "Apollo tape case,"'" Shaffer said. "So Sim Sup says, 'Why am I doing that?' I said, 'Because I'm asking you to.' And he said, 'I got it.'

"So he gave us that case and things really went to hell in a hand basket. The tape was the source for all the CSM systems failure descriptions and data used for training simulations for the flight controllers and flight crews. We couldn't tell where we were in orbit after the launch phase, communication was really ratty, and there were electrical problems, computer problems, etc. I unplugged and ran up to his console and said, 'Tell me quick... what do I do now?'

"The guy looked at me, reached up, unplugged his communications set, got up, and walked out. We never saw him again during a dynamic flight phase." On orbit however, his group was very active via an ad hoc organization called the Mission Management Team.

4. Fifty-six Days in a Can

To start with, I was out in California
in Huntington Beach. And I got this call,
and it was the good Robert Crippen who was calling.
He said, "We had a drawing, and your name was drawn to be
a crewmember on SMEAT." And I said, "What the hell is SMEAT?"

Bo Bobko

SMEAT, the Skylab Medical Experiment Altitude Test, was a full-length simulation of a Skylab mission. The crew selected for the test would spend fifty-six days in a spacecraft mock-up without the benefits of actually being in space. Selection for the mission might seem a dubious honor, but for the commander of the chosen crew, things had been much, much worse.

"June 10, 1969, was probably one of the low points in my life," remembered astronaut Bob Crippen. On that date the future pilot of the first Space Shuttle flight learned that the project to which he had dedicated the past three years of his life was over. The U.S. Air Force had canceled its Manned Orbiting Laboratory program, leaving Crippen and his fellow members of the Air Force's astronaut corps uncertain as to what the future held. Beginning almost four years earlier, a total of seventeen astronauts had been selected by the Air Force from the ranks of military pilots. During that time they had completed training on the NASA-developed Gemini spacecraft, which was to have been used in the Air Force program. They had also undergone training on the tasks they were to perform on the space-based laboratory.

At the time the program was canceled, the members of the corps were excited about the prospect of spaceflight, but now the Air Force would no longer have need for astronauts. The nation's civilian space program, on the other hand, still had an astronaut corps, but that group had become overly crowded as well. The last class of astronauts NASA had selected, a second group of scientist astronauts brought into the corps two years earlier in 1967, had dubbed themselves the "Excess Eleven" (or, in test-pilot terminology, XS-11) when they realized just how low their odds were of being assigned to a spaceflight anytime in the near future.

Crippen said that after the program was canceled, "We sat around, and it seems like for a month afterward, we'd go to the bar every night at probably about 2 o'clock in the afternoon and have a wake. One day, I remember a crew meeting, and we were trying to figure out what we were going to do, and Bo [Bobko] said, 'Why don't we ask NASA if they could use any of us?' And we said, 'Bo, that's the dumbest damn idea I ever heard. They're canceling Apollo flights, and they've got more astronauts than they know what to do with.'

"But long story short, somebody asked. In fact, in some of my talks, especially with kids, I always remember Bo asked me that question, which I thought was dumb. It doesn't hurt to ask, even if you think you know the answer. It really doesn't."

And, seemingly against the odds, the answer was "Yes." The request for the MOL astronauts to be accepted into NASA's astronaut corps made its way to Office of Manned Space Flight associate administrator George Mueller, who was near the end of his tenure with the agency. The cancellation of the Manned Orbiting Laboratory marked the end of a period during which Congress had essentially forced NASA and the Air Force to compete with each other. Now NASA was beginning to make plans for its next crewed spacecraft, the Space Shuttle. Mueller hoped to enlist the Air Force as an ally as it lobbied to make the Space Shuttle a reality. Although NASA already had more astronauts than it needed, Mueller believed it would be in the agency's best interest to try to curry favor with the Air Force by accepting its erstwhile future spacemen into the NASA corps.

Director of Flight Crew Operations Deke Slayton, however, was unwilling to accept the entire group of Air Force astronauts into his already crowded corps. He invoked NASA's requirement at the time that only candidates under the age of thirty-six be accepted, cutting the applicant field roughly in half. Seven MOL astronauts were accepted into NASA's corps as the seventh group of astronauts on 14 August 1969: Maj. Karol "Bo" Bobko, Lt. Cdr. Robert Crippen, Maj. C. Gordon Fullerton, Maj. Henry "Hank" Hartsfield Jr., Maj. Robert Overmyer, Maj. Donald Peterson, and Lt. Cdr. Richard Truly.

Even after NASA hired them, things weren't settled for the former Manned Orbiting Laboratory corps. "We were fired twice the first year we were here," Bobko explained. "They came and said, 'You guys are fired. You're going to have to leave.' It wasn't any joke; they were really serious. I don't know if they called us in all at once or one at a time, but they told us we were fired. Twice." However, each time the astronauts' superiors in Houston gave the orders for the Group 7 astronauts to leave, their superiors' superiors at NASA headquarters gave the orders for them to stay.

"At the time, I think, both Deke and Al were worried about the cancellation of flights," Crippen said. "In fact, Deke was honest when he finally hired us the first time before the firings. He said, 'I don't have any flights for you until the Space Shuttle flies, and it's not even an approved program.' He said that'll probably be around 1980 at the earliest, but he added, 'I've got lots of work you can do.'"

Even though they were allowed to stay, the newest members of the corps sometimes felt like they were second-class additions. "I mean, we were not particularly loved and watered," Bobko said. "When I got here, I was the last guy to ever study the Apollo. I'd go and say, 'Can I get some manuals?' And they'd say, 'Yeah, but they're all out of date.' 'What about classes?' 'No, those have been all canceled.' Now it wasn't that bad, because I'd go over to the simulator, and nobody cared about the simulator. So I'd be over there myself, and they'd let me stay almost as long as I wanted.

"There was a time I felt like I was a cosine wave in a sine-wave world," he said. "We got on board, and they canceled [flights] before we got here; but after we got here, they canceled a lot more. There was supposed to be more than one Skylab, and I don't think it was until after we were told we were coming that they canceled the last two Apollos. And then the Shuttle was supposed to be ready a lot faster."

The ongoing cancellations were already having an effect on the corps when the MOL astronauts arrived. "People were bailing out," Bobko said. "Every crew meeting we went to, they talked about, 'Well, they've canceled another thing.' So the first year was pretty dismal, it really was."

If Crippen and Bobko felt underutilized during their first years at NASA, that was to change in June 1971, when they were selected for a mission—of a sort. "[Pete] Conrad called me into his office, and said 'OK, Crip, we've got this test that we want to run, and we want you and Bobko and [Bill]

10. (*From left*) Bo Bobko, Bill Thornton, and Bob Crippen.

Thornton working with it.' So I said, 'I learned never to volunteer, but it sounds like the best job available.'"

The third member of the group, Bill Thornton, had been selected to the corps on 11 August 1967 as part of the second group of scientist astronauts. Though his path to NASA differed from that of his two colleagues, he had much in common with them. Like Crippen and Bobko, Thornton had come to NASA from the Air Force, where he had been a flight surgeon, among other things. Also like the other two, Thornton had been involved in the Manned Orbiting Laboratory program before coming to NASA, though in a very different capacity. At the Aerospace Medical Division at Brooks Air Force Base, Thornton had been involved in research and development for projects for NASA and decided to submit an application during the second round of scientist astronaut selections.

His qualifications were very good. He didn't have the flight time Bo and Crip had, but he had over a thousand hours of testing (including flight testing) war weapons and missiles (during his first hitch in the USAF as a physicist) and then testing instruments designed for MOL as a flight surgeon. He was awarded a Legion of Merit for this work and accumulated over twenty patents. (Today, his total of over thirty-five patents includes everything from military weapons systems to the first real-time computer electrocardiogram analysis.)

The Skylab missions were intended to pave the way for the sort of long-duration spaceflights that would be needed to send humans beyond the moon and onward to other planets. For a trip to Mars to be possible, NASA would need experience with mission lengths far beyond the fourteen-day record that had been set during the Gemini program. Skylab would be the bridge between the two weeks that NASA had experience with to the months or years that would be needed to go to Mars. The plan was, with the first three Skylab flights, to quadruple the previous record, doubling it once with the twenty-eight-day first manned mission and then doubling that again with a fifty-six-day second mission. The plans called for the third crew to further demonstrate that a crew could successfully complete a mission of that length, rather than increasing the duration any more. (That plan changed, however, when the first two crews demonstrated just how well astronauts could function on long-duration missions, and the better part of month was added to the third crew's stay on Skylab.)

However, the unprecedented length of the missions would mean that unprecedented preparations would need to be made. Attention was focused in two areas of concern: whether human physiology could withstand such long-term exposure to microgravity and whether everything developed and planned would actually work as intended.

Regarding the former concern, in 1967 the President's Science Advisory Committee recommended an expansion of the Biosatellite program, which used animals to baseline the biomedical effects of spaceflight before longer-duration human missions were undertaken. The Biosatellite III mission was carried out in the summer of 1969, sending a monkey, Bonnie, into orbit in a small capsule for what was intended to be a thirty-day mission.

On the ninth day of the mission, controllers were forced to abort the mission and deorbit the capsule because of concerns about the monkey's health. The recovery team successfully recovered it, but Bonnie died hours later. Fortunately any negative side effects of Biosatellite III were minimal for Skylab. There was plenty of evidence that the monkey's death was not directly due to microgravity exposure.

The experiment had at least one positive result for Skylab. Due to the concerns about Bonnie's body mass loss, the microgravity mass-measurement device Bill Thornton had designed while with the Air Force became a high-priority payload for the workshop so that any body mass loss by the Skylab crews could be tracked in flight lest they suffer similar problems.

Crippen, Bobko, and Thornton were selected to participate in a more down-to-Earth and ultimately more meaningful preparation for the Skylab missions: the Skylab Medical Experiment Altitude Test, or SMEAT.

Rather than have the flight crews break away from their busy training schedule for full-length simulations, a surrogate crew was selected to complete a full-duration dry run of a Skylab mission. This SMEAT crew would test out various elements of the Skylab equipment and procedures in a series of trials, culminating in a full-scale simulation that was set at fifty-six days, at the time the longest planned duration of the Skylab missions and the length for which the second and third missions were scheduled.

The first part of the name came from the fact that trying out the medical experiments would be a major focus of the simulation, and the "altitude" referred to the fact it was conducted at the lower atmospheric pressure that would be used on Skylab.

In addition to the qualifying of the medical experiments, many other elements of the Skylab program were to be tried out during the program. The crew was to eat a diet according to the guidelines that had been planned for the Skylab astronauts. Even the interpersonal relationships of the crew sealed in the chamber for almost two months, both with one other and with those they dealt with on the outside, would be a learning tool for the upcoming orbital missions.

Thornton, in particular, was excited about the possibilities SMEAT presented to do some hands-on testing of the Skylab equipment. He had already volunteered his services to the Marshall Space Flight Center in 1967 to help with the design and testing of Skylab equipment. He was determined that it should work on orbit and had expressed dissatisfaction with several of the designs. To him SMEAT was an opportunity to complete development and to test the flight gear as only he could test it—as he put it, "with a forced injection of operational reality." His largest concerns going into the test were the urine collection and measuring system, the food system, and the bicycle ergometer.

The fifty-six days spent inside the altitude chamber would be only a fraction of the time that the three SMEAT crewmembers would devote to the test.

"It was about a year from the time we first started with all the planning and the engineering, and then the training and the preflight stuff, and then the actual test itself, and the writing reports," Bobko said.

The training for SMEAT was an intensive endeavor in and of itself. For example, though they were to be safely on Earth the whole time just a short distance from help, the SMEAT crew went through the same medical training as the Skylab members. Crippen said that the dental training, during which the astronauts learned to extract teeth, was a rather memorable experience. "We'd each done a tooth and done the deadening with the Novocain and all that kind of stuff," he said. "And they had this one kid that had a horrible looking mouth come in, and he needed to have a tooth out. They left Bo and I in there. The doctor said, 'You guys pull teeth.' We said, 'We've pulled one.' He said, 'Go.' He left, and I think I did the deadening, and Bo did the extraction."

Bobko said that the youth was nervous about having the extraction done and was anxious about having to have a shot before the tooth was pulled. "And so 'bedside manner Crippen' here whips around with this needle that's about that long," he said, holding his fingers several inches apart. "But we went through with it," Crippen said, "and he told us, 'You're the best dentists I've ever had.'"

In another memorable incident during the medical training, Crippen broke his hand learning CPR. During training at Sheppard Air Force Base in Wichita Falls, Texas, the SMEAT crewmembers were taught CPR techniques with a "Resuscitation Annie" training dummy. "Back in those days, they always had you whack the person on the chest before you started," Crippen said. "So I whacked the dummy." When he did, the trainers told him he needed to hit the patient much harder than that. "And I did, and I broke my fifth metacarpal! So don't have a heart attack around me."

The SMEAT crew also spent time before the chamber test participating in the engineering design for the simulation. They played an important role in determining how the facility would be configured for the test. Bill Thornton was a stickler for good engineering in the chamber itself. The fire detection and "deluge system" sprinklers for putting out fires were of particular concern to Bill, who had been at Brooks Air Force Base when a serious chamber fire had taken place. The deluge system was tested successfully, but he followed up by tracing the power system to its source, supposedly a bank of specially designed, long-life, high-reliability lead-acid batteries. But these batteries were corroded, and some had been replaced by ordinary automobile batteries. "He raised hell, and the batteries were replaced—with other automobile batteries," Joe Kerwin recalls. "He raised hell again, and eventually the correct batteries were obtained."

The tests took place in a vacuum chamber used to simulate atmospheric pressure at various altitudes, from ground-level value of 14.7 pounds per square inch (psi) down to a space vacuum. For the full-duration test, the pressure would be held at 5 pounds per square inch, which would replicate the atmosphere that would be present on Skylab (5 psi, with 70 percent of it oxygen). The cylindrical chamber had a twenty-foot diameter and was twenty feet high, which allowed for it to be configured with a Skylab-esque two floors. The chamber was outfitted with equipment to replicate the Skylab layout closely enough for a meaningful simulation, though it was far from an exact copy. Bunks in SMEAT, for example, obviously had to be placed parallel to the floor rather than perpendicular as on Skylab. The chamber was outfitted with the medical experiments that were to be flown on Skylab, including the vestibular-adaptation-testing rotating chair, the lower body negative pressure device, the bicycle ergometer, and the body mass measurement device. The SMEAT crew was to use the same toilet facilities as were on Skylab ("Except ours wasn't on the wall," Bobko noted), and their waste output was to be measured as it would be on orbit.

"We had a second deck on the thing, and then we divided up the first deck into compartments," Crippen said. "We had the one sleep compartment where Bo and I had a bunk, and another compartment for Bill, and we had a head compartment, and we had one where all the medical experiments were set up. It was similar but not exactly like the living deck on Skylab. It was comfortable living."

The two bunk rooms were outside the main cylindrical area in a rectangular extension that led to the main airlock. The waste-management compartment was an area partitioned off on the first floor of the cylindrical area. The large open volume of the main area housed the medical experiment equipment as well as the SMEAT equivalent of the Skylab wardroom, a food storage and preparation area with a table. The main room also featured a small access hatch through which items could be passed to or from the outside world. This small airlock was about the only compromise made in SMEAT that was not available on orbit. The second level featured desks at which the three astronauts could work (an additional desk was located on the first level).

Before the full-length fifty-six-day run, the crew conducted shorter tests in the chamber to work out any problems before committing to being sealed in for the full duration. After a "paper simulation" in which the crewmembers went into the chamber and talked through a day's activities, two run-throughs of two and three days were conducted. As with the full-length test, the shorter runs required that the crew go through the process of preparing to enter the lower-pressure environment in the chamber. "We ran a large number of tests where we'd only go in the chamber for a day or so and run these things to wring it out before we actually got in for the long duration," Crippen said. "Otherwise we'd never [have been] able to do it.

"I remember one case where there was this one tech that worked in Building 7," he said. "He was normally one of their chamber guys that were trained to operate the chamber. He and I were doing one run one day. They'd always prebreathe you [require you breathe 100 percent oxygen for three hours to eliminate nitrogen from your tissues and thereby prevent the bends] before the pressure is reduced from sea level to 5 psi in the chamber. We were setting up in the prebreathe room, and only he and I were there, and he got up and took off his oxygen mask and made a phone call to his girlfriend. Sure enough, we got in there and he was on the bicycle, and I was overseeing him. And he started hurting, and they had to take him out and put him in the hyperbaric chamber, 'cause he almost 'bent' himself—well, he did get the bends."

Just as the actual Skylab crews did, the SMEAT crew received small tattoos on their bodies to mark where sensors went for the medical tests in order to ensure the sensors were placed consistently and thus increase the accuracy of the results. According to Bobko, "They came to me, and they said, 'We're going to tattoo you so you know where to put the electrodes.' And I said, 'OK, only after one of you guys shows us exactly how it's gonna look.'" He acquiesced after one of the doctors had the tattoos placed on himself. "He said, 'If I could figure out how to see [behind me], I would have put it on my ass.'" Well over thirty years later, SMEAT and Skylab crewmembers report that their tattoos are still visible.

As any good crew would, the SMEAT astronauts came up with an official crew patch for their mission. The patch, reflective of the crew's plum assignment, depicts Snoopy the beagle from Charles Schultz's *Peanuts* comic strip (a favorite icon of the astronaut office) with an aviator's cap, goggles, and scarf and a rope tied around his neck. Their original idea was to use Snoopy and "put a fishhook in his mouth." The crew contacted Schultz to see if he would be willing to draw Snoopy for their patch. He agreed, but with one change: Crippen said, "[H]e wouldn't put a fishhook, so he did the little noose-around-the-neck thing for us."

Another part of the SMEAT simulation that began before the crew actually entered the chamber for the fifty-six-day test was the premission diet. Just as the actual Skylab crews did, the SMEAT astronauts ate beforehand a diet similar to what they would eat during the mission in order to establish some baseline information with which the metabolic data collected during the mission would be compared. According to Bobko, the "preflight" and "postflight" diets the crew ate were not exactly the same as what they ate in the chamber during the test but were carefully selected to have the same mineral count and nutritional value. The crewmembers had two refrigerators brought to their homes before the chamber simulation: one stocked with the premission food that was all they were allowed to eat, the second was used for storage of waste output, which would be taken back to MSC for analysis. As things worked out, the astronauts got plenty of opportunity to enjoy the preflight diet; the planned twenty-one-day period during which they were supposed to eat it stretched to twenty-eight days when the start date for the test slipped by a week.

Crippen said that he'd certainly had his fill of the prescribed diet after eating the twenty-eight-day preflight diet, the fifty-six-day mission diet and then the postflight diet. "That got to be a pretty long time," he said. "I can remember after we got out that I wanted a hamburger something awful." (Other astronauts had similar experiences. After weeks of preflight diet and almost sixty days of Skylab meals, Owen Garriott made arrangements to have a chocolate milkshake waiting for him on the recovery aircraft carrier when he landed after his mission.)

"We used to give them a hard time about the food," Bobko said, "Like I'd ask them, 'What's your analysis technique?' and I never got an answer. We'd have a meeting, and they'd hem and haw around it, but they never gave it." The SMEAT crew's persistence in challenging the experimenters' dietary planning was to be a vital contribution during the actual test, which led to

11. The SMEAT mission patch.

an important change in the Skylab flight program. Thanks to Bill Thornton's persistence, a one-size-fits-all, relatively low approach to caloric intake planning was amended. "We tried very hard," he said. "I tried to get information from them; we'd say, 'How are you going to do this? We're going to be eating this three months or so; how are you going to do the analysis?' On the first day, obviously, we have outside influence, when does that wash out? You can't average it over the fifty-six days, that doesn't sound reasonable, etc. etc. And I never got an answer."

"I don't think they had an answer," Crippen agreed.

"There were a number of things like that we had questions on that nobody really knew," Bobko added.

Finally, on 26 July 1972, only ten months before the first crew would launch into space, the time came to enter the vacuum chamber. And so the fifty-six-day stay began, and the astronauts were faced with what seemed at the outset like one of the mission's biggest challenges—keeping occupied for fifty-six days. Apart from its terrestrial location, one of the main differences between SMEAT and Skylab was the lack of much of the science package that would make up much of the actual work in orbit. While the SMEAT crew conducted most of the Skylab biomedical experiments, they were obviously unable to conduct the astronomy experiments and Earth resources observations, which depended on Skylab's location in Earth orbit, or the materials science research and microgravity experimentation, which depended on its constant state of free fall. Thus, they were given the full duration of a Skylab mission, without all of the Skylab activities that would fill that duration in orbit. In addition they were unable to share some of the favorite free-time activities of the orbital crews—viewing the Earth and enjoying the wonders of weightlessness.

However, despite not having those Skylab activities to fill their time, the SMEAT crews managed to find ways to avoid becoming bored by their extended isolation. "I think we all worried about that ahead of time," Crippen said, "because it wouldn't be like the guys flying where you had the ATM and all that to do. We worked on trying to find stuff to do. They let us take things in. We built a model, or tried to build a model. We took Russian. We found enough activities where I think we were reasonably busy." (Notes Kerwin: "I have a memo from Crip, April 1971, to 'Skykingdom' [Conrad, et al] asking for things to do. We suggested bridge and ping-pong.")

"We kept up the pretense: 'OK, this is like a spaceflight,' and we communicated through Capcoms, and all that," Crippen said (Capcoms being short for "capsule communicators," the people in Mission Control assigned to talk to the astronauts on a flight). "They said 'We'd like to make it as much like Skylab as possible,' and we did that. We did things like only communicating during AOS schedule." In orbit, a spacecraft could only contact the ground when it was within range of a relay station on Earth, periods known as acquisition of signal (AOS). Using that schedule for SMEAT meant the crew had only the same limited opportunities to talk to Mission Control as the orbital Skylab crews would. A closed-circuit television was used for training classes, and each of the crewmembers was able to use it for two videoconferences with their families during the test.

As would be the case during the orbital program, the SMEAT crew took on some extra work to fill some of the time. Crippen set up regular debriefing sessions during the weekend to help organize the crew's efforts. Just as would be the case on Skylab, housekeeping also filled some of the crew's time. "They [once told] us that things coming out of there were stinking," Crippen said. "And we were very sensitive because it didn't smell bad to us. I can remember, especially after we got the complaint about things kind of smelling that were coming out of there, we'd take Neutrogena soap and rub it down and scrub things around, so we worked hard at trying to keep the place clean."

And then there were the phone calls. As another way to pass the time, Crippen insisted before entering the chamber to begin the test that it be outfitted to make phone calls to anywhere in the country. Bobko recalled the line being a wonderful luxury as his wife used the time that her husband was away to take a vacation through California, and he was able to keep in touch with her as she traveled.

Crippen had a slightly different experience when friend and fellow astronaut Dick Truly arranged a little joke to remind the confined commander just what he was missing out on in the real world. "I remember somewhere around Day 40-something, I got this call from Dick Truly," he said. "I got on the line, and there were two young ladies on the line, and it was the biggest sexy phone call I can remember. I almost came out the door right then."

As it worked out, the premission concern about staying occupied proved to be unfounded. Between their primary SMEAT tasks and the supplemental activities they had scheduled for themselves, the crew not only had no problems keeping occupied but found their schedule so full that they sometimes had to skip some of the supplementary activities they had planned. Work days, six days a week, began at 7:00 a.m. and continued until 9:00 p.m. with breaks for meals.

In addition to managing to keep occupied, the crew also maintained good relationships with one other despite being confined together in a limited space for almost two months. Bobko, though, noted that the question of how they got along after being "shut up" together is really somewhat misleading. "It wasn't something that was a shut-up thing," he said, "because we had worked with each other for damn near a year, for probably eight or nine months or something, before we ever got in there. So any of the crew dynamic had already been worked out. My feeling was that we each had our own little peculiarities, but we understood each other, and we knew what they were, and we accepted them, and we got along."

The same, he said, was true of all of the spaceflight crews of which he was

12. The SMEAT altitude chamber.

a member. By the time the beginning of the actual mission arrived, the crew had worked together in training for so long that the various personalities had already meshed into a team, and any initial problems had been overcome. "I had a woman on one of my flights, Rhea Seddon," Bobko said. "People would say, 'What do you think about taking a woman on your flight?' Well, hell, we'd trained with her for six or nine months. That had all been worked out; the dynamic had been established already."

Despite the eventual monotony that set in by the end of the "postflight" diet period after months of restricted choices, the astronauts said that the Skylab food provided during the fifty-six-day chamber run was not bad at all. "After we got the menus, I don't remember being unhappy with the food," Crippen said.

Bobko, who later went on to command Space Shuttle missions, said that the unique hardware specifications of Skylab were a boon to the program's menu options. Unlike later spacecraft, Skylab had facilities to store frozen food, and unlike previous NASA spacecraft and the Space Shuttle, Skylab did not use fuel cells for power generation.

"Compared to Shuttle, I think Skylab menus were a lot better," Bobko said. "They had the frozen steaks; they had ice cream; they had other frozen things. And, unfortunately, the Apollo having fuel cells, which made water, and the Shuttle having fuel cells, which made water, has kept their food all on a narrow track; they wanted it to be dehydrated to save weight on the Shuttle." Skylab had plenty of lift capability to launch nondehydrated food.

The biggest challenge was setting up the menus in such a way as to make sure that the demanding nutritional guidelines were all met. "The food system was a bit of a problem," Bobko said, "because they wanted us to balance our intake of proteins and minerals every day, which just made selection and consumption and everything else more difficult. That was the difficulty with the minerals and [calories]. Because if you took peanuts, if I remember right, it excluded a whole bunch of other selections because [they] had enough other things in [them] that it really restricted your choices."

As would be the case with Skylab, the crews set up menus for six days, and then cycled through those selections for the duration of the menu, eating the same meals every six days. "The six-day cycle, at least for me, was interesting," Bobko said. "The way certain activities repeated led to some unusual associations." For example, he said, part of his exercise schedule was on the same six-day cycle as the meals; so the same meal—spaghetti—was being prepared every time he did the exercise. "So, it's like, if you're exercising, you know you're going to have the spaghetti smell in the background."

A few of the food items developed for the program, however, were less appealing than some of the others. "I can still remember finding out that Silly Putty and the little pudding that they gave us were in cans that were exactly the same size and looked exactly the same," Bobko said. "So we tried to feed it to one of the experimenters before the test, but he didn't show up to the meeting."

Despite being designed to replicate the Skylab menu as closely as possible, the SMEAT menu did feature one perk that the orbital version did not. Once in every six-day cycle, the SMEAT astronauts were allowed to imbibe a serving of sherry. The original plan had been for the Skylab menus to include a wine selection in each rotation, and a tasting had even been held for the crews to select what they wanted to carry into orbit with them. Medical objections had been overcome, but serving wine on a government "ship" was too much of a break with precedent for the political sensitivities of 1972, and it was removed from the flight diet.

Fortunately for the SMEAT crew, however, by the time the decision was made to remove the sherry from the Skylab menus, the SMEAT menus had already been made out, and it was too late to go back through the process of completely rebalancing the various nutritional factors that would have to be changed if the sherry were removed. "We had it," Crippen said, "and we really looked forward to it."

A more significant disagreement over the menus, however, proved to be of great importance to the Skylab program. The intense scrutiny on diets was not just to make sure that the crews stayed healthy, it was also one of the major biomedical experiments. Since the crews would be setting new spaceflight duration records, scientists wanted to learn all they could about how the microgravity exposure affected their metabolisms. Their dietary intake would be closely monitored, as would their waste output and their body mass, in order to make sure there were no unknown issues that would be a limiting factor for future long-duration spaceflight.

In order to facilitate the close scrutiny of the astronauts' intake, the decision had been made to standardize the intake for all Skylab (and SMEAT) crewmembers so that all crewmembers would consume the same number of nutritional calories each day. One set of dietary guidelines would be established, and all of the astronauts would adhere to it, making it easier to keep up with exactly how much everyone was eating. The astronauts and the investigators had been negotiating the diet since 1969, and when SMEAT took place it had changed from a standard 2,400 calories apiece to a base of 2,000 calories worth of real food (containing all the protein, calcium, and phosphorous allowed) plus up to 800 additional "snack" calories.

Bill Thornton, however, no stranger to medical concerns himself, disagreed with the decision and took it upon himself to prove that the standardized diet was a bad idea before it could be implemented on orbit, where there would be no way to change it. The tall and muscular Thornton, one of the corps' most physically imposing members, believed that setting a uniform standard for all the crewmembers would be unhealthy, that each needed a nutritional plan custom tailored for his own body type and metabolism.

Bill decided to demonstrate the inadequacy of this diet ("2,000 calories plus sugar for a 207-pound man with less than ten-percent body fat") by consuming it as directed. Pretest he maintained his usual extensive exercise regimen. In the chamber he estimated the difference between his usual routine and his chamber activities and made up the difference with cycle ergometry. He gorged on sugar cookies and lemon drops to stay alive.

"Bill like to drove me nuts," Crippen said. "He didn't think the caloric intake they had assigned for the flights was adequate, and he was determined to try to prove that so that they would up it. Bill was exercising on the ergometer. And he exercised on the ergometer, and he exercised. It'd be in the middle of the night and he'd be in there peddling on the thing.

"He finally got to be almost like a skeleton. He got to where I was worried about him. I didn't know how much weight he lost, but it was significant. Somewhere around the thirty-day point, I finally called outside and said either he's coming out or you've got to send some food in. They boosted up what we ended up flying, and I thought it was around 2,500 calories a day. I got irritated at Bill a few times simply because I couldn't get him off the damn bicycle. I thought he was going to starve himself to death. He's a bulldog; but you know, he's a great guy, and that was the only thing that he and I had an issue on—he wouldn't get off that damn bicycle."

And, indeed, the diet was insufficient for Thornton to maintain his body mass. "I was under the impression that a loss of twenty-eight pounds, most of it upper-body muscle, would be enough to convince anyone," Thornton said. As it turned out, it wasn't enough to convince the principal investigator for the mineral balance experiment. Dr. Donald Whedon thought "He overexercised," and his coinvestigator, Dr. Leo Lutwak felt "He only lost body fat." But a lot of discussion resulted in extra food being stowed aboard Skylab for the flight crews. Specifically, the eight hundred calories of "snack" food was now allowed to contain significant amounts of protein, which put many more food items onto the snack list.

Just as the intake monitoring had issues that had to be worked out, so too did the output monitoring. A similar problem occurred in planning the urine-collection system as had with the nutrition-standard guidelines: the designers had taken a one-size-fits-all approach that while wonderful in theory proved to be less wonderful in practice.

The urine system was Thornton's biggest hardware concern. It had to collect and measure twenty-four hours of output efficiently and reliably with very small error, in weightlessness. The contractor had designed a two-chambered bag separated by a "hydrophilic" membrane to transfer the urine into the measurement chamber under enough pressure to activate a complex mechanical displacement indicator. It failed as soon as urine was used to test it instead of water.

"The urine collection burst on us," Bobko said. "They had gone, I guess, into hospitals and figured out what the urine output would be, and it was too low. So two things happened, and one is that if you got up to take a leak at night, you may fill this thing up, so halfway through your evacuation, you had to cross your legs, and you had to [change] the bag." The other thing that happened was that the bags occasionally became overfilled and burst.

Emergency meetings were held. A centrifuge was designed whose centrifugal force would generate enough pressure to transfer urine into new, filterless bags. Thornton was skeptical. He campaigned hard to get the system into SMEAT for test and was the only one of the three crewmen who used it.

There were multiple failures. Seven times the bag broke, usually near the end of a twenty-four hour cycle when it was nearly full. Thornton recalls, "I had only my dirty discarded underwear and a very limited amount of water and soap to sop up a couple of liters of urine into discarded bags and clean up the floor. Then I had to thrust my big hands into a maze of machined parts with sharp edges to dry them, lest they corrode and seize up. My hands looked like I had taken on a bobcat." Crip and Bo joined Bill in telling management it wouldn't fly. A meeting was scheduled, and the three of them collaborated in preparing a rather blunt demonstration of the seriousness of the problem.

"They had the overcans for food, the big cans," Crippen said. "I think it was Bill that was doing this, but we were all complicit. We took one of the big food cans and took a spring out of the tissue dispensers, turned one of the small food cans over and put it down in on the spring, and then took a urine-soaked rag and put it down in there, and sent it out there. So when they opened up, it popped out, to demonstrate that we had a problem in there." The result was a complete system redesign with Dick Truly in charge. Bill suspects to this day that the food can was never opened; Truly just believed his fellow astronauts.

In addition to urine, stool was also sent out to be measured and analyzed. According to Bo Bobko: "We didn't freeze-dry the feces; we didn't have the vacuum as was available in space. We put them in little cans and sent them out. We sent the urine out, but we did the sample first; I think it was thirty milliliters per day.

"Then there was Thornton. I can remember them going to Thornton, and saying, 'Bill, it's Friday noon, and you haven't given us a fecal sample, and we'd like to let all the people go home for the weekend.' And Bill would say, 'Just a minute.' So he turned around, and said, 'You were talking to the wrong end.'

"I can remember them giving us these little cups. I said, 'These little cups, you know—how about something like four times larger?' So they gave us something that looked like a mailing tube. I said, 'You dummies, give us something that looks like an ice cream half-gallon container or something, that we don't have a hard time hitting.' So they did. But there were probably a lot of little things like that flight crews never knew about or cared about.

"They were complaining to us that we weren't sending everything out. Like, they said we weren't sending out all the feces. We said, 'What are we doing with it? Storing it under the boards of the floor?' I remember that time Bill got on the phone with our surgeon, kind of an excitable guy. Bill asked for a private consultation. He got on the phone, and he was saying, 'I've been noticing some strange behavior.' The flight surgeon said, 'Oh, oh, tell us about it.' He said 'Well, you know, these people seem to be paranoid. It looks like we have some paranoid things,' and we have this and that. The flight surgeon was assuming it was us, and he was getting more and more excited. The flight surgeon finally said 'Who is this? Who is this?' And Bill said, 'It's the management.' You don't think of him as a funny person. But when you have things like talking to the flight surgeon about this deviant behavior, you thought about it and laughed about it for days."

In a similar vein, the crew noticed an unanticipated side effect of the lower atmospheric pressure in the altitude chamber: "There was a lot of flatulence," Crippen noted. "We tried to think maybe it was the diet, but I think it was just strictly the 5 psi. It was significant." Common sense supports the latter theory: At the 5 psi of the SMEAT chamber, any given mass of a gas would have three times the volume that it would under sea-level atmospheric pressure. (Skylab crewmembers confirmed that the same phenomenon occurred during orbital operations as well.) Recalls Bobko: "We had a timer, and we were counting. I don't remember how many times it was in a day, but it was a significant number."

The 5 psi atmosphere had more mundane effects as well. The lower pressure reduced the transmission of sound so that during the first few days the crewmembers frequently found themselves shouting, and became hoarse as a result. (On Skylab, an intercom system addressed this problem.) They also found that they were unable to whistle in the lower pressure atmosphere and that sneezes were milder.

The most important part of SMEAT, of course, was the work the crew did in testing out the equipment and procedures designed for Skylab, making sure that everything would function as planned by the time the first crew arrived in orbit. While the problems the SMEAT crew had with the urine collection system were inconvenient for them, to say the least, their inconvenience served a greater good—the consequences of the urine collection problems would have been much greater had they first been discovered in the microgravity environment of Skylab.

One of the most immediate tasks for the SMEAT crew was to begin taking the roughly delineated guidelines that had been developed for Skylab operations and turn them into the finely detailed procedures that would be needed for the astronauts in space. The efforts to refine the checklist were an ongoing process for the SMEAT crew, beginning long before the chamber test and continuing through the simulation.

"We did quite a bit of development on the checklist, because a lot of that was almost nonexistent when we started," Bobko said. "It was in bad shape. So we had to do something; we had to make it operational. A lot of this stuff just wasn't in an operational format." Much of the early checklist, he said, was too short and vague for use in spaceflight. "It was 'Don't do this.' It was all right for training, but it wasn't really good enough to use. So we really worked on that quite a bit. That was part of the engineering and training that took place at the beginning."

Working with the principal investigators for the medical experiments in developing the procedures, Bobko said, had an additional beneficial side effect. The opportunity to witness the SMEAT crew performing the experiments gave the investigators some idea of what they could expect in working with the Skylab crews during orbital operations.

As was the case with the urine collection system, the SMEAT astronauts' use of the Skylab hardware revealed problems with equipment destined for the orbital workshop. Their discoveries meant that the problems could be addressed before the equipment was launched into orbit, where the sort of flaws uncovered during SMEAT would have been devastating for the program.

Bill Thornton's dedicated use of the wheel-less bicycle ergometer, for example, did more than just reveal problems with the dietary guidelines imposed on the crew; it also contributed to breaking—and then fixing—the bicycle. (Though that problem was corrected, the ergometer would present other challenges during its use on orbit, though fortunately all of the crews were able to deal with it in situ.)

Thornton used the ergometer primarily to maintain his normal exercise level. But because he questioned its ruggedness and thought it had not been tested thoroughly, he wanted to put it to that test. He recalled, "After a reasonable break-in period I planned to take it to its 300 watt rating for an hour, but starvation was taking its toll, and I was relieved when it screeched to a stop at about forty-eight minutes. The airlock was used to exchange it for an old, indestructible model. A considerable time later the bike was returned, 'fixed' by restricting it to thirty minutes at 300 watts, now an ordeal with my continued malnutrition. But this time it took only twenty-nine minutes and thirty seconds to destroy the bearings. I shall never forget the look of disgust on Crip's face."

This time, after independent engineering analysis, a different shaft and bearings were installed. The flight unit, however, was further restricted to 250 watts. Fortunately, this proved enough for the mere mortals who flew. Bill still remembers feeling hurt by the subsequent efforts of MSFC management to separate him from his testing role. And he insists that he never actually used it when Crip and Bo were asleep—"maybe when they were watching TV, but not after lights out."

Here are some of the hardware redesigns accomplished as a result of the SMEAT tests:

In the lower body negative pressure device, a seal that was necessary for depressurizing the lower extremities developed a leak and had to be redesigned. In addition, the decision was made to carry a spare seal during the flight program.

The equipment used for measuring blood pressure was discovered to have been miscalibrated, causing it to produce inaccurately high results.

Several problems were discovered in the metabolic analyzer unit. Some

13. Bill Thornton riding the cycle ergometer.

of the measurements it took were found to be substantially higher than they should have been, and the oxygen consumption measurement of the device was discovered to be significantly greater in the 5 psi atmosphere in the chamber than at sea-level pressure. The unit was redesigned to provide accurate and consistent data for Skylab.

The electrode cement used for the vectorcardiogram test was found to cause skin irritation, and action was taken to prevent the situation from recurring on Skylab.

Coagulation problems in samples were discovered to result from the blood sampling techniques used in SMEAT, and additional anticoagulants were added.

The centrifuge used for blood separation was found to be prone to excessive vibration and had to be redesigned prior to the flight program.

Of course, not all the problems the crew experienced were the fault of the hardware. One of the less-coveted tasks for the SMEAT crew was wearing the electroencephalogram (EEG) cap that monitored sleep levels. Crippen initially had agreed to be the one to wear the cap but before the beginning of the chamber test discovered that the salve or jelly that had to be applied to wear the cap caused his head to break out in welts. When Crippen realized that he was going to have to pass on the EEG duties, Thornton volunteered to take it over. But he too was unable to wear the cap. The task was then passed on to Bobko, who with no one left to pass it on to was stuck with it. Though no longer the one who would be wearing the cap, Crippen was still the one trained in its operation and thus had the responsibility of changing the tapes on which the device's data was recorded. Unfortunately due to an error in changing the tape, the data wasn't recorded. No one realized, however, that there was a problem until after the test was over, and the experimenters went back to review the data. "So Bo went through [the test], and there was no damned data," Crippen said. "I guess we got some from the first tape." Even that experience presented new ideas for the Skylab program—for the orbital operations, some of the data was sent down in real time to prevent just that sort of problem.

The SMEAT experiences proved to be invaluable to the Skylab orbital program, and the three men were proud of their contributions to the success of Skylab. "The first mission would have been a lot more difficult for the medical experiments" without the lessons of SMEAT, Bobko said.

Crippen agreed that the breaking-in the SMEAT crew put the medical experiment equipment through on the ground was a key to how well things went when the equipment was used in orbit. "If we'd flown those without running them in some sort of operational situation, I think there would have been a problem," he said.

Thornton praised Richard Johnston, then director of Life Sciences, for initiating a daily logging and review system for medical data which resulted in good status monitoring during the flights—and eventually in a fine document capturing Skylab's medical achievements, "Biomedical Results of Skylab."

Finally, though, the time came to bring the test to a close. Crippen said he was never entirely sure if the test was going to run exactly the full fifty-six days for which the second two Skylab missions were planned. First, he said, he believed that the simulation might be brought to an early close and the crew released from the chamber. But then as the test drew nearer to its conclusion, he wasn't sure if the mission planners might not decide to extend it to continue the experiments.

Bobko said he also wondered whether the crew might have to spend additional time in the chamber. "Near the end, I can remember thinking we may not get everything done, because we just have a lot to do," he said. "Fifty-six days was the target, and they said, 'If anything goes wrong, we can take you out.' But there was a feeling of, we had a purpose, and we had to get to it done. I can remember having some concerns that we weren't going to get all the results that we really wanted to get. And it all turned out; I think that we did. And like I said, I'm not sure that if anybody said, 'You want to go for another fifty-six days,' I would have been ready for that."

Crippen agreed: "I know I wouldn't have. If it were one or two more days to get some stuff done we would have been able to do that."

The SMEAT crew would eventually get their chance to move up from simulated space missions to the real thing. As Slayton had predicted when the MOL astronauts were brought into the NASA corps, their chance to fly did not come until the Shuttle was ready to launch, twelve years after they were brought in. Even before that happened, though, after almost a decade of being on the lower rungs of the astronaut corps ladder, the MOL astronauts saw their situation change in 1978 with the selection of the eighth class of candidates, chosen specifically for the Shuttle program. "We were starting to get into Shuttle before I felt like I wasn't a new guy," Crippen said. By that time, many of the veteran astronauts from the earlier groups had left the corps, and the MOL class played a vital role in the early Shuttle program. Crippen said that he was told that some of the same managers who had opposed bringing in the Air Force astronauts originally went on to feel lucky to have them when the Shuttle began flying.

Ironically, and sadly, though each did get to fly and command Shuttle missions, Crippen and Bobko's careers as flight-status astronauts ended much as they began: waiting on a space launch from Vandenberg Air Force Base that was never to come.

The Air Force had modified Space Launch Complex 6 (SLC-6, pronounced "Slick Six") at Vandenberg Air Force Base in California, once destined to serve as the launch pad for the MOL program, for use with the Space Shuttle. Missions were planned for launch from the "new" Shuttle pad, and crewmembers were selected for those missions, including Crippen and Bobko.

"One of the sad things was, [in MOL] we were supposed to fly from Vandenberg on Slick Six, Space Launch Complex Number 6," Crippen said. "And sure enough, I was up to command 62-A, which was going to launch out of Vandenberg out of Slick Six."

Bobko recalled: "I can remember going out there during MOL and staying in the crew quarters. And then they redid the crew quarters out there to make them a command center. And then they changed it back to crew quarters. I can remember being out there the second time, same place, I don't know how many years later, when they were getting ready to do the Shuttle flights."

The 62-A mission that Crippen was assigned to command was to be the first Shuttle launch from the Vandenberg complex and would have been the first launch of a manned spaceflight into a polar orbit. The flight was scheduled for mid-1986, but it was never to be. On 28 January 1986, the Space Shuttle Challenger was lost during launch from Florida, destroying the vehicle and killing the seven members of its crew. As a result of that tragedy, the decision was made not to launch the Shuttle from Slick Six.

5. A Tour of Skylab

Perhaps the best way to begin a tour of Skylab is to begin where its crews did—on the outside, with a look at the station's exterior.

If a crew in an Apollo Command Module were to approach Skylab with its docking port before them, the nearest module would be the Multiple Docking Adapter (MDA). From the exterior, the MDA was basically a nondescript cylinder, marked primarily by its two docking ports. One of the docking ports, the one used by the crews docking with Skylab, was located on the end of the cylinder. The second, the radial docking port, was at a ninety-degree angle from the first, on the circumference of the MDA.

The other notable feature of the Multiple Docking Adapter was the truss structure that surrounded it and connected it to the Apollo Telescope Mount (ATM), on the side of Skylab opposite the radial docking port. The ATM is easily recognized by its four solar arrays, which had a very distinctive windmill appearance. Between the four rectangular arrays was a cylinder that housed the ATM's eight solar astronomy instruments. Covers over the instrument apertures rotated back and forth, revealing the instruments when they were in use and protecting them from possible contamination when they were not.

Continuing from MDA, the crew would next come to the Airlock Module (AM), a smaller cylinder partially tucked into the end of the exterior hull of the larger workshop cylinder. The Airlock Module was most notable, as the name suggests, for its airlock featuring an exterior door allowing the crew to egress to conduct spacewalks outside the station. While the program that spawned Skylab had been dubbed "Apollo Applications" for its extensive use of Apollo hardware and technology, the Airlock Module was actually a "Gemini Application"—the door used for EVAS was a Gemini spacecraft hatch.

The airlock and all the spacewalk equipment on Skylab were designed for one purpose—to allow the crew to retrieve and replace film from the solar telescope cameras on the Apollo Telescope Mount. "There was no thought of the crews doing repairs or maintenance on other things," Kerwin said. "Little did we know!"

The airlock was partly covered by the Fixed Airlock Shroud, a stout aluminum cylinder that was a forward extension of the skin of the workshop. The aft struts from which the Apollo Telescope Mount was suspended were mounted here. The truss structure included a path, complete with handholds that spacewalking astronauts could use to move from the airlock hatch to the ATM so that they could change out the film.

Finally moving farther past the Airlock Module, the crew would reach the largest segment of Skylab, the cylindrical Orbital Workshop. This was the portion that consisted of the modified S-IVB stage. As it was originally constructed, the most distinctive features of the station were the two solar array wings, which stretched out to either side and which were to be the primary source of electrical power for the workshop. Prior to launch the photovoltaic cells that made up the arrays folded up flat against the beam that would hold them out from the sides of the workshop. These beams, in turn, folded down against the outside of the S-IVB stage in its launch configuration, making the wings much more aerodynamic for the flight into orbit.

After completing their fly around, a crew would return to the top of the Multiple Docking Adapter and dock their spacecraft to the station. A complete tour of the interior of Skylab should begin right there on their capsule. After docking, the Command and Service Module became a part of the cluster. While there were occasions when things needed to be done in the Command Module, they were few. Perhaps its primary use while docked with Skylab was essentially as a telephone booth; crewmembers could float up to the Command Module to find a little privacy for conducting space-to-ground communications with their loved ones at home on a back-up frequency that was not available in the workshop.

Upon opening the hatch and entering Skylab, the crew would first find themselves inside the MDA. Originally planned to have a total of four docking ports around its circumference, the MDA lost three as a result of the switch from the wet workshop to the dry. When the wet workshop cluster, which had to be assembled individually on orbit, was replaced with a facility launched all at once as a dry workshop, the additional ports at which to dock separately launched modules were not needed. Eliminating the three

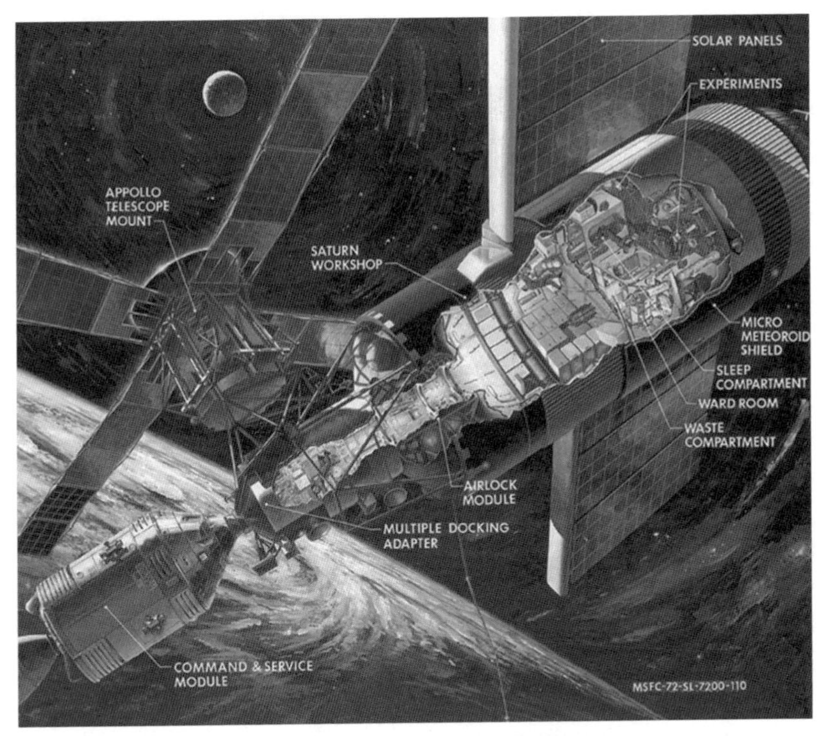

14. A cutaway view of the Skylab space station.

extra docking ports freed up a large amount of wall space around the MDA's circumference, space that was utilized to turn the module into essentially an additional science annex.

The design of the interior of the Multiple Docking Adapter was itself one of Skylab's experiments. The argument had been made that in the microgravity environment in orbit there was no need to follow the same design paradigms that were unavoidable on the ground. There was no need to leave a floor empty to walk on. The ceilings were no more out of reach than walls, and equipment could be placed on them just as easily as on a wall. The MDA was an experiment in designing for that environment, with no up or down. Equipment was located all the way around the wall of the cylinder, allowing more complete use of the available space than would be practical on Earth.

Foremost among the scientific equipment located in the module was the operator's station for the Apollo Telescope Mount, a large flat panel featuring the controls and displays for the ATM with a narrow table in front of it.

The ATM console was arguably evidence of the extent to which the module's designers were influenced by Earthbound thinking. Though care was taken to design the MDA as an ideal microgravity work environment, the ATM console was furnished with a chair for the astronauts to sit in while operating the controls. "We called it the 'Commander's Chair,' because it was Pete's idea," notes first crew science pilot Joe Kerwin. "It didn't survive longer than about the first two weeks of our mission; we then put it away somewhere, and I don't think anyone retrieved it."

Also located in the MDA was the Materials Processing Facility. Included in this experiment was a furnace used to study flammability and melting of solid materials in microgravity. The adapter also housed the Earth resources experiment equipment.

Leaving the Multiple Docking Adapter and heading farther down into Skylab, one would next come into the Airlock Module, the function of which was very aptly described by its name. Joining the MDA and the Airlock Module together was the Structural Transition Section, which connected the larger diameter of the Docking Adapter on one end to that of the smaller Airlock Module on the other. The Structural Transition Section housed extensive systems operation equipment. The Airlock Module provided a way for astronauts to egress the station for spacewalks. Before they could go outside, the Airlock Module would have to be shut off from the rest of the station and then depressurized. Once the atmosphere had been removed, the airlock hatch could be opened, and the EVA crewmembers could go outside.

To prepare for an EVA, all three crewmembers would put on their spacesuits in the larger open area of the Orbital Workshop, where the equipment was stored. The astronaut who would be staying inside stopped short of donning his helmet and gloves but suited up the rest of the way in case a problem occurred. The EVA umbilicals were stored in the Airlock Module, and the ends of these were pulled down into the workshop during this time and connected to the suits of the two EVA crewmen. These provided oxygen, cooling, and communications for the two astronauts who would be going outside as well as tethering them to the station.

Once all three were suited up, the non-EVA crewman would precede the others, move through the airlock and into the MDA/Structural Transition Section. There he would attach himself to a shorter umbilical. With his helmet off, he would be breathing the atmosphere in the MDA, but in the bulky spacesuit, he needed the umbilical for cooling as well as for communications. The EVA crewmen would move to the airlock and close both hatches (helped on the MDA side by the third crewmember). Once the hatches were closed, the Airlock Module would be depressurized by venting its atmosphere into space. The outside hatch would be opened, and the two spacewalkers could venture outside.

Once the EVA was completed, the two astronauts would return to the Airlock Module and close the outside hatch. The AM would be repressurized, and they would open equalization valves in both end hatches to assure equal pressure with the rest of the station. Finally, they'd open both hatches, return to the workshop and doff their suits. The normal pressure regulation system would add gas to the workshop as needed.

The Airlock Module's location in the middle of Skylab meant that a problem with repressurization could mean the end of the mission. If for some reason the module were unable to hold an atmosphere, the third crewman would put on his helmet and gloves and depressurize the Multiple Docking Adapter. The other two would disconnect their umbilicals from the Airlock Module and rely on a reserve oxygen supply in their suits while they opened the hatch between the two modules, and moved into the docking adapter. Once there, they would reconnect their umbilicals in the MDA, and then seal it off from the Airlock Module and repressurize it. If they and the ground were then unable to figure out a way to fix the problem with the Airlock Module, the mission would be aborted. They would leave the MDA for the Command Module and return home.

Continuing deeper into the station, one would next reach the large Orbital Workshop volume. This section was divided into two "stories," with a hole in the middle of the floor of the top story that allowed the crew to move between them.

Like the Multiple Docking Adapter, the workshop was part of the experiment in designing for microgravity. Whereas the MDA was designed without consideration for the direction of gravitational force on the ground, the approach to the workshop design had been to keep in mind that it would be used by men whose brains had long been wired for the one-G environment in which they had lived their entire lives. The "bottom" story of the workshop was arranged with a very definite up and down. Furnishings and large equipment sat on the floor like they would on Earth (with a few exceptions), and the walls functioned more or less the way walls normally do. The upper compartment was more of a hybrid, with variations from the one-G–based design of the lower section.

The area at the top of the workshop was very unusual by spacecraft standards. Traditionally spacecraft design is a field in which mass, and by extension volume, are at a premium, reflecting the challenge of moving anything from the surface of the Earth into orbit. As a result spacecraft tend to be relatively cramped with every inch utilized as much as possible. While modern spacecraft like the Space Shuttle and the International Space Station are roomy compared to early vehicles like the Mercury and Gemini capsules, their designs still reflect the basic limitations in putting any mass into orbit. Skylab had a couple of advantages that made it exceptional in that respect. The availability of the Saturn V as the launch vehicle and the decision to use an S-IVB for the Orbital Workshop meant that it was much less constrained by the traditional mass and volume limitations. Nowhere was that more apparent than at the top of the workshop, which featured an open volume that by spacecraft standards was incredibly large. While the lower floor was divided into separate "rooms," the upper floor, the larger of the two, was not divided. An astronaut could float freely in the middle of this volume without bumping into the walls.

In fact Skylab's designers were concerned that this could present a real problem. They feared that an astronaut could get stranded in the middle of this open volume; without anything nearby to push off, he would have to rely on air currents or his crewmates to push him back toward a solid surface. To eliminate this danger and to provide for easier movement through Skylab, they provided a "fireman's pole" in the middle of the workshop, running from one end to the other. The idea was that the astronauts would hold on to the pole to move "up" and "down" the workshop. The pole, however, proved unnecessary, and the crews found that it just got in the way. It turned out to be quite easy to push off from a surface and glide to one's destination—no pole required. The first crew took it down for the duration of their stay, but at the end, politely restored Skylab to factory specs, reinstalling the pole for the second crew. They in turn did the same—promptly taking it out of their way but putting it back before they left so that the third crew could remove it one last time.

The upper portion of the workshop dome volume was left almost vacant for experiments requiring a lot of volume for checkout, like a Manned Maneuvering Unit prototype. Just below this was a ring of white storage lockers, which the first crew found provided an excellent "track" to enable easy shirt-sleeve jogging and tumbling around the inside circumference of the workshop. Also located in the upper deck were storage of food supplies for all three missions, a refrigerator and a very heavy (on Earth, at least) steel vault for film storage.

A few experiments were also located in this area, including Skylab's equivalent of bathroom scales, the body mass measurement device, which the astronauts used to keep track of how much "weight" they had lost or gained. The upper dome volume was also where the two astronaut maneuvering units were kept. One was a backpack device that was the forerunner of the Manned Maneuvering Unit later used on some Space Shuttle missions and of SAFER, the Simplified Aid for EVA Rescue, used on the International Space Station. (Ironically, a member of the one Skylab crew that did not get to test the maneuvering unit, Joe Kerwin, was a co-inventor of the SAFER unit, while working at Lockheed Martin years later.) The other device was a maneuvering aid that astronauts operated with their feet, rather than their hands.

The upper story of the workshop also featured a pair of airlocks. Too small for a person to go through—only about ten inches square—the two Scientific Airlocks (SALs) were designed for solar physics, astronomy, Earth photography, and space exposure experiments, allowing astronauts to pass materials samples through to see how they weathered the harsh environs outside. The two airlocks were on opposite sides of the compartment from each other; a solar airlock pointed in the same direction as the Apollo Telescope Mount, while the antisolar SAL faced in the opposite direction. (This solar-looking airlock would be an important part of addressing problems that occurred during launch.)

Also located at the top of the dome was Skylab's unofficial "Lost and Found." "Most of us have enough trouble keeping up with our pencils, notes, paper clips, and other small items here on Earth in a largely 'two dimensional' world," Garriott explained. "By two dimensions, we mean that an object may get pushed around horizontally, but it seldom floats away vertically in a third dimension, like a feather might do. But space is different—everything floats away unless it is tethered or tied down. But our eyes and our minds have been trained for years to look only on the tops of surfaces to find lost articles. We may not 'see' a small floating object in space, or may not look in all the more obscure places a lost article may have become lodged.

"But serendipity came to the rescue here," he said. "The very slow air circulation from the lower decks up to the single air intake duct in the top of the dome volume slowly urged all drifting objects to come to it. We found that each morning when we arose, we could find many of our small, lost articles on the screen on the intake duct!"

At the bottom of each of the workshop's two "stories" were floors with an open-grid construction that was a fortunate relic of Skylab's development. During the wet-workshop phase of Skylab's history, engineers looked at whether any of the station's infrastructure could be included in the S-IVB stage while it was being used as a fuel tank up to, and during, the launch. Anything that could be built into the tank would mean mass that would not have to be carried up later, and installation work that the crew would be spared. The catch of course was that it would also have to be something that could withstand the environment of an S-IVB filled with cryogenic propellants, that it could not pose a risk of igniting the propellants, and that it must not interfere with the function of the rocket stage. One item that the engineers decided they could include was the floors of the workshop. However, solid floors could not be used, since they would impede the flow of fuel through the tank. As a result, special floors were designed with a grid pattern that would allow fuel to flow through them.

When the switch was made from the wet workshop to the dry, the grid-pattern floors were no longer needed for their original purpose. However, the design was kept for the dry workshop because it was realized that the grid could serve another purpose as well, solving one of the challenges of life in microgravity. The Skylab astronauts were given special sneakers that had triangular fittings attached to their soles. These pieces would fit into the triangles that made up the floor's grid pattern and lock in place with a small rotation of one's foot. This allowed the crewmembers to stand in place on the floor without the help of gravity.

Finally, one would reach the farthest point from the Command Module, the bottom "story" of the Orbital Workshop. This was the primary living area of the space station and included its bedrooms, bathroom, kitchen, and gym. This area was divided into four major areas: the sleep compartments, the waste-management compartment, the wardroom, and the experiment volume.

Skylab had three sleep compartments, one for each of the astronauts aboard at any time. To make the most of the available space, the beds were arranged vertically in the quarters. Without gravity to keep a sleeper in place, the beds were essentially sleeping bags with extra slits and a vent to make them more comfortable. These were mounted on an aluminum frame with a firm sheet of plastic stretched within it to serve as a "mattress." A privacy curtain took the place of a door at the entrance to each "bedroom." Also in each sleep compartment were storage lockers, in which crewmembers could keep their personal items, and an intercom for communications.

The intercoms in the sleep quarters were among several located around the station, which served a dual purpose—they allowed communication both with the ground and throughout the station. Because of the low air pressure on Skylab, sound did not carry far, which could make it difficult to be heard in other parts of the station.

Voice communications with the ground were carried out in two major ways. The primary means of communication was the A Channel, which was used for real-time conversations with Mission Control. The other was B Channel, which was recorded on an onboard tape recorder and periodically "dumped" to the ground and transcribed. This allowed the astronauts to pass along their thoughts about such things as habitability issues on Skylab, things that were not urgent but were needed for future reference. The crews were given questionnaires about aspects of life aboard the station and would dictate their answers into the intercom on B Channel.

For Project Mercury, NASA had to quickly develop a worldwide satellite-tracking network so that voice communications, data from spacecraft systems, and commands from the ground could be sent and received. Stations were placed in exotic locations such as Zanzibar and Kano, Nigeria—often with help from the State Department—and were staffed by small teams of NASA employees and contractors. There was no real-time communication between Mission Control and most of these stations; data was relayed via leased commercial phone lines, undersea cables, and radios.

Capability of the system was continuously upgraded during the Gemini program. By the time Apollo 7 flew in late 1968, satellite relay of voice and data permitted Houston to communicate directly with the spacecraft; the remote-site teams were called home, and a unique travel experience disappeared. But communication was still only via the transmitters and receivers at the tracking stations.

The system inherited by Skylab was called the "Spacecraft Tracking and Data System." It consisted of twelve stations: Bermuda, Grand Canary Island, Ascension Island, St. Johns (Newfoundland), Madrid, Carnarvon and Honeysuckle Creek (Australia), Guam, Hawaii, Goldstone (California), Corpus Christi (Texas), Merritt Island (Florida), plus the ship Vanguard off the east coast of South America, and sometimes an aircraft (call sign ARIA) used to fill gaps during launch and reentry. As a result, communication between Skylab and Houston took place only in the brief passes over these stations, often interspersed by an hour or more of silence. The crew could tell where they were around the world by Houston's calls—"Skylab, Houston, with you at Guam for eight minutes."

To the left of the sleep compartments was the waste-management compartment. This room featured a water dispenser that was the microgravity equivalent of a sink, a mirror for personal hygiene, and, of course, the space toilet. The Skylab mission required a level of innovation in this area not achieved in previous spaceflights. While the bag-based system used on previous spaceflights for defecation had not been particularly pleasant, there was not really room on the smaller vehicles for a better means of dealing with the issue. For the comparatively short durations of those missions, it was something that astronauts simply had to bear.

Skylab, however, involved both a long-enough duration to merit finding a better solution as well as the space needed to provide one. For urination, the crewman stood in front of the collection facility with his feet beneath straps to hold himself in place. He urinated directly into a funnel with modest airflow drawing urine into individual collection bags, one for each crewman. For defecation, he rotated about 180 degrees and seated himself on a small chair on the wall, rather like a child's potty chair. But here a plastic bag had been placed beneath the seat for each use, which maintained a simple and hygienic "interface" with the astronaut. A lap belt and handholds were provided to allow the user to stay in one place. As with the urine system, airflow took on some of the role that gravity would play on Earth. An innovative feature of the fecal collection system allowed these bags to be placed in a heating unit after mass measurement, then exposed to the vacuum, which dried their contents completely. It was then much lighter and quite hygienic. The dried feces and samples of the urine were saved and returned to Earth for post-mission analysis.

To the left of the waste-management compartment was the wardroom, the station's combination kitchen, dining, and meeting room. (Explained Kerwin: "Why was it called the wardroom? Because the first crew was all-Navy, and they got to name stuff. The wardroom is the officers' dining and meeting room in a Navy ship.") In the center of the room was Skylab's high-tech kitchen table. Its round center was surrounded by three leaves, one for each crewmember. The flat surface of each of the leaves was actually a lid, which could be released with the push of a button. Underneath the lids were six holes in which food containers could be placed, three of which could be heated to warm food. The trays had magnets for holding utensils in place. The table also featured water dispensers, which could provide diners with both hot and cold water. Both thigh constraints and foot loops on the deck provided means for the astronauts to keep themselves in place while eating.

The walls of the wardroom were lined with stowage lockers and with a small refrigerator-freezer for food storage. The wardroom was one of the most popular places on Skylab for spending time—partially because it had the largest window on Skylab, which could be used for Earth- or star-gazing.

The largest portion of the bottom floor was the experiment area, which was home to several of the major medical experiments. The Lower Body Negative Pressure experiment was a cylindrical device, which an astronaut would enter, legs first, until the lower half of his body was inside. After a pressure seal was made around his waist, suction would then decrease the pressure against his lower body relative to the atmospheric pressure around his upper torso. The pressure difference would cause more blood to pool in his lower extremities, simulating the conditions he would experience when he returned to Earth and gravity caused a similar effect.

Also in the experiment volume was the ergometer, essentially a wheelless exercise bicycle modified for use in microgravity. Like its Earthbound equivalents, the ergometer featured pedals, a seat, and handlebars, but it was also equipped with electronics equipment for biomedical monitoring.

The Metabolic Analyzer was used with the ergometer to monitor the crew's respiration. The device itself was a rectangular box with a hose connected to a mouthpiece. The user would put on a nose clip and then breathe in and out through the mouthpiece. The analyzer could not only measure respiration rate and breath volume but also, via a mass spectrometer, the composition of the air he exhaled and thus oxygen consumption and carbon dioxide production.

Another experiment in that area of the workshop was the Human Vestibular Function device, which was basically a rotating chair. With an astronaut sitting in it, the chair could be rotated about the axis of the subject's spine at speeds up to thirty revolutions per minute, either clockwise or counterclockwise. The purpose of the experiment was to test how their vestibular systems (responsible for balance and detection of rotation and gravity) adapted to the microgravity environment. The experiment had been performed with the astronauts on the ground to provide a baseline and was performed again in orbit for comparative results.

Another major item located in the experiment room was only an experiment in the broadest sense—that life on Skylab was all part of research into long-duration spaceflight habitability factors. Because of the way the lower deck was divided and because the shower was a later addition to the station's equipment, the shower was instead located in the larger, open experiment area instead of being located in the waste-management facility, which in other respects was Skylab's bathroom.

Water posed a potential hazard in Skylab. In weightlessness water would coalesce into spheres, which could float around the spacecraft. If they weren't collected, they presented the risk that they could get into electronic devices or other equipment and cause damage. Small amounts of water could be easily managed, but large amounts were generally avoided in spaceflight. To wash their hands, for example, astronauts would squirt water into a cloth and then clean their hands with it rather than putting the water directly on their hands.

The shower provided means for a true spaceflight luxury. In it, astronauts could clean themselves in a manner that, while not quite the same as the way they would shower on Earth, was much closer. They would pull a cylindrical curtain up around themselves and then squirt warm water directly on their bodies using a handheld spray nozzle. Confined within the curtain, the water posed no risk to the spacecraft and after the shower could be cleaned up with towels or a suction device. The crews found the suction it provided inadequate for drying off completely and so used lots of towels. Nevertheless,

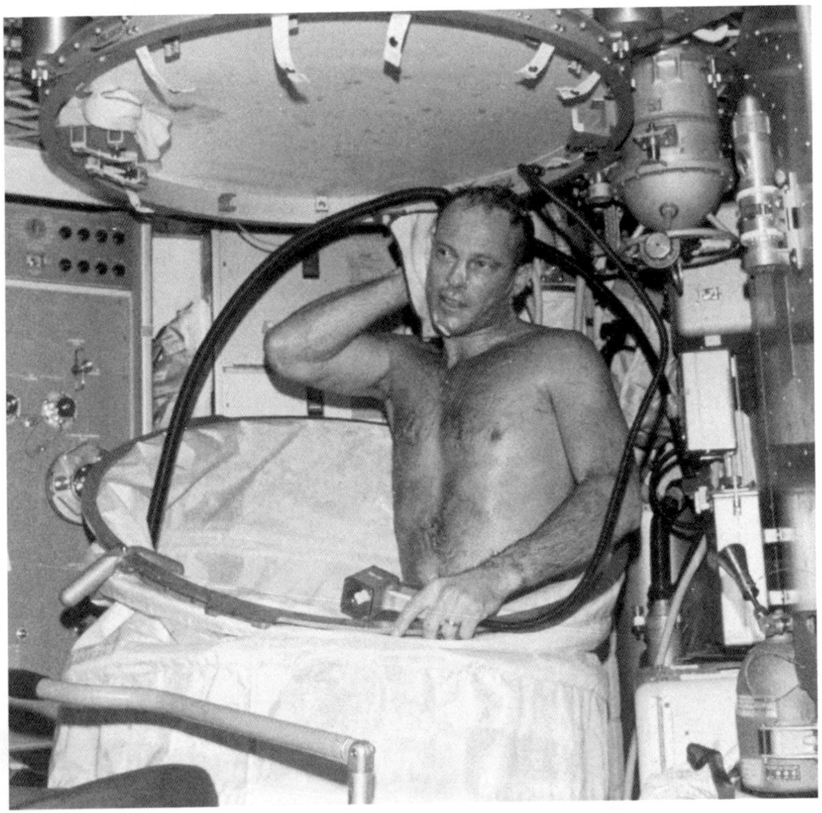

15. Lousma demonstrates Skylab's shower.

at least one crewmember thought this "luxury" was both unnecessary and a gross waste of time.

At the center of this lowest floor of Skylab, the very opposite point from where the tour began, was the Trash Airlock. The S-IVB stage from which Skylab was modified had two tanks that originally would have been used to store the propellant: a larger tank for the fuel, liquid hydrogen, and a smaller tank for the oxidizer, liquid oxygen. The entire manned volume of the workshop was inside the stage's liquid hydrogen tank. The liquid oxygen tank, which was exposed to vacuum, was used for trash storage. Between the two was an airlock that was used to transfer trash into the storage area. The oxygen tank was vented to space, creating a vacuum that helped pull the trash through, but it had a screen to prevent any trash from escaping. The arrangement meant that the waste generated on Skylab was stored safely instead of becoming orbital debris.

6. Ten Days in May

It's interesting when you stop and think about it,
how you get thrust very unexpectedly into an environment that
a few years later is the apex of your career.
You were doing stuff that had never been done before,
and you were successful.

Jim Splawn

"Eighteen days before launch—let's see, that would be April 26—the Skylab 2 crew entered quarantine and started eating our carefully measured flight-type diets," Joe Kerwin recalled. "That meant saying goodbye to our wives and families and moving into a couple of trailers on JSC property. Yes, we missed our families, but the arrangement was efficient, and we were in peak concentration mode. Nobody could come near us without a brief physical exam and a surgical mask. Launch readiness was everything. One of us recalled Coach Vince Lombardi's famous and often misinterpreted quote about football: 'Winning isn't everything. It's the only thing.' He did NOT mean that football was more important than God, country, or family, just that on Sunday afternoon, you should not be thinking of that other stuff. That's where we were. It was a good, team feeling; I remember all six of us [the prime crew of Conrad, Kerwin, and Weitz and their backups, Rusty Schweikart, Story Musgrave, and Bruce McCandless] standing outside the trailer each evening at bedtime, joking as we filled our urine specimen bottles for science."

By launch morning the uncertainties of the Skylab program had practically vanished. All the battles that had been fought to improve the hardware and procedures were over. Chris Kraft, JSC director, had called the first crew in to his office about a month earlier to tell them to knock off trying to change the medical experiments and start working to accomplish them; and Commander Pete Conrad had been able to say that they were already there. Training was over. They felt confident of their abilities and trusted the team. The crew and NASA were ready to fly and had no premonition of disaster. "You could say Fate had us right where she wanted us," Kerwin said.

With only one day left on the countdown for their own launch, the crew watched from the roof of the Manned Spaceflight Operations Building the Skylab station launch on a beautiful May morning. The Saturn V rose slowly and majestically from the pad and disappeared into the eastern sky. The launch looked good. They went back down the stairs to the crew quarters conference room, where the flight director voice loop from Houston's Mission Control was set up so that they could listen to the activation activities as they lunched and did a last-minute review of the next day—their launch day.

According to Owen Garriott, "May 14, 1973, was a beautiful day at the Cape. All of the three planned Skylab crews, along with tens of thousands of space enthusiasts, were in attendance to watch what would be the final launch of NASA's most powerful launch vehicle. It all appeared from the ground to go perfectly with a long, smoky trail headed into the blue sky.

"Jack [Lousma] and I headed back to our usual motel—Holiday Inn, Cocoa Beach—to change into flight suits and head for Patrick Air Force Base where our NASA T-38 was ready for a quick flight back to Ellington Air Force Base, near the Manned Spacecraft Center. We wanted to be home as soon as possible to observe the Skylab telemetry and verify that our home to be was in good shape for human visitation and also to watch the launch of Pete, Joe, and Paul scheduled for the next morning from Mission Control.

"As we were walking out to our rental car for the short drive to Patrick, we noticed the recently appointed director of the Marshall Space Flight Center, Dr. Rocco Petrone, walking along the porch in front of his second story room. 'Looked like a great launch,' we shouted. 'Yes, but don't get your expectations too high. There were some telemetry glitches observed.'

"With no more time for discussion, we headed for our aircraft, hoping that the telemetry issues would be resolved by the time of our arrival in Houston. We could only speculate about what these 'glitches' were, never imagining

16. Skylab was launched as the third stage of a modified Saturn V rocket.

the problems to be encountered and then solved in the next ten days."

That apparently beautiful launch was when things started to fall apart. The problems didn't surface all at once. There had been a "G spike"—a sudden, brief shock to the vehicle—about forty-five seconds after launch as the Saturn booster was accelerating through the speed of sound. Around a minute after liftoff, and very near the time of "max-Q" when the atmospheric pressure on the speeding vehicle is at the maximum, telemetry received in Houston indicated the micrometeoroid shield had deployed prematurely, an anomaly not fully appreciated as the Saturn kept going and deposited Skylab in the correct orbit.

Once the S-IVB stage was in orbit, a planned sequence began to reconfigure it from its launch mode to its operational arrangement. The first deployment action was to jettison the "SLA Panels," four large panels that protected the docking adapter and ATM during launch. That action went well. The next sign that something was wrong began to surface as workshop temperatures started to climb above normal, but the full extent of the problem was still not apparent, and the deployment appeared to be going well. Mission Control proceeded with the next scheduled step: rotating the ATM ninety degrees to face the sun and opening its four solar panels into their "windmill" configuration. That too went smoothly.

Then it was the turn of the Solar Array System, the main solar panels on the workshop itself. First the Solar Array System beams, the solar panel housings, would be deployed to ninety-degree angles from the workshop; then the panels themselves would unfold accordion-style. Fully open they would provide two-thirds of Skylab's electrical power. And after they were open, the thermal/meteoroid shield over which they'd been folded (the "heat shield") could itself be popped up away from the workshop's exterior surface to assume its function of reflecting the sun's heat and breaking up small meteoroids. The beams did not deploy. And the workshop surface and internal temperatures continued to climb. Something was wrong with the heat shield, which should have kept the workshop cool even before deployment. Mission Control went into troubleshooting mode.

By late afternoon they finally realized the truth: the heat shield was gone. That G spike during launch had been the shield departing the vehicle. The anomalous telemetry had been completely accurate. Further there was no response at all from Solar Panel 2, indicating that it was probably gone as well. Solar Panel 1 was showing just a trickle of current, leading controllers to believe it was still present but stuck shut. NASA was soliciting high-resolution photography from other satellites and ground-based telescopes.

While the full import of what had taken place would take some time to pin down exactly, it became very apparent very quickly that something very bad had happened. For those involved in the program, 14 May would be an unforgettable day; many can still recall what they were doing when they learned that what had looked like a perfect launch had in fact been anything but and that years' worth of work was in real danger of being lost. Marshall Skylab program director Lee Belew remembers that he was at Kennedy Space Center that day, having traveled down there to watch the launch. One particular memory that has stuck with him over the years since was being grilled by the national media for answers that were still unknown. "I was interviewed by Walter Cronkite at the Cape, and of course, he asked pretty tough questions."

Phil Shaffer was in Mission Control at Johnson Space Center. He was not actually on duty; he was scheduled for a shift as flight director for launch and rendezvous the following day, overseeing the launch of the first crew. "Don Puddy was the flight director for the workshop launch," he said. "Because I was going to sit down to launch SL-2 [with the first crew] the next day, I was there, more than willing to be a 'gofer' for him. When the telemetry just went nuts and those pieces started coming off, we didn't know what had happened except that a lot of things we were seeing from telemetry didn't make any sense. We certainly hadn't seen anything like that in any of the simulations. We got to orbit, and Don started trying to get the post-insertion sequence to work. Many of the actions he was trying to get done involved equipment that was missing now. It wasn't working, and the instrumentation was so screwed up we really couldn't tell what was going on. Then additional unexpected things began to happen on orbit, began to not work. Probably the best thing I did for anybody that day was start a malfunction list. Puddy didn't have time for it. Two or three hours into the business, Gene Kranz leaned over the console and said, 'You guys better start a malfunction list.' I told him, 'Here it is; it's got forty-seven items on it, or some number like that, things that need to be pursued.' So at that point we were stalled out on the post-insertion activation sequence. And stuff just kept failing, and we could see it was beginning to get hot inside Skylab."

Though things looked grim, Shaffer had a moment that night of being able to view the Skylab in a more positive light. "I remember distinctly, the night of the day of Skylab 1 launch, I knew it was going to come over Houston," he said. "And I went out to look for the 'string of pearls' as it had been advertised—the [booster's second stage] stage, the SLA panels, the refrigerator cover, and the Skylab itself. There it was, a big ol' string of pearls, going across the sky. Outstanding. Beautiful. It was really spectacular. It was a crystal-clear night in Houston, and I watched it for a very long time, almost from horizon to horizon."

As controllers began to determine what the problems meant for the workshop, the first crew realized what it meant for them: they weren't going to Skylab the next morning. Instead they were going back to Houston. Kerwin recalls that his family and friends were having a prelaunch party at the Patrick Air Force Base Officers' Club in Melbourne. "I called my wife and told her the news. 'No launch tomorrow. But might as well keep partying!'" The next morning the crew manned their T-38 jets at Patrick and flew back to Houston. They joined a full-scale battle in progress: the NASA/contractor engineering workforce versus Skylab's problems.

As it happened, the high atmospheric forces near "max-Q" had caused the shield to be torn off the Workshop and drop into the Atlantic Ocean, all unseen from the ground. "Max-Q is a very dangerous, peculiar place, and the pressures are really peculiar at that point and shock waves all over the place," Shaffer said. "If something is going to come undone, that's where everybody says it's going to come undone." And to make matters worse the meteoroid shield also tore one of the workshop solar arrays completely off, dropping it into the ocean as well. The second workshop solar array had a different story. In this case the departing shield caused a metal strap to wrap across the array, which turned out to likely have been a blessing in disguise. On the one hand the strap caught the array and prevented it from deploying properly, leading to severe power limitations after launch. On the other hand, the strap held the array in place during the launch so that it couldn't be ripped off too, a life-saving event for the Skylab science program, which needed the power the array would later be able to generate.

"We knew quickly something was wrong, that's for sure," said Marshall's George Hardy, who had gone down to Houston to monitor the launch and ended up staying there for much of the time before the first crew launched, serving essentially as a liaison between the operations team in Houston and the engineering team at MSFC. "We knew there'd been a failure of the heat shield to deploy, or to properly deploy. We weren't quite sure about that. Because there weren't any extensive strain gauges and instrumentation and things like that on it that, we couldn't pinpoint a structural problem of some kind exactly. But we knew that it wasn't functioning properly. We weren't getting thermal protection. It was a battle for the first days."

That battle was fought on several fronts. First was keeping the vehicle in shape for the crew's arrival, whenever that would be. (The orbit of Skylab

17. The exterior of the Orbital Workshop, stripped of its heat shield, began to bake in the solar radiation.

passed directly over the Kennedy Space Center every five days, so launch opportunities would be at five-day intervals after the original 15 May launch date.)

Skylab's only source of power was the ATM solar panels, and every watt was needed, which meant keeping the ATM pointed straight at the sun. But with the heat shield gone, pointing at the sun was the worst direction for workshop temperatures. It was 130 degrees Fahrenheit inside the workshop. The results of the high temperature could be disastrous: food might spoil; noxious chemicals might outgas from the walls; batteries and other equipment might be degraded. The materials lab at Marshall had conducted a temperature overtest on the substances used inside the workshop to make sure there wouldn't be an outgassing problem if temperature exceeded nominal levels but had only run the test up to over 100 degrees. After the loss of the heat shield, they resumed their testing, this time at higher temperatures.

The flight controllers battled this dilemma for ten days. With no panacea for the conflicting problems of heat and attitude available, those days were filled with constant compromise between the dual concerns. They'd roll Skylab away from the sun, to keep the temperatures from increasing. Power would drop, battery charge decrease, and a roll back toward the sun would have to happen.

The question of controlling the station's attitude was further complicated by its means of attitude control. Skylab had the three large momentum wheels, like large gyroscopes, called "Control Moment Gyros." By ordering the CMG's electrical system to push against the gimbals of one or more of these CMG momentum wheels, it was possible to move the direction in space in which Skylab was pointed, thanks to the principle of conservation of angular momentum. So it was easy enough to change the station's direction. But when this was done, other small forces were encountered (technically called "gravity gradient forces") that tended to drive Skylab's attitude in another direction, perhaps opposite of the one desired. This problem was not so easily solved and required firing the cold gas jets in space to hold the desired attitude. Further complicating the matter, the amount of cold nitrogen gas was strictly limited, and if this procedure were to continue for too many days, perhaps twenty or thirty, all the gas would be expended and attitude control would be lost. So a fix was needed in rather short order to save the Skylab missions.

For ten days mission controllers worked constantly to preserve the delicate balances needed to keep the station fit for when its first crew arrived. Over half the supply of the nitrogen gas for the entire mission was used in these ten days. "It was ironic," Hardy said. "There was a preferred orientation for generating electric power. However, it turns out that most of the time that was the most adverse orientation for the workshop overheating and for drag.

"It was management of the orientation of the workshop on a continual basis, going to one particular attitude knowing that it was penalizing you in some areas. But you had to do that, and you figure out how long you have to stay there and then get back in the other attitude. It was a real balancing act between those three things. The operations people did a fantastic job in that, and of course we loved the great engineering team. There were real questions about whether the batteries would actually survive that kind of cycling. We just started cycling, and I guess we learned a lot of things." But they kept Skylab alive.

The second front of the battle to save Skylab was finding a way to erect a substitute for the heat shield. The spacecraft, its contents, and any crew just could not tolerate those temperatures. And whatever NASA came up with, they had to be quick about it. The longer launch was postponed, the less likely it was that Skylab would remain salvageable.

Engineering teams were formed at NASA centers across the country and told to forget the paperwork for a while. "It was an opportunity for some imagination," Shaffer said. "It was an environment where if you had a good idea, it was really easy to get it executed, 'cause all of the energy was there to do that." The astronauts from the later Skylab missions and the backup crews were sent out to the centers to provide an operations viewpoint for the efforts.

"I remember leaving the launch site and coming back to the Holiday Inn," Jack Lousma said. "I met Ed Gibson and Julie, and they had more word than I did. They were somewhat disappointed and discouraged. I was thinking, at least the Skylab is up there, and even though it's not perfect, there's probably something we can do about it; at least it's up there. I had no other knowledge, and no one else did either. I didn't know what the extent of the damage was, or if I should feel confident that something could be salvaged. But I knew at least it was up there. And that was somewhat heartening.

"I went on to Houston because they hadn't figured out what the problem was for a short period of time. I wasn't there long, because we had to find out what was wrong with the Skylab and figure out what to do about it. I remember coming to work one morning, and Al Shepard said 'I want you to go to Langley and help them develop one of the fixes for the thermal shield.' I didn't go home, I didn't get any clothes, I didn't do anything, I just got in a T-38 and flew there directly. I spent about three days there and worked all day and all night with those guys at Langley to develop one of the concepts that was being proposed. This was an inflatable structure [inspired by an earlier communications satellite design], where it was a very lightweight material that would be shaped in the form of a covering, and it would be inflated when it got up there. It was all this silvery material.

"I think we [would have] extended it out the airlock, but I could be corrected on that. I don't remember there being any external tie downs, or anything like that, that we developed. But anyway they had fabricated one of those very quickly, and started the inflation, and tried to deal with those engineering difficulties, and finally made it work. I introduced my crew comments on it as we went along on how to make sure the crew could actually do it and be able to get it to operate.

"But my conclusion after being there for two or three days was that it was not going to be a satisfactory fix. It was too vulnerable to losing its inflation; the dynamics of its inflation and spreading over the workshop were marginal in my estimation. So all of us rendezvoused down at the Cape directly after that. All of us went to the Cape and met with Pete's crew. I sat and listened to all of the presentations of the fixes that were available, and then I told them that I didn't think that the one I was working on had top priority. They could make up their own minds, but here's what I thought about it. It wasn't one of the ones that went through."

After all the work was done and all the ideas were brought together and reviewed, three solutions seemed feasible. All three ended up being launched, and two were used.

From the Manned Spacecraft Center in Houston, a cloth sunshade that would be deployed by flying the Command and Service Module from point to point around the workshop, while the crew would unfold it and secure it to structure. Installation would be feasible but tricky. This one was not used.

From the Marshall Space Flight Center in Huntsville, the "Marshall Sail" (also known as the Twin-Pole Sunshade), designed to be deployed by a pair of astronauts during a spacewalk after docking with and activating Skylab, using the Skylab airlock. A solid and elegant solution, it was deployed over the original parasol a few months later by the second crew to extend the life of the parasol's delicate aluminized Mylar reflective covering material.

From Houston, a "Solar Parasol" (or "JSC Parasol" or just "the parasol")—a large square of thin nylon cloth attached to four spring-loaded fishing poles and packed in a long metal canister. As luck would have it, there was the Scientific Airlock built into the wall of the workshop on the sun-pointing side. It was designed to allow astronomy and materials experiments in pressure-tight canisters to be pushed right out into vacuum to take their observations. The parasol made some investigators unhappy by monopolizing that airlock for the entire mission. But it played an essential role in saving Skylab, being a quick, safe solution to the heat problem. It was invented by Jack Kinzler.

18. Jack Kinzler (*second from right*) explains his parasol concept.

Jack Kinzler graduated from South Hills High School in Pittsburgh in 1938. He was hired at the National Advisory Committee for Aeronautics' Langley Research Center in 1941 as a journeyman toolmaker. His skills earned him multiple promotions, and in 1961 he was assigned to the Manned Spacecraft Center in Houston as chief of the Technical Services Division. He never got a college degree, but his "equivalent in experience" was worth at least a master's in mechanical engineering.

"Different groups were working on sun shields deployed on spacewalks," Kinzler recalled. "When I realized nobody thought about going inside and doing it the simple way, I thought, 'Well, I'm going to look around some.'"

He found the experiment airlock on the sunny side of the station—he called it a "Sally Port"—in the trainer in Building 8. "So I had one of my techs go down to Houston and buy four fiberglass extendable fishing poles," he said. "I drew up a hub with springs attached to the bottom of each pole. Then I had the sheet-metal shop roll up a tube about eight inches in diameter. I called up my parachute shop and said, 'Get me a twenty-four-foot section of parachute cloth.'

"The machine shop fastened the four fishing rods to my base. I fastened that base to the floor of our big high-bay shop area. We fastened the cloth to the rods and long lines to the tips of each rod. I lowered the big overhead crane to floor level and swung my four lines over the crane hook. Then I called Gilruth, and everybody came over for a demonstration. I said, 'I think I've got something you'll like.'

"So they were standing around thinking, 'What's Kinzler up to now?' I raised the crane back up, letting out excess line 'til I had enough clearance, then let the crane pull all four lines simultaneously. It looked like a magician's act because out came these fishing rods, getting longer and longer. They're dragging with them fabric. They get all the way to where they're fully out, and all I did was let go and it went 'sshum!' So the springs were on each corner (and each spring pulled a pole outward and downward), and they came down and laid out right on the floor just perfectly. And everybody was impressed, I'll tell you. They were impressed! So that concept—my concept—was chosen for the real thing."

Kinzler and his techs gladly paid the price for success, working day and night for six days building the flight version (substituting thin nylon for the parachute cloth) and testing it. Afterward he got a lot of fan letters—and from NASA its highest decoration, the Distinguished Service Medal.

Many others were working on the problem at JSC. Ed Smylie, then chief of the Crew Systems Division, remembered: "We constructed an umbrella that would deploy like a flower petal, rather than like a traditional umbrella. We covered the assembly with a test canopy filled with holes to reduce air resistance during our deployment tests. On the initial test it worked perfectly. I invited NASA management to review our design. Most of senior management from Johnson, Marshall, and headquarters joined us. Upon deployment the frame twisted itself out of shape and failed completely. The managers shook their heads and went away. When I asked my crew what happened, they said they thought the spring was too weak and installed a stronger spring.

"Shortly after, I had a call; I should abandon our design and support Max Faget [director of engineering at Johnson] in the design of the chosen approach. As I recall, the frame was constructed in the JSC fabrication shop under the direction of Jack Kinzler and Dr. Faget. I can remember working with Max laying out the frame on the floor of the shop in the middle of the night. Max turned to me and said, "Isn't this fun?" I had not thought of it that way, but he was right. There was intense pressure, but it was a fun kind of pressure."

Ed shared the fun and the pressure with his key engineers—Jim Correale, Larry Bell, Harley Stutesman, Joe McMann, and others. He set up a clearinghouse called "Action Central." All elements of the center used it for getting things done quickly. One of his branch chiefs took it upon himself to charter a Learjet to move supplies as needed around the country.

"Everybody was working horrendous hours, and towards the end of those ten days one of my branch chiefs was leaving Building 7A one evening, and as he walked out the back door he simply collapsed," Smylie recalls. "He was not injured, but his system had simply shut down.

"The briefing of the Skylab 1 crew was held in the ninth-floor conference room [of the main building at Johnson, then called Building 2] after the crew had been quarantined. Everybody had to wear those little painter's masks over their mouths and noses. It was quite a sight to see the astronauts, senior management, engineers, and industry types—over a hundred of them—crowded together in that room around the huge table, all wearing those little white masks. After an hour or so everyone's masks were getting damp and hard to breathe through, so people started moving the masks a little bit and sneaking breaths. Then Pete Conrad stuck a cigar out the side of his mouth and commenced to smoke it—still keeping the mask on his face. Soon everyone was moving the masks away from their mouths so they could breathe. They started wearing them over their ears, on top of their heads, anywhere—but no one took off their mask!"

While the Houston-designed parasol had the advantage of being easy to deploy and thus providing a quick fix to the heating problem, Bob Schwinghamer, who was the head of the Marshall materials lab, had concerns about its long-term durability. "I was afraid of that because I had run these tests about building the flag, and I knew the nylon wouldn't stand up very long." Schwinghamer was also concerned that the station's TACS thrusters would damage the thin material when they fired. "What if it rips or comes apart totally?"

"For the Marshall Sail, I had used that same rip-stock nylon material, and I had a material called S-13G, which was a thermal-control material, real nice white stuff, very highly reflective," he said. "When you sprayed it on that sail parachute material, it still had high flexibility. It wouldn't flake

19. Seamstresses prepare the Marshall Sail sunshade.

off or anything. So we sprayed it with that, and I ran some ten-sun tests in the lab to see how long it would stand up, and we knew it would last for the Skylab mission.

"Then we had some Navy SEALs in; they packed it. We didn't have a parachute packer; they packed it. That thing was pretty big, it was like thirty by forty, I think. We got some seamstresses in from New Jersey to do the sewing. I remember that stuff was laying all over the floor, and they were just sewing away, and pulling that big sail through there. And we got that stuff all done in ten days, in time to fly."

The failure of the micrometeoroid shield had come as a complete surprise to the Skylab engineering team, which had not foreseen even the possibility of its failure. "The meteoroid shield went through design reviews like all the other hardware did," Hardy said. "There were known design requirements; there were some tests that were done. You couldn't do a test that exposed it directly to aerial loading and things like that, obviously. Nobody predicting a failure or anything like that. But like a lot of other things, after it happened, you go back and look and see where you were deficient in your analysis and some of your designs, but not prior to launch.

"McDonnell Douglas Huntington Beach was the prime contractor for the Orbital Workshop, which included the heat shield. But other contractors were involved in design reviews and things like that. I don't know, there could have been somebody out there that was expressing some concern that wasn't taken into account. But if there was, it was almost after the fact, because I had no knowledge of that."

While the rescue effort was underway, an investigation was started into the causes of the failure. The board of investigation was chaired by Bruce T. Lundin, director of NASA's Lewis (now Glenn) Research Center and presented its findings on 30 July. The board determined that the most probable cause of the failure was pressurized air under the shield forcing the forward end of the shield away from the workshop and into the supersonic air stream, which tore the meteoroid shield from the workshop. The report stated that this was likely due to flaws in the way the shield was attached to the workshop, which allowed in air.

The failure to recognize these issues during six years of development was attributed to a decision to treat the shield as a subsystem of the S-IVB, based on the presumption that it would be structurally integral to the tank. As a result, the shield was not assigned its own project engineer, who could have provided greater project leadership. In addition, testing focused on deployment, rather than performance during launch. The board found no evidence that limitation of funds or schedule pressure were factors and that engineering and management personnel on Skylab, both contractor and government, were highly experienced and adequate in number.

Resolving one final issue was absolutely essential to Skylab's long-term success. It was to answer the questions "Is one of the solar array wings still there; what's wrong with it; and how can we get it deployed?" Related to these were the further questions of how much the crew could do with ATM power alone and how they would do it. Even if it were possible to deploy the remaining solar array wing, living off the ATM power would be necessary when they arrived and until (and if) they got the solar panel open. The crew spent a lot of time with the flight planners, going through all the checklists, and marking them up for low-power operations. General conclusions were:

> With care and frugality, the crew could live in the workshop and do some of the experiments.
> They should use lights sparingly and turn them off when they moved and not make coffee or heat food.
> The solar physics and Earth resources work should be minimized. (The medical experiments were pretty low power users and OK.)
> And so forth.

There was serious doubt that Skylab could go three missions like that or that it would be worth doing. One or two failures would put it out of business. The science return would be badly hurt. They had to get that solar panel out to save Skylab.

Some images had been obtained. They were blurry but showed that the solar panel was there. What was holding it shut would have to await inspection after crew arrival. But the holdup must have to do with the ripped-off heat shield. Freeing the array would require a spacewalk. And the team had to select a suite of tools for the job right then without waiting for the inspection. Engineering did a superb job of outlining possible situations they might meet and finding a collection of tubes, ropes, and cutters to make up their tool kit.

And soon the decision was made that NASA couldn't possibly be ready to launch on 20 May. The flight controllers said they could hold on another five days. So they aimed for 25 May, the next fifth-day opportunity to launch.

"I was in the flight operations management room at JSC at the launch on May 14 and spent about the first week down there," George Hardy remembered. "After the first ten-twelve hours of meetings and looking at data, everybody was walking around with their heads down. I'd just come out of a meeting with Gene Kranz. I felt we'd bought the farm; we'd lost the mission. And Kranz said, 'No, no, we'll figure it out. We'll figure out some way to get this thing done.' That's the sort of guy he was, still optimistic.

"The next seven to ten days was remarkable. I stayed down there with the first-line Marshall engineering support group. And, of course, we stayed in touch with what was going on at Marshall. But I got to see firsthand what the JSC folks were doing too. It was quite remarkable.

"We were getting all kinds of suggestions from the public. I still have, tucked away somewhere in shoeboxes or somewhere, letters that I got from people all over the world that had suggestions about what we might do.

"The other thing I remember, Chris Kraft had a meeting every morning in his conference room, and he accepted me in that conference room just like I was one of the JSC guys. We had the best working relationship you can imagine. There was no 'invented here or there.' Both centers worked aggressively to get their sun shield out, and both of them worked; both of them served their purpose. Those were good days, great days.

"That was something, the first seven days. It was actually ten days, but there was a movie about that time called '*Seven Days in May,*' and Jim Kingsley and I joked later about that. It turns out at that point in time we'd worked on Skylab about seven years, and we almost lost it, so we decided we were going to write a book called *Seven Years in May.*"

About a week after the launch, Hardy left JSC to return to Marshall and oversee his team there. He recalled that Rocco Petrone, who in January of that year had been brought to the center to serve as its new director, was very involved in the discussions about the status of Skylab. "Dr. Petrone would come by every morning and get a briefing on what was going on," Hardy said. "He'd usually call sometime during the day. And he'd come by every afternoon. And every afternoon we'd have a briefing for him on every problem, all the plans for the next day and everything. He'd come by like six o'clock in the afternoon, and sometimes those briefings would last until midnight. You had to stay on top of it. At that particular time, initially, he hadn't been [at MSFC] but a very short time, and his wife wasn't here, and he was living in an apartment, so he had a lot of time. He'd come over there and eat Kentucky Fried Chicken, or whatever it was we'd ordered for that night, and stay with us. A lot of times he'd stay five or six hours."

Lee Belew had a similar recollection: "We worked day and night. I mean, absolutely day and night. With Rocco whipping us all the time. He was something else."

Bob Schwinghamer was also at Kennedy Space Center for the launch. "I was down there on the fourteenth," he said. "I had the family along for that launch. That was a big deal, Skylab. Right away we started hearing that the temperature was not the way it should be; and they concluded that something had happened with the micrometeoroid shield and everything.

"I hung around the control room as long as I could down there. I finally went back to the motel and told everybody, 'We've got to get out of here, and we're going to leave very early, about three o'clock in the morning,' and it was about nine or ten then. We drove all the way back, nonstop. And I got in sometime the next day—I don't know exactly when—went over to HOSC, which is where our control center was, and for the most part stayed around there. We started trying to figure out what might have happened, and what should we do."

Like most at NASA's manned spaceflight program at that time, Schwinghamer has memories of long hours of hard work, and as with most at Marshall then, Rocco Petrone features prominently in those memories. "It was ten days and ten nights," Schwinghamer said. "I didn't go home the first three days. And then after that, Rocco would come in about ten o'clock at night all the time, eleven o'clock, 'What kind of progress are we making?' After we started building the sail, I had three teams for around-the-clock coverage on the sail, and I had a big chart in my office, and I had a guy that his responsibility was to log in the results: When did you paint? Have you inspected? All the steps were there, and they were logging them in as they occurred.

"And Rocco would come in at ten or eleven o'clock at night, and he'd say, 'I'll be back later.' Well, hell, you didn't know if later was in an hour, or two hours, or tomorrow morning. So the first three days, my deputy Gene Allen and I stayed all night. Finally on the third day, we looked at each other, with dark circles under our eyes, and I said, 'Gene, I can't keep this up.' He said, 'I can't either.' I said, 'I'll tell you what let's do, let's go on two twelves, and a little bit extra if it takes it, fifteen, maybe, or whatever. But nominally two twelve-hour shifts.' Well, we did that after the first three days, and it worked out pretty good. Somebody was always there when Rocco showed up. Oh, he was tough on us. Oh, he was tough. But that's what it took."

Schwinghamer recalled that during the preparation of the sail, the process laboratory at Marshall insisted that they should be responsible for spraying the protective coat on it. He warned them that it would probably be more difficult to apply than they were anticipating and said that his materials lab could do it. "They said, 'Ah, we can do anything,' And I said, 'All right go ahead, spray it.' So I went over there, and it was about six o'clock, seven o'clock in the evening, and I was standing there, and it was the damnedest mess you ever saw. All of a sudden, somebody behind me says, 'Did you guys ever make any flight hardware in this place?!' It was Rocco, and he was madder than a hornet. It was a mess. He said, 'Bob, you get everybody that has anything to do with this sail into your office; at eight o'clock tonight we're going to decide how this gets done.'

"So I called Matt Sebile, the lab director in the process engineering department. I call him on the phone and he said, 'Well, I'm cutting the grass, I'd like to get finished.' I said, 'I tell you what, I can't tell you what to do; I'm just a division chief. But if I were you, I'd get my butt over here just as fast as I could.' And he did, and he had ratty old Bermuda shorts on and dirty tennis shoes. At that eight o'clock meeting, Rocco says, 'All right Schwinghamer, the sail's your responsibility. J. R. [Thompson, head of crew systems at Marshall], you're responsible for the deployment and how it gets used. Now get out of here, and get it done.' Of all things, my dad came down to visit us from Indiana right in the middle of that. I said, 'Dad, no fishing this time.' He always used to come down, and I'd take him fishing one time."

Despite the busy schedule, Schwinghamer did get to talk to his father, once, during the visit, while he "was home shaving one morning, and taking a shower." His father questioned why his son was spending so much time at work. "He said, 'Are just you doing that?' And I said, 'No, everybody's doing that.' He said, 'I can't understand that.' He was the superintendent of a chair factory up there in Indiana. He said 'You know what, I couldn't get my people to do that.' I said, 'Everybody's doing that, not just me.'"

While Petrone loomed large in Marshall's efforts during those ten days, he made at least one contribution of which he was utterly unaware. One evening Schwinghamer found himself in need of some material used in the sail, which was only produced at the Illinois Institute of Technology in Chicago. Obviously simply ordering the material would have taken far too long; he had to have it right away. As happened often during the around-the-clock work, it was already into the evening when Schwinghamer discovered he needed the material.

"I wanted to get [the center's] Gulfstream to go up and get that stuff so we'd be there at seven o'clock the next morning and could bring it home by noon the next day," he said. "Of course, everybody was gone, and you couldn't get the normal approvals. I tried to call some people at home to get approval to use the Gulfstream. I finally conned somebody—I don't remember who,

I don't wanna know—who would give me enough paper that I could go out there and say I need to take the Gulfstream. And we did that.

"And I kept thinking, 'Rocco Petrone's the director, if I get caught doing this, I'm done for.' That would have been a little too far. But since it was at night, there was no call or request or requirement for the airplane, and we got by with it all right. Boy oh boy, that would have been very bad. I can tell you some center directors we had that I could have talked my way out of it, but I couldn't have done it with him."

The only-semi-authorized use of the center's Gulfstream was practically by the book though, compared to another vehicle use Schwinghamer was involved in during that ten-day period. When he ran out of another material he needed one evening, Schwinghamer called Tom McElmurry in Houston for more at around 9:00 or 10:00 p.m. "I called him; I said, 'Tom, I've got to have some of that damn material. Can you bring me some?' And he said, 'I think I can.' He got here at midnight, I'll never forget. He said, 'Bob, I'm not flying back to Houston tonight. I'm going to get a hotel room. Can you get me a car?'

"I thought, 'How am I going to get a car now?' And then I thought, 'Oh, the lab director's got a car, and I know where they keep the key.' The secretary always left the key under the desk. I said, 'I'll get you the lab director's car.' My thought was, he's going to take off at six in the morning, and Heinberg wouldn't come in until seven thirty, and I could get that car back, and he'd never know it was gone. But I worked all night, and I was getting pretty woozy. By that time in the morning, I didn't think about the damn car any more. The car was sitting out at the Redstone strip.

"Heinberg comes in—the secretary told me this story—he looked through the Venetian blinds, and said, 'Where the hell is my car?' She said, 'Mr. Heinberg, it's in your parking place.' He said, 'You just come over here and look, and the car was gone.'

"So then they called all around, and they had the MPs running around looking for the car with the license number and all that crap. And they finally found the car out there at the Redstone airstrip, and somebody out there, said 'Yeah, Schwinghamer did that.' So he called me up about ten o'clock that morning, I'll never forget. And he said, 'Schwinghamer, you know what, you're totally irresponsible." And he hung up. And I thought, 'Oh, I'm fired.'"

There were several other occasions when the exigencies of the situation meant that protocol was waived in favor of expediency. "We had very strict rules about handling flight hardware," Schwinghamer said. "You couldn't just put that in the trunk of your car and drive off. This guy that worked for me had this big Ford van with just two seats in the front. I got so tired calling transportation and then having them show up four or five hours, or the next day, later. And I thought, 'The hell with it.' So every time we had to move [something], I'd get him and we'd put it all in the back of that big Ford van and move it."

On one occasion, though, he had "unofficially" transported some flight hardware for a qualification check, and someone came back and told him it wasn't working. "And I said, 'What are you talking about? We checked that thing out ourselves here and it was working fine.' So I went up there and said, 'Nothing happened here, did it? I mean, when we moved it?'

"I looked around real carefully, and one corner of that cube was scrunched, and it had green tile in it. And they had green tile on the floor up there. I said, 'All right, this metal got yielded. I suspect it fell on that corner.' Finally one guy broke down, and he started crying. He said, 'I did it, I didn't mean to.' I'll never forget, he was one of the best techs we had. I really hated that. I said, 'Just forget about it. You get this thing going again, fix it, and I won't say anything.' I felt so bad, 'cause he was one of the best techs we had and that would have gotten him in a lot of trouble. And I had a little personal interest too: I thought, next thing you know, they're going to start inquiring about how this stuff gets moved around. We were in it together, I do believe. He fixed everything up, and it worked fine.

"One night I needed something in that same building, and I got in before they closed everything, but I stayed too long, and then I couldn't get out," Schwinghamer recalled. "I had what I needed; it was two small parts. And I couldn't get out. I thought, 'If I call these damn security guys, they'll be here in two hours or something. I'll climb the fence.' It was one of these fences with that overhanging barbwire. I didn't think about what the hell I'm going to do when I get up to the barbwire. I cut a big gash in my butt, and I fell off the fence and fell on the ground. It was about eight feet high. And just when I hit the ground, two headlights came on. These darned security guys drove up and slammed up the brakes and jumped out. One of their names was Miller; I knew him well, 'cause he had nailed me for speeding four or five times. He walked up there, and I'm lying on the ground, saying

'Oh, my . . .' And he said, 'Hell, I should have known it'd be you, Schwinghamer.' He didn't do anything; he let me go. Boogered my butt too."

"We didn't let anything deter us in those days," Schwinghamer said of the construction of the sail. "A lot of funny stuff happened on our way to the Skylab. We did all kinds of stuff in those days. I can remember I was getting in hot water all the time. And it was day and night, I'm telling you what. It was really something. Oh, we did all kinds of stuff like that at that time, but we got her built. And they got it deployed, and the temperature came down real nice."

A total of seven of the sails were built; two of which were set aside as flight hardware, and the others were designated for testing. "I even had one left and gave it to the Space and Rocket Center out here," he said.

During the ten days they were grounded, the astronauts of the first crew were still under quarantine and continued to eat their flight diet so as not to ruin the metabolic balance experiment regardless of when they launched. But a trip to Huntsville was necessary, to inspect hardware and discuss repairs with Skylab engineers and to try out the Marshall Sail twin-pole sunshade in their water tank.

"That spacesuited water tank run was a memorable experience," Joe Kerwin said. "There was an air of quiet intensity. We were all old hands at this, donning the suits, checkout, entry into the water were quick and easy. A preliminary version of the sail was ready, and tools were available. We worked our way through the deployment process, noting omissions and improvements. We liked what we saw. It was late in the evening when we emerged, and we didn't know that reporters had gotten wind of the exercise and were clamoring to get in and take pictures. Word came to us, we'd done enough, go back to the motel and home tomorrow; let the backup crew take over. So we returned to the motel, and who should show up but Deke Slayton, carrying a couple of large pizzas and a bottle of wine. Best pizza and wine I ever tasted. Morale was good. Deke knew how to lead. And the metabolic balance experiment didn't suffer. That was the only time we broke the meal regimen."

Marshall's Neutral Buoyancy Simulator, large enough to hold full-scale replicas of space hardware underwater, had already played a vital role in the Skylab program in the wet workshop versus dry workshop decision. Now it would make yet another key contribution. "It really proved out to be very,

20. Jim Splawn (*left*) shows Rocco Petrone (*center*) and Bill Lucas the mount of the sunshade poles.

very beneficial," said Jim Splawn, who was the neutral-buoyancy tank manager at Marshall at the time. "Because once we had the difficulty at the launch of the Skylab itself heading toward orbit, it really proved its worth because of all the hardware we had to assemble underwater."

Like so many others involved in the effort to save Skylab, Splawn had been at the SL-1 launch and was already putting plans in motion even before boarding the NASA plane for the return flight to Marshall. "Once we knew we had a problem, I got on the phone with the crew back here in Huntsville and said, 'Hey guys, we've got to start thinking about how we may help with the repair—design of tools, the mechanics of the procedures we go through. We need to start thinking about that, not tomorrow, but right now. And just 'think outside the box,' as we say today."

The two biggest tasks were to contribute any ideas as to how the problems with the orbital facility could be fixed and to begin thinking about how they would train the flight crews on the procedures and the tools that would be needed to make the repairs. "And that started within probably an hour or so of knowing we had a problem," he said.

Once development of the Marshall twin-pole sunshade began, the water tank played an important role in qualifying the design. Test hardware would be fabricated and tested in the tank, problems would be identified and corrected, and the cycle would begin again. "With us being there right next door to the shop areas, we had the run of the entire shop to make whatever we needed," Splawn said. "There were times when the shop would make a welded piece, and we'd grab it and run to the water, and when we would put it in the water, it would sizzle because, luckily, with the close proximity we were just working that fast."

During that period the tank essentially operated as a twenty-four-hour-a-day facility. "Our wives, our families brought changes of clothes to us; they brought hot meals to us. I have no idea how many hours we went without sleep. I think at one point I personally went about thirty-four, thirty-six hours. But it was simply because the adrenaline was flowing; you had a mission, and you felt that it was just absolutely your job to do everything you could to make it happen. And we were no different from dozens and dozens of other people at Marshall that were pursuing an answer to the problem that we had. So it was very rewarding to see it all come together eventually and put us in business." Other key members of the MSFC neutral-buoyancy team included Charlie Cooper and Dick Heckman, and astronaut Rusty Schweickart spent a large amount of time working with the team to provide crew operations input.

For Splawn his work on Skylab, and particularly the effort to save the station, was the highlight of his career. "When we were doing flight crew training, we never dreamed that we would end up doing an exercise like this, to have a salvaging kind of 'Let's fix the Skylab for the crew so that they can go and spend their days on orbit and do the things they had been trained to do.' We never dreamed we'd end up in that kind of posture. But the way it worked out, that really is a highlight of my career, even beyond the flight crew training, which for us was probably what we thought would be the apex of our careers. But that ten days, obviously, topped it.

"For the launch of the last crew, we took all of our staff—all the divers, secretary, everybody—to the Cape and got to see the last launch," he said.

At the same time that the work was going on to figure out how to erect some sort of sunshade on Skylab, another effort was underway to work out how to free the station's remaining solar array so that it could deploy and provide power. With an attitude that was typical of people involved in the

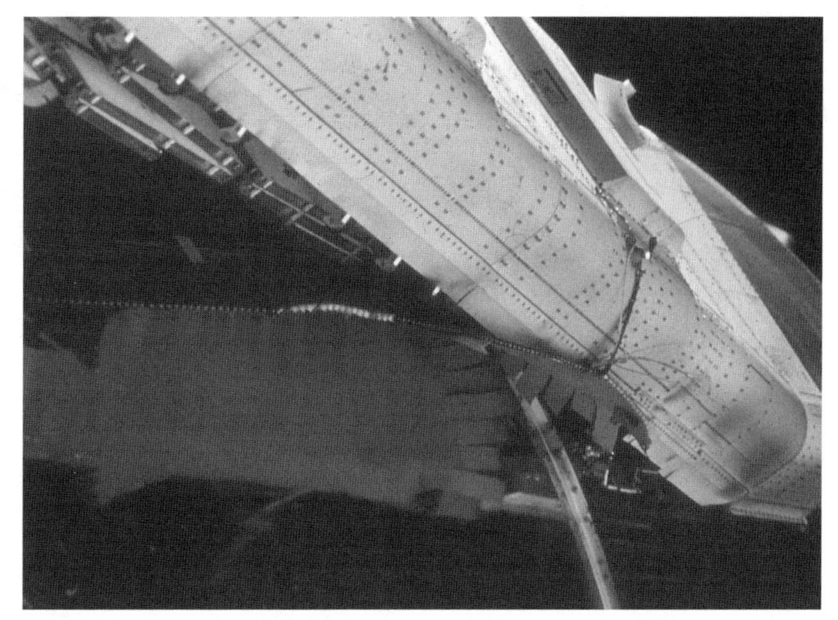

21. When the first crew arrived at Skylab, they discovered that debris had prevented the solar array (white beam at top) from deploying from the workshop (black cylinder at bottom).

project, the engineers took a task that was considered impossible and promptly began working to figure out how to make it happen.

"Six weeks or so prior to that, maybe it was a little longer, we had done a failure modes and effects analysis, which is kind of typical," said Chuck Lewis, a man-systems engineer at Marshall at the time (not the same as the Chuck Lewis at Mission Control in Houston). "And one of the things that we and JSC did together was a study of what sort of contingency plans were feasible. One of the issues, believe it or not, was if one of the solar arrays malfunctioned. And the answer to that was, 'It's not possible [to do anything]. Can't get there. Don't even need to worry about doing any analysis on that.' The message was, 'You're not going down the side of the S-IVB. It's not possible to get down there, because there are no handrails.' That was the killer in the notion of whether you could do anything about this."

Just weeks after the task had been deemed impossible, J. R. Thompson, who went on to become the director of the Marshall center, and later NASA's associate administrator, called in Lewis and others after the Skylab had arrived in orbit and controllers had learned that the unfixable problem had occurred. "J. R. Thompson called three or four of us up to the conference room up there," Lewis said. "He had the whole vehicle on one drawing spread out on a table. And he said, 'OK, guys, I know we've said we aren't going to do this. But we're going to do it. We're going to figure out how to get down there and see what's going on.'"

Work on solving the task began even before the initial imagery revealed what exactly had happened. "J. R. was jumping the gun a little bit, not knowing what the details were. Essentially telling us, 'Start thinking hard about this. We don't know what's going on yet, but start thinking about it.'" A "war room" was set up to deal with the situation, with communication lines established with other NASA centers and with the contractors who would be able to provide expertise on the equipment either involved in the problem or needed to fix it.

"We were sitting in that office," Chuck Lewis said, "and at that point we were basically going on a thirty-hours-on and eight-hours-off schedule; you worked till you dropped—which was about thirty hours—and then went home. I spent the first or second night under the conference table in the room under the one-G mock-up in 4619. My wife brought up a sleeping bag, and I sacked out the first night under the conference table, and J. R. Thompson went and slept in that advanced concepts module, which later turned out to be the precursor to Spacelab. And we were all growing beards and had sunken eyeballs and everything else.

"We were all in contact with an awful lot of people. Because at this point the environment had changed from 'Who are you, what's your need to know, who do you work for, and why should I give you this information?' to 'What can I get for you?' Managers were walking around asking what we needed—money, people, airplanes, whatever—which was a 180-degree reversal from the typical NASA mindset.

"It seemed like everybody in the world had ideas for solutions. And everybody was working on something. So we had the MacDac [McDonnell Douglas] contractors and our NASA guys sitting in the same room across the aisle from one another. Everybody was keeping track of what was going on. It was the day after launch because everybody was working—Boeing and MacDac and Martin and a bunch of people were working overnight to put concepts together.

"We were being called constantly. The phone was ringing off the wall; everybody had an idea. Every now and then Thompson would come in, and he'd have some reporter calling him on the line, and the typical question was, 'What are you guys doing?' And Bob's answer typically was, 'It's what we're not doing that's easier to answer. Ask us what we're not doing, and that's an easy answer. We're doing everything."

Relatively early in the discussions, the idea had arisen of conducting a "stand-up EVA," in which the first crew would fly their Command Module around to the stuck solar wing, open the spacecraft's hatch, and one crewmember would stand up through the open hatch and attempt to free the array. They got in touch with their counterparts at Johnson Space Center, who said they had been thinking the same thing. "We'd been sitting there with a mock-up on the table, playing around with toothpicks and strings," Lewis said. "We kinda brainstormed that together and thought what we need to do is figure out something to take up there that will let you poke and prod and maybe pull on something."

By the morning after the SL-1 launch, someone had found a lead on just such a tool. "About nine o'clock in the morning, one of the MacDac guys poked me on the shoulder and said, 'Chuck, we just learned about a company in Centralia, Missouri, called A. B. Chance Company,'" Lewis said. "They made what they called hot poles for linemen to use from the ground to reach up and activate breakers and stuff like that. They were fiberglass collapsing poles.

"[The Mcdonnell Douglas engineer] said, 'We found out they also have a number of detachable tools that go on the end of these poles. It might be that if you call these guys, there might be something useful there.' Anyway, the MacDac guy gave me the phone number for this A. B. Chance Company, and I said, 'Sure, I'll give them a call.'

"So I gave the guys a call, and I talked to the first person who answered and kind of explained the situation. I said, 'This is Chuck Lewis, and I work for Marshall, and you probably know that we're having a little trouble with the last launch, and we think you guys might have a tool with some end effectors on the end that we might be able to use.' The answer I got back was, 'Well yes, we do, we have a lot of those, and I can send you a catalog right out.'"

Of course, with the crew launch at that point scheduled just days later on 20 May, a mail-order catalog wasn't going to be anywhere near fast enough. Lewis explained the situation and was directed to the next person up the ladder at the company, who told him that the same- or next-day delivery he needed just wouldn't be possible because of an airline strike.

"I said, 'That's OK, no problem, no problem. Let me talk to the next guy up,'" Lewis said. "He said, 'OK, I'll let you speak to the product manager.' So a guy came on the phone, and his name was Cliff Bosch."

Bosch told him that A. B. Chance did carry the sort of tools that it sounded like NASA needed, took his phone number, and promised to call back in fifteen minutes with an inventory of possible solutions.

"So I let him go, and about that time, the same guy that had originally clued me about the A. B. Chance company bumped me on the shoulder again and said MacDac had been working on an automatic, pneumatically operated nibbling tool that you could pull the trigger, and it would nibble through," Lewis said. "They'd been working on that all night, and [the president of McDonnell Douglas] wanted to bring it out, and he was headed to Huntsville, flying his Aerocommander, and if the guy from A. B. Chance really came up with anything, he'd land at the airport in Centralia and pick it up.

"So when Cliff called back, I asked him how things went, and he said, 'Oh I haven't had so much fun in a long time. It's the first time I've ever done this; I walked through the stock room with a box in my hands, and I just picked up stuff off the shelves—now how do I get it to you?'

"And I said, 'Well, do you have a little airport there in Centralia?' And he said, 'Yeah.' I said, 'Well, the president of MacDac is going to pick it up in his Aerocommander and bring it to Huntsville.' And he said, 'Hell no, you guys don't know how to use this stuff; you need somebody there to tell you how to use it. Has he got an extra seat?' So we negotiated back and forth, but the upshot was that Cliff came to Huntsville."

By the time Bosch arrived at Marshall, first-crew backup astronauts Story Musgrave and Rusty Schweickart had joined Lewis's team in working on the solution to the stuck solar array. When Bosch arrived in the high-bay room with the full-size Skylab mock-up, he and the team all sat in a circle on the concrete floor, and he began showing them everything he had brought in his box. "We all sat down cross-legged on the floor, and it was kinda like opening presents at Christmas time," Lewis said. "He had all kinds of stuff. He must have had thirty or forty pounds worth of end effectors, and two of the extendable poles."

One of Bosch's tools, in particular, struck the team at Marshall as having potential. "The one thing they had that was really neat was this scissorslike

22. Ed Gibson (*from left*), Rusty Schweickart, and Pete Conrad participate in discussions about saving Skylab.

cutter that they used to clip electrical cables, that was kinda like the tree trimmer," Lewis said. "The guys at Marshall re-engineered that in about a day and a half to provide some extra mechanical advantage because they did the analysis on the strap—we'd figured out the material by that time—and we knew what sort of load it was going to take for those jaws to get through that and whether they were going to be able to pull on it. So they put a double-pulley arrangement in there that was not there originally."

Of course modifying a tool for spaceflight and actually being able to fly it are two different things. "They had to get that through stress test and everything else," Lewis said. "And how it usually goes is it takes forever. Well, in this case what they had done was to set up stations, just like college registration. Station one, there was a desk out in front of the office, and somebody would be there with a rubber stamp. Put your drawing down, stamp it, take it over to station two, stamp it. And of course, while all of this was going on, we still had managers going through, saying, 'What can we do to help? What do you need, money?' So we had a lot of support."

(That support wasn't just limited to people within NASA, either, as Steve Marks, who was working at the time as a NASA aerospace education specialist, recalls. Marks was involved in an effort to explore the untapped potential of television for NASA education and was transporting some equipment for that project to Marshall. Driving late at night, he was pulled over by a police officer for speeding. However, when the officer saw the NASA logo on the van, and Marks explained that he was delivering equipment to Marshall, the officer promptly sent him on his way—without a ticket.)

While the team at Marshall was going through the goodies that Bosch brought, another team took the poles to look into how they could be stored in the equipment bay of the Command Module. (Ultimately, the collapsible A. B. Chance poles would not be used, and NASA would instead re-create a modified version consisting of segments that could be assembled.) By that afternoon, the group working that issue at Kennedy said that they believed they had figured out how to store poles in the Command Module, but they wanted to actually see what they were working with to be sure.

"And J. R. came striding in; he strode into our meeting circle on the floor," Lewis said. "We were still sitting cross-legged on the floor of the high-bay area, and he said, 'Where's this guy from Centralia, the guy with the tools?' Cliff raised his hand. J. R. said, 'Your plane leaves for the Cape in forty-five minutes, can you be ready?' And he said, 'Yeah, give me a chance to call my wife.' He hadn't talked to her since he left that morning for work. He'd had a glass of orange juice, that was it. So he called his wife, and he said—at least, this is what he reported back—he said, 'I asked her to pack me a bag, and send it on a plane to Florida. I tried to explain what was going on, and I promised I'd explain it when I had a chance.' I think it was the Gulfstream that he got on and took his tools, with the exception of the flight items that our guys were still running through the machine shops.

"My recollection is, it was about a week and a half before he got home," he said. "To me, that's just a wonderful story, and it's such a wonderful example of what was going on around the world."

During the ten-day period between the troubled launch of the Skylab on 14 May 1973, and the launch of the first crew on 25 May, three completely independent repair methods had been conceived and built by the NASA team and tested and practiced by the flight crews who would deploy them.

"I consider these ten days the crowning achievement of NASA's real-time performance in their near half-century of existence—only matched in the recovery of the Apollo 13 crew a few years before," Owen Garriott said. "With Skylab, however, all NASA centers and thousands of civil servants and space contractors were involved in saving the program.

7. "We Fix Anything"

"Launch day arrived, ready or not, as days do," Joe Kerwin said. "It was a beautiful, quiet morning at the Cape. We went through our checks and soon were standing on the platform at Pad 39B, waiting for ingress and looking out over the peaceful ocean, with sea birds flying below us. The Cape was practically deserted; all the guests had long since gone home. The families would see this launch on television.

"I was the last crewmember to be inserted into the Command Module because I had the center couch and would be in the way of the others. There was plenty of help strapping in and making the oxygen and communications connections. And a friendly handshake and pat on the shoulder, and the hatch was closed. Communications check. Countdown continues; here's the right place in the checklist. Pete's on the left, in charge; he has the abort handle. Paul's on the right, the Command Service Module systems expert. I'm in the middle, computer backup and navigation. We've done all this a million times (two million for Pete), and it's all going well.

"About ten minutes before launch, Pete said on the intercom, 'Guys, Mission Control needs something to cheer them up. What can we say at liftoff that'll do the trick?' We discussed it a little bit and Pete made up his mind. Liftoff came, and amidst the noise and shaking, as the tail of the Saturn IB rose above the level of the launch gantry, Pete made his first voice call: 'Liftoff. And the clock is running.' And his second: 'Clear the tower. And Houston, Skylab Two, We Fix Anything, got a pitch and a roll program.'

"One of the longest, busiest days of our lives was underway."

Mission Control had reason for needing cheering up on the first crew's launch day, 25 May 1973. The team felt the clear need to get the astronauts to Skylab before it was too late to save it. There were launch constraints: one was that calling a hold too late in the count required detanking then refueling the booster, making a launch the next day impossible. And a new

23. The "milkstool" used to raise the Saturn IB boosters for launch on the Saturn V–fitted launch pad is visible in this photo of the SL-2 launch.

problem had cropped up. The mission operations computer began experiencing overload problems. When that happened it needed to be taken offline and reinitialized. It must not be offline during launch. And the cause of the overloads was unknown.

The launch flight director, Phil Shaffer, had several intense conversations with his computer supervisor, who assured him that he could bring the computer online for liftoff and that it would stay online throughout launch. Phil had a decision to make. He recalled: "At T minus six minutes the launch director at the Cape came to me for a 'Go for Launch.' At that time the mission operations computer was down and being brought up. The last status check was at T minus three, and if we went down, then it would preclude a next-day launch. I gave a 'Go' to the Cape. And then the computer did come online, and it performed nominally 'til the end of the first stage burn. At staging the computer overload problem just disappeared."

A similar problem would occur at the end of the last mission when because of memory problems the mission operations computer was dropping out every ten to twenty minutes. Mission Control found that they could reinitialize between station passes and keep coverage seamless that way, and it worked. It provided a pair of bookends for Skylab: two problems assessed and overcome by the flight control team.

Despite having been through the launch procedure countless times in simulation, the two astronauts making their first spaceflight found the real thing rather exciting. "The liftoff and ascent were of course quite an experience for us newbies," Paul Weitz said. "There is a programmed activity with the booster's first stage called 'Propellant Utilization Shift.' When we got the PU Shift, the thrust dropped off dramatically as far as I was concerned—my first thought was that we had lost an engine. Pete, of course, said something to the effect 'Cool it, rookie.'"

They reached orbit without incident, and gloves and helmets came off. Kerwin, who had "average" susceptibility to seasickness, took a planned scopolamine/Dexedrine capsule before leaving his couch; Conrad and Weitz pressed ahead with rendezvous. They were due to arrive at Skylab in about eight hours, and there was a lot to do. The space beneath the crew couches was a sight to behold, a sea of brown cloth and rope securing all the strange equipment that made up the Skylab repair kit. That under-the-couch volume was normally kept clear for launch. In case of a very early abort, the Command Module would be pulled away from the booster by its Launch Escape Tower, and the parachutes would deploy and land the spacecraft offshore. If the winds were unfavorable, a landing on the beach was possible, a hard landing, which would cause the couches to "stroke"; that is, to crush compressible aluminum material inside their struts and move down a foot or so, cushioning the shock to the crew. But that space was needed for the three sunshades and all the poles and cutters that were selected for repair. So NASA just made sure the east wind at altitude was not strong enough to blow the vehicle back onto the beach.

Upon reaching orbit, Kerwin disconnected his center couch and slid it underneath Conrad's. Then he spent much of the eight hours before rendezvous untying and rearranging all the gear to check it out and have at hand what might be needed on Mission Day 1. After a series of maneuvers, Conrad spotted Skylab out the window and expertly executed the braking maneuvers that brought the Command Module beside it.

Once in place, Pete flew the Apollo spacecraft slowly around Skylab. It all looked pretty much as expected, but the sight of all that gold-colored Mylar where the heat shield was supposed to be—already turning a discolored brown in spots from the intense ultraviolet sunlight—was a little alarming. It *looked* hot.

"After orbital insertion, we started our catch-up with the OWS [Orbital Workshop]," Paul Weitz said. "Joe got the first look at it through the telescope. After rendezvous we did an inspection fly around. The ATM and its solar array had deployed normally. One solar wing on the OWS was missing—it was cleanly gone, with no apparent damage to the OWS. Broken wires, cables, and mechanical attachments protruded from the base of the array like tendons on a broken turkey wing. The other array was partially deployed but was held close to the OWS's surface by a piece of debris from the missing meteoroid/heat shield that had wrapped itself up over the top of the solar array wing. All the rest of the meteoroid shield was gone. We took photos and downlinked TV to the ground.

"It was a frustrating time for me because my job was to take the photos with a thirty-five millimeter camera with a telephoto lens, and to get the TV imagery. Well, both of these devices had relatively long lenses, and there was not much clearance between the window and the couch support structure. It was difficult in zero-G to get a good stable picture, but I guess it turned out OK."

After completing the fly around, Pete maneuvered to the Multiple Docking Adapter's centerline docking port and performed a trouble-free soft dock. The Apollo docking system had two parts. The Command Module had a probe, a device that looked like a diamond-shaped metal-frame skeleton of a box standing on one corner with a cluster of small latches at the opposite corner—the capture latches. The four corners in the middle of this box were hinged and motorized. Commanding the motors to straighten those corners, would make the probe get longer and skinnier—it would extend—and commanding the motors to bend the corners would retract the probe, pulling the capture latches in. In the Multiple Docking Adapter's docking port was the other half of the mechanism, the drogue—a concave metal cone with a hole at its apex just big enough to allow the three capture latches to push in, then snap out like door latches. The procedure was to extend the probe, then drive the vehicles together until the capture latches engaged the hole in the drogue, retract the probe, pulling the two spacecraft together until the twelve big main latches touched, engaged, and made an airtight seal.

For this interlude Pete engaged the capture latches but didn't retract the probe; the Command Service Module just swung lazily in place by the end of the probe while the crew had a bite to eat and planned the evening's activities with Houston. A full hard dock wasn't desirable at this point because of the likelihood that they'd undock again shortly. The docking system needed to be dismantled and reset after a hard dock.

The team agreed that an EVA would be done that day. It looked as though there was a chance that if the crew pulled on the stuck solar panel cover, it might come free. They had a tool in their kit with which to make the attempt—a curved metal "shepherd's crook," which they could attach to a five-foot aluminum pole. Everyone felt fine, no motion sickness; and NASA had not yet passed the mission rule prohibiting EVAs (except in an emergency) until the fourth day of a mission to allow time to adjust to weightlessness—just to make sure astronauts don't vomit into their helmets. So they prepared the shepherd's crook, put their helmets and gloves back on, and Pete retracted the capture latches and undocked.

The next step was to let the air out of the Command Module. Pete suffered an ear block during that depressurization but insisted the crew keep going. He flew back around to the offending solar panel cover, and Paul opened the side hatch. Paul got to be the shepherd; he glided halfway out the hatch, crook in hand. Joe's job was to hang on to his legs to keep him from going out all the way. He maneuvered the crook into the gap between the solar panel cover and the side of the Orbital Workshop and pulled with all his strength.

"I positioned the crook under the end of the wing and gave a mighty heave." Weitz recalled. "The wing did not move, but it pulled the CSM toward the OWS. Now, the hatch opened to the left, which blocked half of Pete's field of view. So I am yanking on the pole; the CSM is being pulled in; and much to my amazement, in zero-G I was even moving the one-hundred-ton lab. I could see the cold gas thrusters on the OWS firing to maintain attitude, and Pete is mumbling and cursing in his attempts to maintain some semblance of station-keeping.

"Pete decided to give up on the shepherd's crook and to try the branch loppers. Joe changed out the end equipment, and I tried to cut through the material that was holding the wing down. We could not get a satisfactory grip on the debris that would allow me to cut through it, so we in exasperation decided to call it a day."

After the unsuccessful attempt, Kerwin helped Weitz get the pole and himself back into the spacecraft. They closed the hatch and headed back to the station's axial docking port to dock with Skylab again, this time for the long haul.

Joe Kerwin recalled the docking: "Fate had another bear trap to fling in our path. When Pete approached the docking port to soft dock, the capture latches would not engage. He tried it again, with a slower approach. Then he tried it with a faster approach. He tried the first backup docking procedure in the checklist. No joy. Then the second backup docking procedure. Still no joy. Suddenly there was a grimmer problem than the solar panel. If we couldn't dock, we would have to come home. With nothing accomplished.

"Pete backed off and kept station with Skylab and talked it over with Houston. It was close to midnight back home, and the crew had been awake for twenty-one hours, but Pete's flying was as smooth as ever. And there was one more procedure in the book, labeled laconically: FINAL DOCKING ATTEMPT. It required what amounted to a second spacewalk. Could we do it?

"Back up three months. We're in the Command Module trainer, going over a few procedures with our instructor, Jake Smith. We'd finished everything on the list and were ready to go home. Jake said, 'Guys, there's this third backup docking procedure we've never gone through. It's never been used. Why don't I just talk you through it once, so you can see which wires to cut.' We were OK with that, and fifteen minutes later we had filled that training square. God bless you, Jake!"

Gloves and helmets went back on. Once more they brought the Command Module down to vacuum. Pete's ear was still blocked but was not too painful. This time they removed the hatch to the docking tunnel. Joe made the changes to the docking probe per the checklist, put the hatch back in place, and they repressurized. Paul then did the rewiring in the right-hand equipment bay. The idea of this procedure was to remove the electrical interlock that prevented the main latches from actuating unless the capture latches were secure. Pete would drive the CSM into the drogue and just keep thrusting while he commanded the probe to retract. Then if the crew were lucky, the main latches would close on contact with the Skylab half of the tunnel.

Everything was checked. Pete stood in for the last attempt. The checklist said it would work in ten seconds or never. He closed in and made contact and kept the forward thrusters firing. Probe to "Retract." "We counted, ten, nine, eight, seven, six, five, four, three, two—and a machine gun went off in our faces," Kerwin said. "That explosive rattle was the main latches engaging. We were staying!"

As he checked the tunnel between the two spacecraft, Weitz discovered that eleven of the twelve latches had captured during the hard dock. The last one engaged manually with no problem. "There was a great collective sigh of relief onboard and in Mission Control," Kerwin said. "We equalized pressure, opened the tunnel hatch, snacked again, used the waste management system, and so to bed. And we slept like babies. No one had time to even think about space motion sickness."

Reflecting concerns that the equipment issues that had caused problems for docking could also cause further complications later, Pete's final comment as the crew went to sleep in the Command Module at 1:30 a.m. had been, "Now that we're docked, I'm not sure how we get *un*docked."

On Day 2, Houston's first call came at eight minutes after nine. The first words from Skylab were, "We're all healthy, Houston." Whether they were likely to stay healthy would be determined by a couple of tests scheduled after breakfast. The issue was whether the atmosphere in Skylab was safe to breathe.

The walls of the workshop, originally built to contain liquid hydrogen, were insulated with a thick layer of polyurethane foam covered with fiberglass. The foam was manufactured with toluene diisocyanate, a toxic material. Testing showed that the polyurethane would begin to break down from heat at a temperature of about 390 degrees, releasing toluene and other nasty products into the air. Had that happened? The estimated temperature the foam had reached was 350 degrees on the skin side and 160 degrees on the interior. But that was just an estimate.

Dr. Chuck Ross, the Skylab 1 crew surgeon, recalled: "This was one of our biggest concerns about Skylab's condition—that breakdown of the insulation would release a 'toxic soup' of poisonous gases into the atmosphere. We worked the problem hard under the leadership of our chief toxicologist, Elliott Harris. We identified and procured gas-sampling tubes which would measure the levels of toluene and adapted them to suck gas through Skylab's hatch equalization valves. That was so that we could do a 'sniff test' before anyone entered Skylab. We procured two activated-charcoal [filter] masks for the crew to use while first entering the OWS. And we prevailed upon the control team to evacuate all the gas out of Skylab and refill it—I think they did that twice before the crew arrived."

Now was the time to conduct those tests. First the crew verified that pressure had held steady in the short tunnel between the Command Module and the Multiple Docking Adapter. Then the Command Module hatch was opened, and the docking probe removed and carefully set aside for future investigation. Next Paul Weitz broke out a sniff-test sampling tube and drew a sample of air from the Multiple Docking Adapter to test for toluene. The test was negative as was a previously planned test for carbon monoxide (a routine precaution that would be repeated periodically throughout the Skylab missions). At about 11:30 a.m. Paul removed the second hatch and moved into the Multiple Docking Adapter, eagerly followed by the rest of the crew. This was the crew's first exposure to moving around in a large volume—and about ten minutes after entry Pete said to the Capcom, "Tell the docs we didn't need our motion sickness pills."

Pete did a thorough inspection of the docking probe and drogue. Houston made several suggestions about what to look for. Pete replied, "All right. And what's your opinion on if we had to undock, how'd we go about doing it? Do you think we could get the capture latches to cock?" Houston said they were thinking about that.

Removal of the probe showed that two of the three capture latches had opened but the third was stuck down. Pete also noted a scrape along the side of the drogue, probably due to one of his more forceful docking attempts. After cycling the latches several times and noting that the same one hung up repeatedly, Pete stowed the probe for the time being.

While Joe started activating Apollo Telescope Mount systems, Pete and Paul worked on starting up power and ventilation to the workshop. Paul had the first fans turned on by 1:30 p.m. Wearing a charcoal mask, he made a quick trip into the workshop to check the airflow gauges and general condition and turn on the rest of the fans. The air recirculation system contained charcoal filters that would absorb toluene if it was present. Returning to the Multiple Docking Adapter, he told Houston, "Okay, on our fairly quick inspection the OWS appears to be in good shape. It feels a little bit warm, as you might expect. From the three to five minutes I spent in there I would say subjectively it's about—it's a dry heat. I guess it feels like 90 to 100 degrees in the desert. Hank, I could feel heat radiating from all around me. . . . I had the soft shoes and the gloves on, and nothing I touched felt hot to me."

Power was a major consideration (and would remain so for the next two weeks).

Houston: ". . . and to help our power situation, I guess we'd like to get the OWS entry lights turned off there while we're eating lunch—after you complete the sniff."

Paul: "I thought I turned them off when I came out, Hank."

Houston: "Okay, you may have."

Pete: "Yes, he did."

Paul: "You think I'm not power conscious?"

The toluene sniff test was now repeated for the workshop and was negative. The crew floated back into the Command Module to eat lunch, and at about a quarter past four Pete and Paul entered the workshop again, this time without masks, to prepare to deploy the parasol.

Pete: "How do you read, Houston?"

Capcom: "I'm reading you loud and clear, Pete."

Pete: "Okay; I got you on the speaker box—ow. Yes, I got my hot gloves back on again. The speaker box is about 130 degrees."

The parasol had been packed in a rectangular aluminum experiment container about one foot square and five feet long. Now Pete and Paul carried it from the Command Module down to the Scientific Airlock on the sunny side of the workshop, the same side of the cluster as the Apollo Telescope Mount and so normally always pointed at the sun. This location was a lucky

24. The parasol deployment mechanism fitted to the solar airlock.

one for it was centered on the side of the workshop that needed to be shaded. (It wasn't so lucky for a couple of experiments designed to use it.)

One end of the container was inserted and locked into the airlock to make an airtight seal. Then the metal Scientific Airlock outer door was cranked open, exposing the parasol to vacuum. At this point, Pete and Paul retreated to the Multiple Docking Adapter for a cooling-off break and a drink of water. Meanwhile Joe was in the Command Module setting up a TV camera to catch a glimpse of the deployment looking aft through a window down the length of the cluster.

Now a series of seven metal rods were inserted one at a time through a seal in the free end of the container and slowly pushed, carrying the apex of the folded-up parasol out into space. To the parasol's apex were connected four fishing rods to which in turn the nylon parasol material was fastened. The rods had been telescoped; as they extended fully, each section locked in place. Then the crew released a brake holding springs designed to push the fishing rods out and down until they opened fully and locked into place, covering the workshop.

Considering the haste in which this rig had been designed, built, tested, and loaded into the Command Module for flight, it's a wonder it worked at all. It did work, but it required a lot of sweat and some ingenuity to get it laid out flat.

Pete (at four minutes after 8:00 p.m.): "OK, Houston, we had a clean deployment as far as rods clearing and everything, but it's not laid out the way it's supposed to be. . . . The problem seems to be that the folds in the material have taken too much of a set. . . . We're open for suggestions."

Capcom: "Roger. First off, we'd like to get Joe to tell us what he saw out the window. We'd like to know if all the rods are approximately the same plane."

Joe: "Well, we don't think so, Houston. . . . The front legs, that is the forward ones closest to the Command Module, came out smartly. . . . It looks as if they actually went over center a little bit, then bounced back. The back ones did not come out, it looked like, all the way—didn't come to ninety degrees. They went slowly, and they just kind of drifted to a stop."

Capcom: "Okay. What kind of an angle do you think they made with the plane of the first two rods?"

Joe: "It's your guess, but I guess thirty degrees, something like that."

Capcom: "OK. We would like for the CDR and PLT to go back in the workshop and pull her in, and we want you to complete the procedure down to Step 43 so you've done a full retraction."

Paul and Pete got the parasol pulled in, and noted that the rods that had been exposed to vacuum were nice and cool as they came back in.

Paul: "Rod B is gathering frost as it lays here in the fiery workshop."

Finally, they pushed it out a couple of feet and immediately pulled it in again. That straightened out the aft rods pretty well.

Houston: "We're looking at two flight plans tomorrow. We're taking a tentative look at one that doesn't consider anything in the Workshop. The alternative is going as planned. We'd like to get your opinion on this. . . ."

Pete: ". . . we spent the better part of two or three hours down there; and every time we'd get hot, we'd come up and take a rest. Now, if the temperatures are coming down, I would like to stick with our original flight plan, and we'll start activating it down there."

Houston: "Okay. Our best estimate, Pete, is we'll be below 100 degrees in there by tomorrow morning."

25. The sun-shield parasol, as deployed by the first crew of Skylab.

Pete: "Well, what do you think it was in there today?"
Houston: "We guess about 125."

It was going to be pretty hot work. But the Multiple Docking Adapter was about sixty degrees—jacket weather. And the crew knew they'd be eating, sleeping, and cooling off "upstairs" for the next several days.

Now Houston's priority was to get Skylab back into solar inertial attitude—with the Apollo Telescope Mount pointing directly at the sun. This was both the coolest attitude, assuming the parasol worked, and by far the best for generating electricity from the Apollo Telescope Mount's solar panels. Skylab had a very accurate system for pointing directly at the sun. It had a cluster of Fine Sun Sensors, which could measure the difference between Skylab's attitude and the sun line within a tenth of a degree. But the error had to be less than ten degrees to start with; outside of that, the sun sensors themselves couldn't find the sun. Houston was struggling now to get the cluster pointed within that ten-degree cone.

Houston, at 9:40 p.m.: "We're going to take a look at the temperatures, and we think they're coming down. We're prepared to command solar inertial here over Hawaii."

At 9:45 p.m., Houston: "Skylab, Houston. You're on your way to solar inertial now."

At 11:15, Paul: "We're not in solar inertial, you know."

Houston: "Roger. We assumed that we should be close to solar inertial attitude. We're not solar inertial mode; we'll be working that ourselves."

Paul: "Well, you're not even very close. . . . Do you know where to go?"

Houston (laughter): "Probably not."

Pete: "Well. I'm looking out the window, and it looks as if you need a plus rotation about Y and a plus about X. And I'm not sure of the magnitude, but about 10 degrees or more."

Houston: "Okay, what we're going to do is put in a plus Y rotation of forty degrees, and a plus X rotation of fifteen degrees. If we don't hack it this time, we'll probably suggest turning it over to you. . . ."

Kerwin recalled: "Well, to make a long story short, they did turn it over to us, and we got it done. Pete and Paul looked out the sun-side window in the aft portion of the Multiple Docking Adapter, and when they agreed on an attitude correction, I'd put half of it in the Apollo Telescope Mount's computer, and it would execute the maneuver. That way we got closer and closer without overshooting. When we all agreed we were well within ten degrees, we switched the mode to solar inertial. The ATM said, in computer language, 'Oh, there you are, sun!' and finished the job. When all was steady, we built our final 'how to find the sun again' tool. When we were pointed precisely at the sun, the sun-facing window threw a bright oblong spot of sunlight on the opposite wall. We just carefully surrounded that spot with gray duct tape and went to bed. To my knowledge, no one ever had to use it again. But it was there if they did!"

On Mission Day 3, bedtime having been three hours late, wake-up was an hour and a half behind the preflight schedule—still behind and with a lot of activation challenges ahead, but getting closer.

7:24 a.m., Pete: "Hey, what time would we've gotten up this morning, if it was a normal wakeup?"

Houston: "Okay, we had you scheduled for about 1500 Zulu" [11:00 a.m. Houston time].

Pete: "I meant—we should have gotten up at 1100, right?"

Houston: "Roger, that's the nominal time, Pete. But we were going to let you sleep late, since you didn't get to bed until late last night."

Pete: "Okay. Well, we're slowly trying to work our way back to the normal schedule. . . . Say, what's your cooling look like?"

Houston: "Looks like we've been dropping about a degree an hour. We've got a lot of our measurements back on scale now. We're showing some duct temperatures around ninety-five, ninety-eight degrees. . . ."

Breakfast was in the Command Module again. The wardroom hadn't been activated yet, and it was too hot anyway. But they were eating well and enjoying it.

Pete: "I'm feeling pretty spunky. Got a good night's sleep, just had a little sausage, a little scrambled eggs, and I'm working on my jam and bread, with a little coffee, goes pretty well this morning."

Houston: "That sounds good to me. I haven't had my breakfast yet."

Pete: "Sorry about that."

As the crew began to get serious about workshop activation, little problems and confusions popped up. The crew had their heads down trying to get the tasks done, and Houston was starved for information about where they were on the timeline—and in the vehicle. And all this was complicated by the sporadic nature of communications.

Houston: "Skylab, Houston through Honeysuckle for six minutes. CDR, are you in the Command Module now?"

Pete: "No, sir. I'm in the wardroom."

Houston: "Okay. . . ."

Pete: "What do you need?"

Houston: "Well, we've got a couple of things we'd like to get done up there, if it's convenient for you to take a break."

Pete: "Yes, I'm on my way. I'll be there in a flash."

And here's Pete later in the day:

Pete: "The other thing we just finished spending a little time on was—we thought we had the urine system all rigged up and it wouldn't work. I was the test subject and I had a big failure. And we went back and regrouped on it and we've concluded that you have to have a fecal bag in the thing. You have to have the fecal bag in a certain way or you just don't get enough vacuum through the urine system, enough flow, to pull urine down the urine tube. But the other two guys have been working on that down there for a while. They say it works OK."

Houston: "Roger. Copy."

And later.

Houston: "Skylab, Houston. We've got a dump maneuver starting now. You're going to have to wait on that star tracker procedure. Wait until next daylight."

Pete: "Okay, Henry. Nobody was quite sure who was supposed to do it, so it didn't get done."

Houston: "No sweat, don't worry about it."

Temperature reports were frequent throughout the day.

Pete (at 11:00 a.m.): "Hey, Houston. The biggest thing I can notice is the grid floor is beginning to cool compared to yesterday. Some of those other lockers are bigger heat sources, but everything generally seems cooler in here, although still reading off scale high on the OWS temp gauge."

Paul: "Henry, as Pete mentioned this morning, it's hot over by water tank one on that side of the Scientific Airlock. And just for information, there's a lot of the metal-to-metal fittings that don't fit too well at 130 degrees, like they did at 70. Had a pretty tough time getting the wardroom hose on water tank one, which is still hotter than a two-dollar pistol. . . ."

Paul (at 1:00 a.m., just before bedtime): "Tonight we dug out the personal hygiene kit spares container and found that every tube of Keri hand cream in there had ruptured. I don't know why. . . . Also, about two-thirds of the toothpaste has burst. And some of it has been cooked to the point where it is very, very thick and unusable any more. . . ."

But not everything was troublesome. As Pete had implied the previous evening, getting around the huge Skylab cluster was proving to be a joy.

Pete (at 11:17): "Hey, I'll tell you there is no problem adapting, and you can go anywhere you want. You may get out of control a little bit en route, but you don't bang into anything hard. And if you just take your time pushing off, you can go anywhere you want in the vehicle. Just super fast!"

A few days later, Pete dislocated a finger while doing acrobatics around the ring lockers, but it slipped back into place easily. Dr. Kerwin applied a tongue depressor splint for the day, but Pete removed it after a couple of hours, and all was well. They mentioned it to Dr. Ross as a nonevent.

Pete's evening report, delivered at about 9:00 p.m., showed that they were making progress toward blessed routine but were still behind the plan. Here's a sample:

Pete: ". . . Today, Day 3, let me give you the urine volumes—CDR 210 ml, SPT 160, PLT 200. [Those volumes were for a partial day.] Today, we have no body mass for you. We have no exercise to give you. We'll cover item Echo [medical status] on the private comm. tonight—although there isn't anything to report. We're in good health. And let me read you today's food log. 'The CDR ate everything today except corn. The reason for that was that the bag failed when inflating with water, and I got corn all over everywhere. Now for yesterday, Joe calculated that I should have taken—and I did take—two calcium pills.'"

He repeated the food items skipped and supplements taken for Paul and Joe.

Houston: "Okay. We'd like to talk about the Flight Plan. How we saw things was perhaps—the first thing tomorrow is to just pick up where we left off and work on through, with a couple of exceptions. There will be a press conference tomorrow . . . and there will be a trim burn at about 01:07 [Greenwich Mean Time]—about a twenty-nine second burn time."

Pete: "Okay, twenty-nine seconds, roughly."

Houston: "That's affirmative."

Pete: "I figured I pushed so hard trying to dock the other night I almost deorbited the thing."

Houston (laughter in background): "It's still there."

And so began the tradition of evening status report, always friendly, relaxed, and informative, just keeping everybody floating in the same direction. The food and water data allowed the doctors to calculate supplement pill needs and have the information to the crew in the morning; the crew had a glance at tomorrow's plans; and Houston could appraise where the crew was and how they felt about it.

So ended the third day.

Mission Day 4, 28 May, was a day of firsts:

First time the men were up on schedule—at 6:00 a.m. Houston time.

First breakfast in the wardroom.

First time for Joe to draw blood (theirs and his own), centrifuge it and freeze it for return.

First runs of the major medical experiments, lower body negative pressure (experiment number M092) and exercise tolerance on the bicycle ergometer (number M171).

Last, and maybe least, the first mention by Houston of the Indianapolis 500 to Pete, a racecar driver himself and a big fan.

Houston: "Skylab, Houston. We've got you stateside for 16½ minutes. Good morning."

Paul: "Hi there. Our hands are full of bloody medical equipment, but we'll recover, I think."

Pete: "Hey, Bill. Joe just drew all three of us. That went very smoothly . . . CDR just finished shaving. Breakfast is cooking, and I think with a little luck at all, we might get on to a good routine. This is our first real crack at the post-sleep checklist, so we'll get a good chance to see how long it takes us. . . . If it weren't for the fact that we have such a spectacular view out the wardroom window, which we didn't open until yesterday evening, late, I'd think we were back in Houston simming."

Back on Earth, Rusty Schweickart and his backup crew were working on a spacewalk to pry loose Solar Panel 1. For several days they were still gathering information and comparing notes and opinions; a firm plan hadn't yet taken shape. On the evening of Day 3, Paul had made a lengthy recording on Skylab's tape recorder answering questions about the standup spacewalk he'd done on Day 1. Rusty and the engineers wanted to get a very clear idea of how the strap was dug into the Solar Array System (SAS) beam fairing so they could duplicate it on the underwater mockup and come up with a solution.

Later that day Rusty had a lengthy air-to-ground conversation with the crew. Pete opined that if a guy "had a crowbar he could pry with his feet against the Solar Array System beam and pry it right off of there. . . . Paul got that claw under it, but he couldn't provide enough leverage. The claw is so short and the pole was so whippy that he couldn't provide enough leverage to pry it off of there . . ."

Rusty: "Roger. Copy. Do you think that you could have gotten the cutter (the limb lopper) around that strap down at the base, or did you try to get it around the strap?"

Pete: "No. I don't think that cutter would have done it. . . . I'll tell you what's in the back of my mind right now. We have a pry bar in one of these tools, and I'm going to figure out a way to tether a crewman so when we go out and do our thing on Day 26, it's worth our while to see if we can't whinny around there."

At this point, Conrad was still planning on attempting to free the stuck solar panel during the film-retrieval spacewalk scheduled at the end of the crew's mission.

Rusty: "Roger, Pete. For your information we're already working in the water tank trying to see what we can do along those lines. We've looked at the pictures—and we have determined that the tool we have on board will cut that strap. And we're trying to determine if there is a place that you can get at the strap with the cutting tool. . . . If the strap is cut loose, do you believe that there's anything else holding the beam down?"

Pete: "Not from the outside, Rusty, but. . . ."

A lengthy technical discussion followed. In summary, the answer was "maybe."

Another item that would probably hold the solar panel cover (the Solar Array System beam) down even if the strap were cut was an internal damper, part of the hinge mechanism designed to prevent the SAS beam from deploying too quickly. The wing's designers were pretty sure that the cold temperatures aloft would have frozen the damper, and that somehow considerable pulling force was going to have to be exerted on the Solar Array System to break it free. Rusty's team went away for four days after that talk and worked hard on a procedure. The crew would start working it again when the ground had a proposed solution. In the meantime, life had to continue on the injured space station.

Living in Skylab was already generating a lot of trash. But Skylab had a very convenient place to put the trash and an elegant way of getting it there. At the bottom of the lower deck of the workshop was the Trash Airlock (TAL), which was used to put trash into the S-IVB stage's large liquid oxygen tank. To use it, a crewman opened the upper door and inserted a filled trash bag.

Then he closed and locked the upper door and opened the lower door into the liquid oxygen tank. Air in the Trash Airlock escaped immediately, but the bag had to be pushed out with a plunger. Once that was done, the lower door could be closed and the Trash Airlock refilled with air, ready for reuse. Jamming of the Trash Airlock would be a huge problem: there just wasn't any other place to put all the food waste, packaging, used urine bags, and so forth. It couldn't happen. But it nearly happened.

Pete: "We almost thought we'd jammed the trash airlock yesterday. The bag that had the UCTAs [worn under the spacesuits on launch day] in it really expanded. And we were just flat lucky to get that one out of there."

Houston: "Don't scare us like that."

Pete: "Listen, don't scare you: it scared us worse than it scared anybody else. So we're taping our plastic urine bags. Just as a rule, I think we'll tape things, and we're only going to use about half the volume of the TAL, just to be on the safe side."

At ten minutes before 2:00 p.m., Houston called.

Houston: "Skylab, Houston for the CDR."

Joe: "Go ahead, Houston. He's listening."

Houston: "Thought he might be interested to know that the Indy race is in a hold for rain. However, the sun has come out and it looks like they might get a race off about fifteen past the hour. We show you'll be going pretty close to Indy in about seven minutes. Why don't you take a look at the clouds? If it looks good, drop the flag on them."

Pete: "If I don't get a chance to see it, then you all pass my word up there that I wish them all the best of luck—to all my friends that are driving."

Temperatures in the workshop were still about ninety-five degrees as the crew prepared for the first major medical runs. And Lower Body Negative Pressure was expected to be a bit stressful. It was a clever simulation of gravity's effect on blood distribution and thus on how hard the heart had to work to maintain blood pressure. Here's how it worked: the subject inserted himself feet first into a metal cylinder (think of a slender garbage can) up to the waist, wrapping an airtight rubberized cloth seal around his waist. Then air was pulled out of the can, reducing the pressure around the legs and lower abdomen. This would cause blood to pool in the lower half of the body, as though the subject were standing erect in gravity. (A familiar example of the effect would be "parade ground faint.") Blood pressure and heart rate would be monitored to see how the heart responded to this loss of available blood. This would be an indirect test of whether the astronaut subject was going to have trouble after return to Earth at the end of the mission. Since the amount of blood in circulation was known to decrease in space, it was thought that a gradual decrease in tolerance might be seen.

Kerwin, in view of the high temperatures, recommended that on the first run only the smallest increment of pressure be tested—30 millimeters of mercury, about ½ pound per square inch, and that the exercise experiment likewise be restricted only to the first and second levels. Houston concurred with restricting Lower Body Negative Pressure but asked that Paul Weitz, the first subject, try to go the full protocol on the ergometer. At 7:00 p.m. Kerwin reported how things had gone.

Joe: "Okay. [Lower Body Negative Pressure] was interesting. We ran the whole run, 30, 40, 50, because the numbers looked okay as we went. He had at least twice the increase in leg volume that I've ever seen before, but his figures [blood pressure and heart rate] were normal.

"Then we went to [the exercise experiment], and as a lot of us had suspected, we've got a significant mechanical efficiency problem in riding the bike. . . . We terminated the run with a little under three minutes to go, both for that reason and because of an obvious thermal problem. It's just too hot in there to go 200 watts on the bicycle. And while we will run M171s, pending resolution of the . . . problems, I'm going to strongly recommend against running at the top step."

The design of the ergometer had received a lot of attention. How did one ride a bicycle in zero-G? Your feet wouldn't stay on the pedals, and your behind wouldn't stay on the seat. Without restraints of some kind, you'd float away whenever you pushed. The first insight was to use special shoes to attach the feet to the pedals. And since Skylab's triangular grid pattern was already being used for foot restraints, that design was adopted for the pedals. It worked well.

To restrain the rest of the body called for a more complex device. The astronauts had asked Dr. Story Musgrave, Kerwin's backup as Skylab I science pilot, to tackle the problem. Story had worked with the ergometer designers at Marshall and come up with a sophisticated answer: a padded waist belt with adjustable straps to attach it to the floor and a shoulder harness with adjustable straps as well. There were also adjustments available for the handlebar, the seat height, and the seat's fore and aft position—like the driver's seat on a modern car. All the crewmen had worked out with this harness and found their favorite adjustments, and a large cue card was created with each person's numbers. What could possibly go wrong?

A lot. At the lowest workload things went pretty well. But when Paul increased the resistance and pushed harder, the waist belt's straps began to dig into his thighs, cutting off circulation to the legs. Restrained by the harness, he couldn't rise up off the seat like you do on Earth when you're pedaling rapidly. His legs hurt too much to finish the run.

At day's end, Pete reported to Mission Control that they were still not caught up.

Pete (8:18 p.m.): "We've been running pretty full blower. And all these extra goodies have been coming up. . . . So we need to do some catching up. . . . And it looks like, one of these days, we are going to have to halt for about a couple of hours anyway to GI the place if we're going to keep it clean." ["GI" was slang for soldier.]

A little later, he asked Houston to set up a private communication the next day between himself, the flight director, Chris Kraft, and Deke Slayton. He said, "It's not anything other than I want to just talk to them. It's no emergency or anything like that." And it wasn't. Pete just liked direct contact with the bosses to make sure they knew we were OK and to give them his slant on the mission. He was a great communicator.

He was good at relaxing too—better in this case than the somewhat harried science pilot:

Pete (8:30 p.m.): "Hey, Joe, do you know where the binoculars are?"
Joe: "They were in A-9 last time I looked."
Pete: "We're catching up with some satellite down here. You ought to come down and look out the window at it."
Joe: "I'm busy."
Pete: "OK."

Mission Day 5 was the day the Apollo Telescope Mount experiments got powered up for the first time. Kerwin saw the first beautiful, sharp images from the H-alpha camera, the extreme ultraviolet camera, and the camera that showed continuously what humans had before seen only during rare total eclipses—the solar corona. Still to come were X-ray images. He recorded video, used the TV downlink to show the ground real-time images, and began the slow process of learning how to use the views to recognize and interpret active regions, to spot flares early in their brief violent lifetimes, and to watch for other solar phenomena. The real solar physicists on Earth envied him this visual feast.

Pete at lunchtime: "Hey, we're all congregated in the head, all for different reasons. Why don't you go ahead and slip us the news?"

Houston did—mostly who was making what speeches on Memorial Day.

This afternoon both Conrad and Kerwin got to tackle the bike and the Lower Body Negative Pressure. Pete tried the hardest—he was determined to come back from this mission in good shape—and his heart rate exceeded 180 as he tried to ride his three minutes at 75 percent of maximum. He couldn't do it, and he experienced a couple of irregular heartbeats—premature ventricular contractions—while he was trying. Kerwin couldn't do it either. The harness problem was not going to go away; it was time for an onboard campaign to solve it. As to the Lower Body Negative Pressure, the heart rates, and symptoms on both Pete and Joe didn't look all that different from preflight—yet.

While the crew was pondering the harness problem, the doctors in Houston were seriously concerned. They had committed to twenty-eight days in orbit—despite the Biosatellite monkey's fate and the disconcerting death of the cosmonauts—on the basis that exercise tolerance and Lower Body Negative Pressure testing on orbit would show that the crew remained fit and able to deal with reentry and landing. But already the crew seemed unable to reach the desired exercise levels. Conferences were held. The medical people didn't see Pete's irregular heartbeats until Day 8, due to the complexities of getting experiment data sent to the ground and distributed. When they did, their concerns went up another notch.

Meanwhile the crew pressed on. Power conservation still came first.

Pete (7:30 p.m.): "Hey Joe, did you say we can't use the event timer [on the Apollo Telescope Mount]?"

Joe: "That's right."

Pete: "Is that for power considerations only, or is there something wrong with it?"

Joe: "The power."

Pete: "OK."

Joe (9:00 p.m.): "Weitz, you left your blower on." [In the waste management compartment.]

Paul: "How could I?"

Joe: "Well. . . ."

Paul: "What can I say, dear? I'll try harder. . . ."

They increased their skill in zero-G operations:

Pete (9:05 p.m.): "We have put to bed, once and for all, the question of 'Can you run around the water ring lockers?' I have just made ten trips around the water ring lockers, and the SPT has made five; which means he owes Ed Gibson a steak dinner, and Dr. Faget was right." [Max Faget, the brilliant engineer who had designed the blunt reentry shape of the Command Module, had bet the crew could do it.]

And they began to get personal with the onboard voice recorder:

Joe (10:18 p.m.): "Hey, sweet, lovely B Channel. Your lonely SPT, who is hungering for human companionship, would like to report the serial numbers of the—damn!—tubes for tomorrow's blood letting."

Pete and Joe: (Laughter)

At evening status report, Houston gave the crew a thermal forecast.

Houston: "The average internal temperature has shown a 5-degree drop over the last 24 hours. The magnitude of the drop per day is slowing down. . . . There may be a small portion—less than 10% of the parasol that is not doing its job as a radiator. We will probably stabilize out in the neighborhood of 80 degrees. . . ."

Pete: "Okay. I know where that 10% is. You can run your hands around the wall and find it real easy."

Houston: "Be advised—Indianapolis—you guys are going to be overhead at 12:36 Zulu, almost right overhead. That's 7:36 in the morning local, so if the weather's clear you ought to be able to look straight down and watch the cars warming up."

Mission Day 6 began following the first night all three slept in their sleep compartments. They said it was still a little warm, but they slept well and felt spunky. Houston read up the news, including these bits:

And the Texas legislature has voted to restore the death penalty in certain cases. . . .

The engagement of Princess Anne, daughter of Queen Elizabeth II*, was announced today in England. The Princess will marry a commoner, Cavalry officer Lieutenant Martin Phillips. . . .*

Again the Indianapolis 500 was postponed until hopefully this morning. . . .

People in nearby Dallas are concerned with a mysterious ooze called 'the blob' which first appeared about two weeks ago, up through a suburban back yard.

There was going to be good news and some more trouble for Skylab this day. The good news came with the first run of the M131 Vestibular Function experiment at 9:30 a.m.

M131 was not a crew favorite. It was a complex medical experiment designed to explore the function of the balance system in weightlessness. The principal investigator was Dr. Ashton Graybiel of the Navy's School of Aerospace Medicine in Pensacola, Florida—a distinguished, experienced scientist and a wonderfully gracious gentleman. A couple of times, he even lent his red convertible to crewmembers who'd come to Pensacola for testing.

Several tests were part of M131, and most of them were easy and interesting. But one of the tests was designed to find out what happened to people's susceptibility to motion sickness in space. And to do that, he had to—well, make them sick. The subject would strap himself into a rotating chair and put on a tiny blindfold consisting of two small eye cups on an elastic band (the crew called it "Minnie's Bra," referring to Minnie Mouse). Then the observer would start the chair rotating at a rate selected before flight, and the subject would begin to move his head slowly forward and back and side to side at a steady pace. On the ground, this action would result in unpleasant symptoms after about seventy head movements. Graybiel had defined a symptom complex called "Malaise III"—which meant you usually threw up. The crew objected violently to this, and he backed off to "Malaise IIA," which meant pallor, sweating, stomach awareness, then nausea. If you stopped right away you wouldn't throw up, but it was not fun. Minnie's bra had to be small so that the observer could detect the subject turning pale.

It was agreed that this was an important test. There'd been quite a bit of motion sickness reported in the Russian space program. (Second crew science pilot Owen Garriott recalled: "We were told that the Russians were beginning to wonder why only they experienced space sickness, when Americans were apparently immune! There are several reasons, probably related to the volume available for crew mobility, but that's another story.") There were also several suspected cases in NASA's Apollo program, including most notably Apollo 9's Rusty Schweickart.

The Space Shuttle was already being debated in Congress; it called for an astronaut to land the orbiter manually, a task involving considerable skill; no one wanted to trust it to a motion-sick pilot. NASA was even slightly averse to using the term. The researchers had begun to call it "Space Adaptation Syndrome." The crew still called it DSMS for "Dreaded Space Motion Sickness." And the Skylab crew didn't want to compromise their safety either. For the first mission, they requested that the commander, Pete Conrad, be excused from the Motion Sickness Susceptibility test. That way he'd always be ready to fly home if an emergency took place aboard Skylab. Management readily agreed, and so did the ever-cooperative Dr. Graybiel.

Today would be the first test. Paul Weitz, the lucky one, was scheduled to be the subject, Kerwin the observer. Paul had not had motion sickness in the first days of the flight, and he didn't relish getting it now—even though the first series of head movements would be made with no rotation. He passed with flying colors. He made 150 head movements, the maximum allowed, with none of the symptoms of motion sickness. He had some sensation of rotation, "as if my gyros were being tumbled," but felt fine.

What was going on? On Day 7, Kerwin underwent the test with rotation, (seven and one-half revolutions per minute was his number), and went 150 head movements with no symptoms. It was decided to increase the revolutions in steps. On Days 12, 16, 20, and 24, Joe and Paul were tested, until they were spinning at the maximum safe value of thirty revolutions per minute. No symptoms occurred. Both the crew and investigators were amazed. Pete had been let off the hook for nothing. It seemed that in zero-G, once you were adapted, you were immune to space motion sickness. This was pretty good news for the Shuttle program.

The other big Day 6 event was the first "EREP Pass." EREP stood for "Earth Resources Experiment Package," Skylab's complex of cameras designed to take multispectral, high-resolution, three-dimensional photographs of ground and ocean locations of scientific interest. The crew would call them "targets," but some NASA folks thought that sounded too military.

Skylab normally pointed its upper side, and the Apollo Telescope Mount solar experiments, directly at the sun. It retained this attitude on both the day and night sides of the orbit for two reasons: it required minimum fuel, and it also kept the solar panels perpendicular to the sun whenever the spacecraft was in sunlight—for maximum electrical power. Since at that point Skylab had only the Apollo Telescope Mount solar panels, energy was marginal.

What did this have to do with Earth Resources? In order to point those Earth cameras directly at the planet, Skylab had to maneuver out of solar inertial attitude into local vertical. In local vertical mode, the side of Skylab opposite the ATM was pointed straight down at the Earth at dawn. Then a gentle pitch rate was begun, which rotated Skylab gently nose down, keeping the cameras pointed directly at the Earth for most of one day-side half orbit.

As the locations of interest passed beneath the spacecraft, the various cameras would be operated, often in conjunction with low-altitude photography by aircraft and sometimes observations by scientists on the surface. That done, Skylab would rotate back to solar inertial attitude.

The pass was scheduled for between 3:00 and 4:00 p.m.—daylight over the United States. The local vertical maneuver was initiated a little after three, and by 3:30 Pete and Paul were busily photographing sites.

Pete: "Ready—do MODE AUTO."

Paul: "I found it. How about that? Looking through a little hole in the clouds."

Pete: "What did you find?"

Paul: "I found the site. . . . Okay. On White Sands and Tracking."

Skylab sliced rapidly across the United States from northwest to southeast at four miles a second. With a lot of good film in the can, they shut the cameras down and returned to solar inertial attitude about 4:00 p.m. As they did, the crew noticed a BATTERY CHARGE light that shouldn't have been on. The bad news was about to happen.

Houston (5:40 p.m.): "CDR, Houston . . . Could you do a little favor for us? We'd like to get regulators 6, 7, 8 and 16 off the line."

Pete: "Okay."

Houston: "Roger, copy. And also, for information, we're going to be powering down the EPC [the ATM's Experiment Pointing Control system] to conserve power. And we'll also be turning down the airlock module's secondary coolant loop."

Pete: "Okay. What'd we do, run lots more [power] than you thought?"

Houston: "Negative . . . but a few of the batteries went down. . . ."

Houston (5:45 p.m.): "Skylab, Houston, we'd like to hold up on the M131 run 'til we get into daylight."

Paul: "What for, Hank? We're halfway through the OGI [Oculogyral Illusion, one of the vestibular system tests] Mode now."

Houston: "Well, we've got—we've got a power problem here."

They sure did. What had happened was that several batteries reached a state of charge of less than 45 percent during the Earth Resources pass, and their controllers took them off the line—they weren't recharging. The next hour was spent turning things off and trying to get the batteries to recharge. By 6:50 Houston was able to report, "As we go over the hill here, it looks like the electrical power system is at least stable now. The batteries are coming up. So we'll see you at Vanguard. . . . And Pete, the Indy's over now. It got stopped by rain in 130 something laps, and Gordy Johncock was the winner."

Pete: "Very good."

The next day's EREP pass was canceled. Houston realized that there would be no more full passes until and unless the workshop solar wing was freed. Work on that continued with determination. But morale remained high.

Houston (8:30 p.m.): ". . . One of the things we were wondering is, have you learned to ride the portable fan yet?"

Pete: "No, that's next. We've got to master the front and back flips while running on the water ring lockers."

Mission Day 7 started out on a relaxed note.

Houston: "Skylab, Houston. I've got some sad news in this morning's paper that the blob is dead. I'm sure that Joe will be glad to hear that. And they killed it with nicotine. (Laughter.)"

Joe: "I'd like to be the blob."

Houston: "Getting to feel it now, huh?" [Not smoking.]

Joe: "You guys are all going nuts down there."
Paul: "We're going nuts up here too; the CDR thinks he can fly."
Houston: "The Astros won a game yesterday, 4 to 1."
Joe: "Hurray! Are the Cubs still in first place?"

That morning Pete took the Sound Pressure Level Meter out of stowage and toured Skylab, taking measurements. It was a quiet spacecraft. The levels in the workshop averaged between forty-five and fifty decibels. The Multiple Docking Adapter returned a reading of fifty-three (the level of a quiet office), and the noisiest compartment was its aft end, the so-called Structural Transition Section leading into the Airlock Module, where the pumps and fans registered sixty-two.

The other noise characteristic of Skylab was due to its atmospheric pressure, five pounds per square inch, one-third of an atmosphere. Maybe you've hiked high enough on a mountain to notice that up there where the air is thin, sounds are diminished. Skylab was at the equivalent altitude of Mount McKinley, above twenty thousand feet; the silence was eerie. One had to raise one's voice to be heard by someone ten feet away. Between the workshop and the Multiple Docking Adapter voice communication was impossible; the crew used speaker/intercom boxes to communicate. The good side was that you could play your own music at, say, the Apollo Telescope Mount control panel (each man had his own cassette tape player) and not interfere with the music down in the medical experiment area, where by agreement the subject got to pick the tape.

All three crewmen tried and tried again during their daily exercise periods to conquer the bicycle ergometer—adjusting the harness tighter or looser, changing the angle of the floor straps, raising or lowering the seat. Nothing worked. Weitz attempted M171 again on Day 7, but the problem hadn't been solved.

It was still taking too long to do things; the crew was still behind the timeline. Calibrating the Body Mass Measurement Device had Kerwin an hour and a half behind by noon. Houston wanted Weitz to reinstall an experiment. He said, "Okay. Is that in my flight plan?" Houston said yes and Paul said, "Oh, Okay. I hadn't read that far ahead yet. I'm still trying to catch up. Sorry."

Pete expressed their frustration this evening:

Pete (6:30 p.m.): ". . . You guys are slipping things into the pre and post sleep activities every day, without adding the time. We've already given up shaving in the mornings, and we do it at night after 0300 [10:00 p.m.]. And your time estimates for small activation tasks have turned out to be completely wrong. . . . Handovers take time. . . . I had alarm clocks going off in my pocket, and if you'll look back over my plan, I've been whistling all over this spacecraft today."

Neither crew nor schedulers had yet caught on to the secret. The first time you did something complicated in weightlessness, it took twice as long (or longer) than it had in training. The second time was faster. By the third or fourth repetition, you were back to the preflight estimate. But things were getting better. And the next day was Day 8, the first scheduled "day off."

Pete (8:15 p.m.): "I would like to add one thing, Dick. I think tomorrow, rather than a day off, it's going to be a field day. We've got an awful lot of cleaning up to do in here."

Later, Paul had a request. "Say, Dick, awhile ago we asked for the coordinates of the Pyramids . . . and tomorrow's our day off. I'd also like the coordinates of Mt. Kilimanjaro, if you can find them."

Houston said OK and good night.

Pete: "Yes, we're all shaved and we're leaving for the party. . . . Good night, Dick!"

Unburdened by medical or solar physics duties, the crew did spend much of Day 8 (Friday, 1 June) cleaning up and restowing. They also did some sightseeing out the window, three noses pressed to the glass, three pairs of legs out in different directions. As an aid to where they were, they had a map of the world marked with latitude and longitude lines and pasted onto both sides of a big piece of stiff cardboard with slick plastic rollers at each end. Stretched over the map was a continuous piece of clear plastic, marked with a curved line representing Skylab's orbit, inclined fifty degrees to the equator; and on the line, short cross markings at intervals of how far Skylab would travel in ten minutes.

On request, Houston would give them the exact time and longitude where Skylab had last crossed the equator from south to north—its ascending node. They'd slide the plastic overlay to match, and following the line from there in ten-minute increments, they'd see where they were and what was ahead. For example, set the slider to cross the equator at 160 degrees west longitude, due south of Hawaii. The orbit line showed that Skylab would cross the U.S. west coast at San Francisco, speed over southern Canada north of Montreal, leave America between Newfoundland and Nova Scotia, fly southeast over Spain and right down the African continent out into the Indian Ocean east of the Malagasy Republic, up into the Pacific between Australia and New Zealand, and back close over Hawaii—all in ninety-three minutes.

During that hour and a half, the Earth would have rotated one-and-a-half twenty-fourths of 360 degrees, or twenty-two and a half degrees, which at the equator is about 1,350 miles eastward. So the ground under Skylab's windows would be different each revolution. So would the time of day, the lighting and the weather—a thousand permutations to look for, and the greatest world tour imaginable. Hundreds of frames of film were used on clouds, ocean, islands, mountains, hometowns.

Paul (11:30 a.m.): "Hey, pass on to the people that we are sure glad we came up with this big Earth slider map we've got. It's been the most used single piece of gear on board."

But a good deal of thought was also going into a TV special for Houston and America. At ten after two:

Houston: "Skylab, Houston, with you for about fifteen minutes."
Joe: "Okay. Let us know when you get the picture."
Houston: "It's good now."

Weitz cranked up the volume on his cassette recorder, and the strains of "Also Sprach Zarathustra" by Richard Strauss—the theme from the movie 2001—filled the air. Pete, Paul, and Joe floated up from the experiment compartment into the upper workshop doing their best imitations of swimmer and movie star Esther Williams. They twisted and rolled; they flew from wall to wall; and as the climax of their show they ran around those ring lockers in an exercise they dubbed—what else—the "Skylab 500."

Joe: "Pete's got a couple of free style maneuvers here. The difficulty of that one was a 1.6. Here's a 2.2 (laughter). He didn't get many points for that one . . . that was a new one even on us. That's it. Can we show you anything?"

Houston: "Story wants to know if you've gotten around to a handball game yet."

Joe: "No, not yet."

Houston: "We've just had offers from Ringling Brothers Barnum and Bailey and Kubrick both, if you can bring that show down to Earth and do it."

Pete: "Everybody's adapted super well. We all got to talking about what's going to happen to us when we get back to Earth, because the first thing we're going to do is dive out of our beds in the morning and crash on the floor."

On the afternoon of Day 8, the crew had a surprise call about the stuck solar panel from Deke Slayton, the Mercury astronaut who was the corps' "big boss."

Deke: "Okay, Pete, this is Deke. I'm sort of playing the middle man between you and Rusty. He's over at Marshall trying to work some procedures on this thing, and he has a couple of questions. . . . As a starter we'd like to know if there's any daylight anywhere between that strap and the beam."

Pete: ". . . I'd say a half to three-quarters of an inch."

Deke (after more conversation): "Okay. We'll keep working the problem down here and keep you advised. You guys are doing great work. Hope you're having a nice day."

Pete said they were. It's always nice when the boss is happy with your work.

Deke: "When you have another spare minute you might pull out that wire bone saw—that's Rusty's favorite tool. And try it on something around there; you'd be surprised how well that beauty works. And I guess that's still his favorite choice to solve your problem."

Pete: "Okay; we'll do that. We also talked about the possibility of us putting the suits on inside the vehicle and seeing how much purchase we can get on—something around here like a food box, you know . . . and see how well we could hang on. . . ."

Deke said OK and signed off.

The field day gave both crew and Mission Control teams a chance to catch up and apply some lessons learned. At 7:30 p.m. the crew reported that once again, most of the coffee on the menu wasn't drunk. "Coffee isn't going over too well in the subtropical climate, you can see," remarked Paul. Later, Pete had another first to report.

Pete (9:05 p.m.): "We've had one through the shower, one in the shower and one waiting for the shower."

Houston: "What does the one that went through it think of the shower?"

Pete: "He's clean and sweet and smelling good right now. That's Commander Weitz. [One might notice the crew usually got Weitz to try something first.] And we've got Joe in there right now, and we're timing him to see how long it takes. It takes quite a while to sop the water back up again."

And so to bed.

Saturday, 2 June, Day 9, was Conrad's birthday (his forty-third), and he talked briefly with his wife and children that morning.

The crew's increasing efficiency and Mission Control's increasing experience with scheduling finally dovetailed; the crew finished on time. In the evening report, Pete said, "There were no flight plan deviations today. And we thought today's flight plan was excellent." Did the crew's performance have anything to do with adaptation? Kerwin thought it did. He remarked, after the flight: "I didn't get motion sick early, just a little less appetite. But Day 8 was the first time I woke up in the morning and said to myself, 'I really feel great today.'"

A little extra time for Earth gazing was appreciated:

Pete: "I just got some good pictures of Bermuda with the 300 millimeter camera for the guys in the tracking stations down there. . . . And it's a lovely day down there. And with the 300 millimeter, the girls look very nice on the beach."

Houston: "Come on, Pete, you haven't been up there that long. Your eyes aren't that sharp yet."

Joe: "He even thinks Weitz looks good."

That evening Houston gave an update on the EVA plan. "There's a big management meeting scheduled on Monday to evaluate all the work that Rusty's been doing down in the tank and to formulate several options. And we'll send up the results to you for your evaluation. Then we'll mutually settle on an EVA plan and go from there. There is no EVA planned for Tuesday." Pete said OK. They were getting closer.

One more modest first was recorded that evening:

Pete (10:00 p.m.): "We broke out the first ball tonight about 20 minutes ago. We had a little pitch game going, the three of us. And then we turned it into a kind of football game—ricocheting off the walls and throwing a few passes. So we're working up a few dynamics and orbital mechanics for the ball."

Houston: "I can't say I've really got a little bet going, but there has been a discussion going as to whether you can really throw the ball straight the first time. Did you?"

Pete: "Yes, it goes straight as an arrow."

Houston: "Amazing. We always thought you'd throw it high without the gravity there."

Sunday, 3 June could truthfully be described as a normal working day in space—with the overtone of increasing attention to the forthcoming EVA.

Houston (6:15 a.m.): "Skylab, Houston. Do we happen to have anybody in the area of the STS [Structural Transition Section] panel?"

Joe: "Don't be coy, Houston. What do you need?" [They wanted a switch position checked.]

Houston (7:20 a.m.): "And for the CDR, I'm informed that you now hold the record for more time in space than any other man around, namely Shakey."

Pete: "Holy Christmas! You mean I finally passed Captain Shakey? I can't believe it."

Houston: "I think you've got him beat by a long way before this thing's over."

Pete: "Send him my regards while he's off on his tugboat." ["Shakey" was Jim Lovell, an old friend of Pete's and commander of Apollo 13.]

At 9:00 a.m. Joe played the Navy "Church Call" from his tape cassette for Houston. And later, he recorded on B Channel his evaluation of handholds: "Okay, the Workshop dome and wall handholds are adequate for their jobs, maybe even give them a 'very good.' Their job is not to hand-over-hand it—you never do that around this place, unless you are carrying a large package. You ordinarily fly from one location to the next, and all you need when you get there is something to grab onto." The science pilot was obviously getting used to flying; a couple of weeks later he wrote his wife a poem about it.

On Sunday evening, the Capcom was Story Musgrave, Kerwin's backup and very good in a spacesuit.

Story: "We're planning on an EVA this coming week to deploy SAS panel number one. . . . Next evening we'll send up some procedures for you and also talk them over with you real time to a limited extent tomorrow. Tuesday evening we'll have maybe two or three revs [orbits, more or less] discussing the procedures with you, including probably a TV conference. . . ."

Pete: "A TV conference. OK. You guys happy you worked something out over there, huh Story?"

Story: "Yes, it's looking pretty good. . . . It's basically a five-pole extension with a cutting tool on the end of it and grabbing hold of the strip on the end of the SAS wing, tying down the near end to the fixed airlock shroud, and this will give you an EVA trail going out there."

Pete: "Very good. We aim to please. We're more than happy to do anything we can."

To make it easier to follow the story as it unfolds, here's an overview of the EVA plan Rusty's team had developed: Once the airlock's EVA hatch was open, Conrad (outside) and Kerwin (inside) would assemble five five-foot-long aluminum poles into one twenty-five-foot pole—long enough to reach from the nearest accessible point (the A-frame) to the SAS cover and the strap that held it down. On the business end of the pole would be a telephone company limb lopper (referred to by the crew as the "cutter"). Two ropes from the cutter's handles (pulling one closed the jaws, the other opened them) would be strung back along the pole to the near end. Kerwin would egress the airlock and proceed around to the A-frame; Conrad would hand him the pole, then follow him around.

Kerwin would maneuver the pole so that the cutter's jaws slipped over the offending strap, then pull the right rope to close the jaws. The jaws would bite into the strap but not yet cut it. Kerwin would tie the near end of the pole to a nearby strut, forming an "EVA trail" down to the SAS. Now Conrad would move down the pole, hand over hand, until reaching the SAS. He would carry another rope, known as the "BET" (short for "beam erection tether"), which had two hooks on its end. Kerwin would hold on to the other end of the BET.

Being careful to avoid cutting his suit on sharp edges, Conrad would fasten the two hooks into ventilation openings on the SAS, down past its hinge.

26. From Skylab's airlock (A, hidden in this picture behind the Fixed Airlock Shroud, the black band at the far end of the workshop), an EVA–path provided access to the sun end of the ATM (B). However, there were no translation aids going toward the solar array (C).

Kerwin would then tie his end of the BET to a beam as tightly as possible, so that it would lie nearly flat along the surface of the OWS. Kerwin would close the cutter's jaws all the way, severing the strap.

Conrad would then wriggle his body between the BET and the workshop surface and "stand up." Kerwin would try to do the same up at his end. Together they would put tension on the BET, exerting a pull force on the SAS beam. The engineers figured it would take a couple hundred pounds of force to break the frozen hinge and free the solar panel.

On Mission Day 11 the crew produced a swing and a miss and a home run. The swing and miss occurred during another Earth Resources pass—not with the full maneuver so as to preserve precious battery charge—but a valuable pass, and it went well. The only problem was that after the pass, Pete, reading the checklist, told Paul, "Close the S190 window cover." And Paul said, "It's already closed." And then both of them said, "Oh, dear, [or some word like that]—we forgot to open it." The S190 experiment included a set of six cameras which used a high-quality optical window. To keep the window pristine, it was protected by a cover at all times except during a pass. On this pass those cameras took photos of the back of the door instead of the Earth.

They immediately confessed to Houston and constructed a new cue card written in big black letters with that and other key steps on it and posted it over the Earth Resources switch panel. Its title was "Skylab I EREP Dumbs—t Checklist." They never forgot again.

The home run was the breakthrough on how to ride a bike in space. On Days 9 through 11, the crew perfected the secret to riding the bike. The secret was taking the harness, wrapping it into a bundle, and throwing it away. Then they just hung on and pedaled. The body tended to pitch forward because the handlebars didn't extend back far enough, so some arm fatigue resulted. They found that letting their rear ends float away from the seat and their heads press against a couple of towels duct-taped to the ceiling as a headrest helped ease the strain on their arms. And they didn't actually put the restraint harness down the Trash Airlock. They were tempted but ended up returning it to its stowage locker, where it would come in handy in a different role on the third mission. The breakthrough was complete; exercise to full capacity was now possible.

But on the ground, medical management was coming to some very conservative decisions. Dr. Chuck Ross, the crew Flight Surgeon said, "The Skylab medical team was conditioned to be concerned about irregular heartbeats in space because of their occurrence on a couple of Apollo missions. On Apollo 15 isolated premature ventricular beats had been observed on both lunar surface crewmen [Dave Scott and Jim Irwin], and Jim had had a series of coupled irregular beats during return from the moon. There was little the doctors could do during the flight. Postflight analysis led to the conclusion that loss of fluids and electrolytes during the strenuous lunar surface EVAS had caused the problem. [There was a suspicion that the men's habit of taking long runs on the beach before launch had also cost them some electrolytes.] Extra potassium was added to the Tang for the next flight [John Young, the commander, complained about the taste] and several drugs were added to the medical kit to treat severe heart rhythm disturbances if they occurred. None did.

"But when, on Day 8, the team saw Pete's premature ventricular contractions from his Day 5 M171 run, their concern increased. The Skylab I medical team was headed by Dr. Royce Hawkins, a 'by the numbers' man with little tolerance for risk and a stickler for procedure. Dr. Chuck Berry, the chief astronauts' physician for the Apollo program, had left for a headquarters assignment and was not involved day-to-day in Skylab. And the senior physician at JSC, the very well qualified Dr. Larry Dietlein, was ill.

"During the day both Pete and Joe had described the modification to the ergometer to Houston; and Joe recommended letting the crew run the exercise experiment again at the flight-plan levels. But Royce decided that he could not risk a serious arrhythmia occurring in an unmonitored crewman during exercise. And he was prepared to give a no-go for the upcoming EVA unless he received assurance that the crew could exercise safely. That was the background when I was directed to relay Royce's decisions to the crew."

Late on Day 10, the crew had received a teleprinter message directing them to eliminate the top levels on their bicycle ergometer exercise runs. The teleprinter was a typing device with heat-sensitive printing on a three-inch strip of paper that could be up to thirty feet long. Messages arrived every morning with the day's plan for each crewman, procedural changes, instrument settings, and so forth. And at the Day 11 medical conference, the crew flight surgeon, Dr. Chuck Ross, reluctantly relayed another decision to the crew. From now on, no free exercise was allowed. All use of the ergometer was to be fully instrumented with the twelve-lead electrocardiogram and had to take place when Skylab was flying over the United States, so that the surgeon could watch the heartbeats in real time. This procedure would make it much harder to schedule exercise and less would be accomplished, but the doctors were not going to take a chance. And Pete's permission to perform the EVA was to be conditional, depending on his ergometer run on Day 12.

Conrad took immediate action. He requested another private teleconference with Dr. Kraft and told Chris in the most positive terms that the ergometer modifications had solved the problem and the crew had to have free exercise. He got the decision changed, as long as the ergometer run on Day 12 with him as the subject went well. It did. That reversal and its result removed a potential roadblock to the imminent spacewalk and had a lot to do with allowing Skylab II and III to proceed to set new duration records.

Why was Conrad so passionate about exercise? Kerwin explains: "Both before and during the mission, Pete told us the story of his first try at joining the astronaut corps, along with the Original Seven. During medical testing at the Lovelace Clinic, he received a surgical examination that he considered to be unnecessarily rough and brusque—and called the offending doctor on it that evening at the Kirtland Air Force Base Officers' Club bar. Pete was disqualified medically from that selection as being 'not psychologically adapted for long duration space missions.' He was selected with the second group, and you could say that his subsequent astronaut career had as at least one of its goals to prove that doctor wrong. He wanted Skylab to be a success, and he wanted to walk off that spacecraft after twenty-eight days in good shape. It was, and he did."

Houston was true to its word. At 9:33 and 25 seconds on Monday evening, Rusty started his discussion. Three sets of data came up that night on the teleprinter while the crew slept, to review Tuesday and a lengthy, detailed discussion took place Tuesday evening. The crew went through a dry run in the workshop Wednesday morning with TV coverage of part of it so they could show their equipment to the folks on the ground and discuss how to use it. Wednesday evening, the crew would start the EVA prep, getting all the setup tasks they could out of the way—so that they could get a nice early start Thursday on the spacewalk. And Rusty assured them that if they needed more time, delaying until Friday was OK too.

Rusty then went into a detailed walk-through of the plan, using a diagram of the workshop previously sent up on the teleprinter. The crew mostly listened. When Rusty got to the part about standing up under the rope to break the damper and pull the beam hinge free, Pete remarked, "OK. I hope it don't pull too hard, or we're going to get swatted like by a fly swatter." Rusty replied, "No—we've done it a lot, Pete, and it's kind of fun as a matter of fact. You'll enjoy watching it come up."

The astronauts didn't go outside cold. They had a very detailed discussion Tuesday evening with Rusty and Ed Gibson and then stayed up late preparing the equipment. On Wednesday morning, they did an "EVA Sim" in the workshop, the best dress rehearsal they could do inside of what they were going to do outside. They cut and spliced lengths of rope, sewed cloth containers ("Pete did the sewing; he was a real sailor," Kerwin noted), connected aluminum poles together with the cutters on one end and a place to tighten the rope on the other. Kerwin put his spacesuit on—minus the helmet—and practiced moving the twenty-five-foot pole around and grabbing something with the cutters. (The floor triangle grid was used as a test target.) The dress rehearsal was as they say Broadway dress rehearsals often are—messy and filled with surprises but very productive. They discussed the details on the air to ground.

Joe: "Can we use the discone antenna as a handhold?" [That was a large radio antenna which stuck straight out from the workshop near the astronauts' work area.]

Capcom: "You can put something like—four feet up—forty pounds or so. Okay, we also—I don't know if you've had a chance to practice with the bone saw, but we've got identified for you a piece of 7075 aluminum inside, and that was the launch support bracket. Feel free to cut through it. The only precaution is—you want to have a vacuum cleaner sitting right on top of it, so you don't end up with aluminum chips."

Pete: "OK."

Rusty: "Pete, let me continue on the strap . . . after you cut the strap, we expect, because of the frozen damper situation, that the beam may rise about four degrees before you really put any tension in the BET . . . also, a recommendation after cutting the strap and when you get down there to play the human gym pole game, getting under the BET to push up on it . . . it's very important that when the beam first starts to give, and you can feel it in the BET, you want to slack off so as not to put any additional energy into the beam coming up."

Pete: "OK."

Rusty: "Just for safety reasons, it's a good idea, when you feel something break, to just stand back and let it go."

Rusty: "We see the doctor getting into his suit. We wonder if he's going to try to go out today?"

Joe: "No. I want to get a halfway feel for the difficulties of handling that 25-foot pole."

Rusty: "Okay. We think that's a good idea, Joe."

Pete: "Tell me another thing. Just how do you get yourself under the BET?"

Rusty: "Okay . . . after the beam is free, what you do is, using that rope as a trail, you just move back above the hinge line and just work your way underneath the line. It's not that tight. There really is no problem. And as soon as you get underneath it, as you begin to raise up, you put compression on yourself. And so it's quite natural to be able to stand up and the rope holds you down nicely against the beam fairing."

Pete: "OK. Now the pin by the discone antenna. If you are looking aft, minus X, the discone antenna is on the right about nine inches away, sitting in an angled channel, isn't it?"

Rusty: "That's very good, Pete."

Pete: "Okay; we can see it from the window—the STS window." [The Structural Transition Section—the lower portion of the MDA—had four small windows.]

Rusty: "Ah, that's great. We never even thought about that. . . . A precaution . . . at the base of the discone antenna there are two coax connectors which provide the signal path, and Joe wants to be careful not to mash those connectors."

Joe: "Okay, about those connectors. It's obvious that there's a risk that they'll be damaged or broken off. . . ."

Rusty: "We recognize that, and all we're asking for is reasonable prudence on your part, Mr. SPT. . . . One thing for you, Pete. The folks down here have looked at the optimum place to put the vise grips on a flange of the PCU [the spacesuit's Pressure Control Unit in the chest area of the suit]. . . ."

Pete: "Joe and I figured we'd put them on the blue hose."

Rusty: "Okay. We really didn't have any use for those vise grips out there, Pete. We figured they were just a pretty generally useful tool. . . ."

Pete: "Yes, we agree."

And back and forth for nearly three hours. From the transcript, it's clear that they felt a little skeptical about their chances. Near the end of the sim, Joe said: "That's right. I guess we'll know better when we see it, but our initial impression is that we've got a fifty-fifty chance of pulling it off. And then even if we don't, we'll have a fine reconnaissance for you and some real good words on techniques and possibilities for another try later on."

Rusty: "Right, that's just the way we figure it except we'll give you a higher probability."

Pete: "Yes, well. Just let me caution you. There is no doubt in my mind, as you mentioned, that we could get involved like we did in Gemini 11. And if we do a lot of flailing around out there, I'm sure that we can run out of gas pretty easy. So I think you'd better figure if we're unsuccessful in the first hour and a half we're probably not going to get the job done . . . but we'll give her a go tomorrow. I'm pretty sure we understand everything. We're going back and smoke them over and talk about it some more. And I think the biggest thing depends on Joe being able to get the pole hooked on to something. There's number one, and two—either cutting it or me cutting it or however that works. And let's hope there isn't something else holding it besides that strap."

Rusty: "Yes, sir. . . ."

The day proceeded with normal activities, EVA preps and last-minute decisions. Pete decided not to try bringing the TV camera outside on the EVA—too much complexity, another long cable to contend with. (As a result, there are no good pictures of the EVA.) The doctors on the ground let Kerwin take the night off from wearing his electroencephalograph sleep cap, so he'd get a good night's sleep before the EVA. Everybody—on Skylab and in Mission Control—went over all the checklists one more time with the usual last-minute changes. Pete's comment: "You got 500 guys down there keeping three of us busy." And a little later: "It's like the night before Christmas up here. The suits are hung by the fireplace with their PCUs in place, just waiting to go."

Power was important. The additional power load imposed by the EVA had to be balanced by turning things off. Here's how Houston summed it up:

Houston: "And all this [the shutoffs] comes to 1,100 watts. And then the things that are required for your EVA—all your lights, SUS [Suit Umbilical System] pumps, tape recorder and converter, primary coolant loop and PCU power, comes out to about 887, and then VTR [videotape recorder] is another 125 for a total of 1,012."

Joe (petulantly): "OK. We noticed that little note not to use the food heaters for lunch tomorrow. I'll have you know that we've only been using the food heaters for one food each day, and that's the evening frozen item."

And here's a sample of the news for that day, 6 June, as read up to the crew by Capcom:

I'll start off by saying on this day in history, 1944, we landed in Normandy. President Nixon's made several new appointments this week. . . . General Alexander Haig will retire from the Army to become Nixon's assistant in charge of the

White House staff. Haig, as you recall, was former assistant to Henry Kissinger. . . . Vice President Spiro Agnew spoke to the U.S. Governors at the National Governors Conference at Stateline, Nevada. Agnew told the audience that he was 'available for consultation.' . . . In Paris, Henry Kissinger resumed secret talks with Le Duc Tho, Politburo member from Hanoi. The two representatives are seeking ways to halt continued violations of the cease fire in Viet Nam. The Senate Watergate hearings continue to be televised during the daylight hours. . . . A bill has passed the House to raise the minimum wage from $1.60 an hour to $2.20 an hour next year. . . . Brigette Bardot announced that she will retire from film making. "I have had enough," she was quoted as saying. Some baseball scores from yesterday: Philadelphia 4, Houston 0, Dodgers 10, Chicago 1. . . .

Talking about Watergate on the air-to-ground was a sensitive business for the Capcoms. Weitz said, "Good sense of proportion. Good night, you all."

"The day of the crucial spacewalk dawned bright and clear," according to Kerwin. "But since we had a bright, clear dawn approximately every ninety-three minutes in Skylab, there was nothing special about the way this one looked. There was a slight air of unreality again—sort of like launch day; you know what to do, but you don't know what's going to *happen.* Both the spacecraft and the control room were quiet and businesslike. There wasn't any hurry. We had the checklists and were methodically working them off, staying a half hour early."

Pete: "Houston, CDR."

Houston: "Go ahead."

Pete: "Oh my gosh, is this Rusty?"

Rusty: "That's affirmative."

Pete: "You better give us—what's the earliest time we can start, Rusty?"

Rusty: "Okay. You've got a sunset right around 1410. Hold on, I'll get an exact time."

Pete: "Okay. I'm not sure that we'll make that but there's—we've kind of got a leg up on things and just depends how fast it goes. Otherwise we'll cool it to the right time."

Rusty: "Okay, we understand. And we're sort of semi-prepared for that. Let me give you an exact time here, Pete. Okay. The prior sunset time is about 1403. And Pete, for positive ID purposes we'd like just a word of confirmation that you'll be playing the role of EV-1 today and that Dr. Kerwin will be playing the role of EV-2."

Pete: "That's Charlie." [Playing off the commonly used "Roger" confirmation.]

Rusty: "Charlie, Pete Conrad."

The crew didn't make the one-rev-early start time for the EVA; there were problems with the coolant loops aboard Skylab that kept Paul occupied and held the two EVA astronauts back. But they had time.

Joe Kerwin said, "Paul helped us don the suits. It seemed harder than usual to get snugged in and the zipper zipped. On the ground we were usually able to zip it ourselves; up here much pulling and tugging was required. It wasn't till much later that Dr. Thornton explained to us we'd grown a couple of inches taller.

"Helmets and gloves were secured with a series of satisfying clicks, and checked. The oxygen we were now breathing smelt cold, metallic, and good. Moving to the airlock had been a terrifyingly clumsy task when we practiced it in the big Huntsville water tank. Here it was easy, pleasant, a cakewalk. We'd pulled our long umbilicals out of their storage spheres in the airlock, 'down' to the workshop where the suits were stowed for donning. Now we carefully pulled the excess behind us as we floated 'up' and in; everything had to fit in those tight quarters."

Paul went ahead of them. The Skylab airlock was right in the middle of the cluster's layout. Aft was a hatch into the main workshop; forward were the Multiple Docking Adapter and, right on the end, their Apollo Command Service Module taxicab. Paul had to move to the forward side of the airlock; once it was depressurized there'd be a vacuum between the workshop and the safety of the Apollo, and no way for him to cross it. Before he left his two crewmates, he made sure they had all the gear they'd need: pole sections, cable cutters, ropes, and spare tethers. Gray-taped to the front of Joe's suit was Rusty's favorite tool, the wire bone saw from the dental kit in its cloth container, just in case. Suit integrity checks were performed, and at a quarter past ten it was time to get on with it. It was Pete's prerogative to open the depressurization valve and let the air out of the airlock. Even that wasn't routine:

27. A rare photograph from the solar array deployment spacewalk.

Pete: "Houston, you may be interested in knowing that on the AIRLOCK DUMP valve, a large block of ice is growing, on the screen [a small mesh screen to keep debris out of the depressurization valve]."

Houston: "That on the inside, Pete, or on the outside?"

Pete: "On the inside. Must have enough moisture in the air, Rusty, that as it hit the screen, it froze. That's what's making the lock take so long to dump down."

Pete finally scraped off some of the ice with his wrist tether hook, the airlock got down to 0.2 psi, and the hatch was opened at twenty-three minutes after ten.

Pete went out first and got his boots into the foot restraints just outside the hatch. Joe started assembling the twenty-five-foot pole, pushing it out to Pete as he did so. At 10:47 Houston was back in communication, and Pete gave them a status report: "We have five poles rigged swinging on the hook. And we're just intrepidly peering around out here deciding how far around Joe can get in the dark. Now, the pole assembly went super slick. I had a little juggling problem getting the last longie with the tool on it . . . but she's all rigged and ready to go hanging on the hook here."

Inside Paul was having trouble getting a source of cooling water to his suit. (He was suited just in case Pete and Joe, when finished and back inside, couldn't close the hatch.)

Paul: "Hey, Pete?" [To Don Peterson, the Capcom.]

Don: "Go ahead, Paul."

Paul: "Yes, I'm ready to start working on getting some cooling water, if you think you got a way."

Don: "OK, P. J., we do have a way to do that for you. Are you ready to copy?"

Paul (who is suited, helmet off): "No, I'm not. Can you just tell me?"

Don: "Yes, okay, forget that. Are you ready to listen?"

Paul: "Yes."

But, outside, Pete and Joe were admiring the view.

Pete: "Look, there's half a moon— "

Joe: "You can see the lights, you can see the moonlight on the clouds."

Pete: "Oh, I can see the cities, yes."

Joe: "Horizon to horizon."

Don: "Hey, can you guys stop lollygagging for just a minute so we can get a word to Paul?"

The next ten minutes were a medley of Pete and Joe pulling out fifty-five feet of umbilical for each and solving the resulting tangles, interspersed with Paul talking to Houston about cooling pumps.

The route to the solar panel was uncharted territory. To get there, the crewmen would have to move underneath the Apollo Telescope Mount struts to reach a point at the edge of the Fixed Airlock Shroud where those struts were anchored, a point the crew called the "A-frame." From there they could look straight aft at the stuck Solar Panel 1. And that was as close as they could get. There were no handholds, no lighting, no EVA accessories out there.

Pete: "I tell you, you're going to get worn out doing the things that require you to go there. Do it. Well, that's a big snarl down there. I hope it all comes out right."

Joe: "Now I suggest you take that loop in your hand and put it up over your head."

Pete: "No. How did we do that? . . . Okay. That all right?"

Joe: "Yes. And it all goes behind you."

Houston: "Joe, are you going through the trusses or up over the top? You should be going through them."

Joe: "Through, through. I'm right on the MDA surface."

Houston: "OK."

Joe: "I'm looking at Paul through the window right now. The other window, Paul. Hi there. I've got one hand on the handrail, one hand on the vent duct, and I'm looking at the discone antenna."

Pete: "Do you see the pin?"

Joe: "I'll tell you—no, the base of the antenna is pretty dark."

Houston: "The next thing you'll be doing when you get enough light is to go up and hook your chest tether into the pin."

Joe: "Understand."

Houston: "And for your information, you've got seven minutes to sunrise."

Skylab sailed out of communications range at eleven minutes past eleven. When Houston regained contact over the United States at nearly 11:30, Pete and Joe were well into daylight. Their struggle with the aluminum strap, the twenty-five-foot pole, and Newton's third law had begun.

Getting into position had been easy. Kerwin was floating beside the discone antenna, loosely tethered to the pin at its base (an eye bolt shaped like an upside-down U). He held the twenty-five-foot pole in both hands and had it pointed aft, right down the left edge of the Solar Array System, with the cutter's jaws tantalizingly close to the plainly visible villain of the piece, the aluminum strap. Conrad was just above and behind Kerwin, holding with one hand to one of the sturdy beams that supported the Apollo Telescope Mount. All Kerwin had to do was move the jaws (each about four inches long) over the strap and clamp them tight.

When Houston came back into communications range over the United States fifteen minutes later, Conrad and Kerwin were still struggling. Here are some excerpts:

Pete: "Okay, Houston, we're out there. We have the debris in sight. There looks like enough room to get the cutter in, and I'm trying to help Joe stabilize. And Joe, you're way past it, it looks like."

Joe: "I don't think I am."

Pete: "Yes you are. Come—come towards me."

Joe: ". . . See, I've got it tethered, and that prevents me—"

Pete: "You're battling the tether. [The tether securing the pole to the structure, to prevent them losing it if they lost their grip.] . . . I'll re-tether it for you. Can you hold the pole?"

Joe: "I've got the pole."

Pete: "Got it. Now you're in business."

Joe: "I'll tell you, holding that on there is going to be a chore. Goldang it. Wait a minute. . . . If you could hold one foot, man, I could use both hands on this." [Whenever NASA uses a word like "goldang" in an official transcript, the actual word used may have been a different one.]

Rusty: "Okay, we're reading you. Understand you're having trouble maintaining your position in order to hook it on the strap. Can you give us a little more detail?"

Pete: ". . . We're working the problem. Bunch of wires in the way. Gosh, that prevented you from getting it that time."

Rusty: "Okay. The only thing I can say is that in the water tank we stood almost parallel with the discone with our feet down by the base, and used the discone as a handhold."

Joe: "Yes, I'm doing that. It's not a handhold I need, Rusty, it's a foothold."

Pete: "I tell you, Rusty, it looks like if we ever get it on the strap we've got it made. Because I can see the rest of the meteoroid panel, and most of it's underneath and looks relatively clear."

The pair kept on struggling. But when Houston went out of contact at 11:44, they still hadn't clamped the strap. They called a halt for rest and thinking. And while he was resting, Kerwin looked down at the base of the discone antenna, at the eye bolt where his chest tether was fastened. And he had an idea.

The chest tether was six feet long and adjustable. Kerwin said, "What if I just double the tether? Instead of hooking it on the eye bolt, I run it through the eye bolt and back to my chest? And tighten it a little so I can stand up against the tension, sort of a three-legged chair, two feet and the tether?"

Pete helped run the tether through; Joe clicked it onto the ring on his chest and stood up. He was as stable as a rock. Three minutes later the jaws were closed on the strap. Houston came back at 11:54.

Rusty: "Skylab, Houston, we've got you through Vanguard here. Sounds like you got it hooked on somewhere."

Pete: "Yes we do, and now all we're trying to do is straighten out the umbilical mess before I go out."

Rusty: "Great."

Pete: "I don't think we'll have to move the cutter. We've got it in the thinnest spot. All right, you ready?"

Joe stabilized the pole and Pete went out hand over hand, his legs out sideways to the left, his end of the Beam Erection Tether tethered to his right wrist.

Pete: "Let it come over the end first. Don't pull it all loose. That a boy. Bye."

Joe: "Take your time; I want to feed this rope behind you."

Pete: "I'm going to tighten the nuts on these pole sections on the way. . . . Every single one of them has backed off." And there was a lot of straightening of umbilicals. Houston unfortunately was going out of communications range again—with things looking up, but the issue still in doubt.

Rusty: "Okay . . . we're going to have an hour dropout before we pick you up again at Goldstone. That'll be at 1803 [three minutes after one Houston time.] And you have about thirteen minutes of daylight left. And no big sweat."

For a moment it appeared Pete's umbilical, now pulled out to its full sixty feet, wasn't long enough. But he and Joe straightened it out and it was. Pete could only get one of the two hooks on the BET fastened to the Solar Array System beam; the opening for the other was just too far away to reach. Joe tightened and tied the near end of the BET; it now stretched from the hook on the SAS beam to the A-frame strut. Pete carefully inspected the jaws of the cutter; they were perfectly positioned.

Now Joe positioned his body parallel to the pole and pulled on the jaw-closing rope with all his might. The jaws closed, the strap was cut, and Pete was a bit startled as the SAS beam jumped out a few inches, then stopped. This was just as predicted. Pete inspected the area; the beam was free, the damper was frozen, and the hinge would have to be pulled open.

Pete carefully backed away, six feet or so toward Joe. Carefully, the umbilicals were pulled back and straightened, with care to make sure that one didn't get pinched. Carefully, both men moved their bodies under the BET, feet toward the solar panel, face toward the workshop surface.

Kerwin recalled, "Pete gave the word and we both pushed away with our hands and got our feet under us. We pushed and straightened up. Suddenly—I almost remember hearing a 'pop,' but I know I couldn't have heard one. I guess I felt the pop. The rope was loose, and we were free in space, tumbling head over heels and floating away. Then I got hold of my umbilical, and pulled myself back down till I could grab a strut and turn around. And so did Pete. And we saw the most beautiful sight I've ever seen—well, almost. That solar panel cover was fully upright—ninety degrees from the workshop—and steady—and you could see the three solar panels inside it beginning to unfold. Touchdown! When I think about it now, thirty years later, I can still feel that glow."

Rusty (three minutes after one): "Hello there. We're listening to you. You're coming in loud and clear. And we see SAS amps."

Pete: "All right. I'll tell you where we are. We've got the wing out and locked, the outboard panel and the middle panel are out about the same amount, and the third one is not quite. Got the main job done."

Conrad and Kerwin spent quite a while inspecting the area, tidying up their ropes and discussing everything with Houston. Back at the hatch, they stopped to push Pete's umbilical back into its sphere inside the airlock—about the most physical work they'd done on the EVA. Then Joe got a reward. He got to move up the EVA trail to the sun end of the Apollo Telescope Mount to inspect the doors, pin one open, and replace film in one of the cameras. It was a treat to work where handholds and footholds were plentiful—and a real treat to stand up above the ATM with sun overhead and the Earth spread out below, beautiful as ever, its roundness apparent. It was a "king of the hill" feeling.

The hatch was closed a little under four hours after egress.

Rusty: "Okay, I got some good news for you. First of all, everybody down here is shaking hands, and we wish we could reach up there and shake yours. That was a dandy job and everybody was very pleased. And secondly, we're saying press on with the normal Post EVA Checklist where it says 'EAT.' Go ahead and have a nice one and just cool it."

Pete: "Yes, roger. When we have time this afternoon we'll debrief the EVA. I can tell you where the differences are between the water tank and up here. That's why it took us longer."

Rusty: "You got the job done. We don't care."

Pete: "Well, we got the job done only for one reason, and that's because Joe asked for the end of the long tether to double it up to get himself anchored. If he hadn't been able to anchor himself we wouldn't have been able to do it."

Later in the afternoon, this exchange took place.

Houston: "Skylab, Houston, how do you read?"

Joe: "Well, the PLT is shaving and the CDR went by and said, 'You've been a good boy this week, Paul; you can have the Command Module tonight.'" [The orbital equivalent of a paternal loan of an automobile to reward a teenage son.]

Houston: "Roger, copy. Everyone listening up?"

Paul: "Yes."

Houston: "Okay, I got a message I'd like to read up to you. It's to Skylab Commander Conrad. 'On behalf of the American people, I congratulate and commend you and your crew on the successful effort to repair the world's first true space station. In the two weeks since you left the Earth, you have more than fulfilled the prophecy of your parting words, 'We can fix anything.' All of us have a new courage now that man can work in space to control his environment, improve his circumstances and exert his will, even as he does on Earth. Signed, Richard Nixon.'"

And that is one of the legacies of Skylab.

The crew thought their day was over, but it wasn't. At about 8:00 p.m., when they were doing final cleanup chores and looking out the window, Houston had cheerful news: "We're showing them [the solar panels] all three 100 percent, and we're starting to command you back to solar inertial at this time."

But an hour later, there was this call:

Houston: "Okay, let me get with you guys on a problem we've been watching here, which is the secondary coolant loop. [It] got very cool during your EVA, and we can't seem to get the devil warmed back up. . . . As you know, the primary loop we can't use because of the stuck valve . . . what we're looking into now is what critical items we can turn off tonight so that we don't have to be waking you up."

The crew rogered, and signed off as usual at about 10:00 p.m. The men were tired. There wasn't much window gazing this night, just quick trips to the bathroom and a few minutes' reading in "bed."

But they were just getting into deep sleep when Houston was back.

Houston: "Hey, sorry to bother you guys, but this coolant loop is getting away from us. It's down to two degrees below freezing now. And we're going to have to get you up and work on it until we can get the thing warmed up. . . . It may freeze up in the condensing heat exchanger, and that's an intolerable situation. Sorry to do this to you guys. . . ."

Pete: "No, we want to keep the show running, pal. Don't worry about that."

It was an ironic situation; in a spacecraft plagued by heat, an essential system was threatening to freeze. And what saved the situation were those warm workshop walls that the parasol didn't quite cover.

Here's what was happening. The airlock coolant loop consisted of two circulating "loops" of fluid driven by pumps. In the interior loop the fluid was water. It flowed through pipes in metal "cold plates" to which electrically powered devices were attached, cooling them, and during spacewalks, through the EVA umbilicals to the crew's suits and into plastic tubes in their undergarments. After being warmed by the astronaut's bodies, it flowed through a heat exchanger where its channels were in contact with those of the exterior loop. The exterior loop contained a water/glycol mixture, antifreeze, with a low freezing point. It warmed up in the heat exchanger, taking heat from the water, then radiated the heat to space in an external radiator.

The problem was that the airlock and Multiple Docking Adapter had been cold throughout the mission, and much of the electrical equipment had been turned off because of the power shortage. When Pete and Paul did their spacewalk even more equipment had been turned off, enough to cool the loop further despite the heat transferred from the astronauts. The external loop's temperature dropped dangerously. If it dropped below the freezing point of water, the glycol wouldn't freeze but the water would, rupturing the heat exchanger and making the loop inoperative.

The crew scurried into the workshop, found umbilicals and two of the liquid cooling garments (LCGs) normally worn under their spacesuits to circulate cold water around their bodies and hooked them into the coolant loop. Then they taped the LCGs to the warmest part of the workshop wall they could find, near Water Tank 1. They turned on the pumps, flowing water into the loop, warming it. They covered the LCGs with clothing to prevent heat from being lost into the atmosphere. They and Houston powered up every piece of equipment on that loop. They were thankful for the power newly available.

And it worked. Temperature at the heat exchanger rose to forty-one degrees in just under an hour. The crew stayed up until midnight, to make sure Houston had it under control, then went back to sleep, weariness mixed with relief. Houston promised not to wake them until 8:00 a.m.

The coolant loop emergency marked the transition between the first and second halves of the Skylab 1 mission. With reveille on Day 15, both crew and team were relaxed and confident, schooled in their roles and determined to "get back on the timeline."

Systems were powered up. There was hot water for the coffee. Hot showers began to appear on the flight plan (but only once a week). Skylab talked to Houston about the possibility of scheduling another spacewalk to erect a better sunshade, the so-called "Marshall Sail." They decided to leave that to the second mission; but the crew unanimously made the decision to substitute Weitz for Kerwin for the end-of-mission film retrieval EVA on Day 26. All three had trained, and spreading the experience around would be good for the astronaut cadre.

Suddenly it was the second half of man's longest space mission; everyone was now thinking ahead to its conclusion. Houston negotiated a plan to shift the crew's workday several hours earlier, starting with Day 21. The big shift was made necessary by a nominal landing at dawn in the Pacific, and crew and ground agreed it would be easier to take it in little steps. The steps weren't that little. They shifted earlier by two hours on Mission Days 21 and 22, then by four and a half hours on Day 28, their last night aboard

Skylab. Pete and Paul each took a Dalmane sleeping pill that "night," but nobody got much sleep.

Days settled into a routine. Had there been a murder on Skylab, and had Hercule Poirot wanted to check everyone's whereabouts on the crucial day, he'd need only to refer to the air-to-ground conversations and the telemetry data showing what was on or off. Medical experiments were "down below," and the subject got to pick which music tape to play. More busy passes at the ATM, including usually an evening pass. Earth Resources passes daily for six straight days. Somehow the crew started keeping up and getting ahead. On Day 17, Pete actually settled into his sleep compartment at 9:30 with a book. "Yes," he told Houston, "We ran out of things to do." Houston answered, "You better be careful, Pete. I saw three guys reach for 482s [a task form] down here to start scheduling."

They began to look for activities not in the flight plan.

Kerwin, on Day 18: "I have my hobby up here. I have my do-it-yourself real doctor kit. Right now I'm staining slides."

Conrad: "He's working on my throat culture or something."

Pete invented new games involving the blue rubber ball.

Pete: "We're working on a new game up here, Houston. It's called 'get the rubber ball back to you.' Try it off the water ring lockers first."

Houston: "Which ball you using, Pete?"

Pete: "The blue rubber one, but—it gives up energy awful fast. It kind of poops out after four or five bounces. What we really need is one of those super balls."

Paul, on Day 19: "Roger, Houston. We're pretty busy right now. The CDR is trying to break the PLT's world record of thirteen bounces around the ring lockers. . . . Don't ask for the rules. It's extremely complicated, involves orbital mechanics and everything."

Houston: "Just be sure it's only the world's record that you break."

On the evening of Mission Day 22, the crew gathered around the wardroom table for ice cream and strawberries around bedtime, and somebody said, "It's been Day Twenty-two up here forever!" Now that routine had taken over, just a little boredom had crept in. They were starting to think about coming home.

The very next day, Houston discussed with the crew the possibility of their staying aboard one extra week, to complete additional experiment runs. Of course Conrad said, "You betcha, Houston—we're ready!" But all three were just a bit relieved when the idea was dropped, as NASA gained confidence that the second and third missions would take place. This crew was ready to smell the sea air.

As the flight went on, Pete developed an unfortunate addiction to the butter cookies. He was exercising hard and needed the extra calories. And the butter cookies were "homemade"—done in a NASA kitchen to the recipe of Rita Rapp, a wonderful food system specialist. It got so bad that on Day 23 he asked Houston specifically to assure that there were plenty of butter cookies aboard *Ticonderoga,* the recovery aircraft carrier.

One evening the three decided to check out what it felt like to navigate around a big spacecraft in absolute darkness. They turned off every light in the workshop and covered the big wardroom window, then waited for the spacecraft to fly into night.

"It was really different!" Kerwin recalled. "I never had a sensation of falling till we did that. But you were absolutely clueless about where you were and where anything else was. It was scary. I just clutched my handhold and didn't want to move. It was my first real sensation of fear in space. And others have reported similar feelings. I remember Ken Mattingly talking about emerging from the Command Module hatch on the way home from the moon on Apollo 16, to retrieve film from a camera in the Service Module. Neither the Earth nor the moon was in sight; space looked like an infinitely deep black hole. He just wanted to hold on to something."

All three crewmen had their twentieth college reunions going on. All of them asked that greetings be relayed to their classmates.

Most evenings the Capcom would have time after the evening report to give the crew a brief news report. There was plenty of unrest on planet Earth:

Houston: "The Texas wheat crop is expected to be the third largest in history, but it's in danger because of our fuel crisis. . . . They're limiting gas at lots of places to ten gallons per fill-up. . . . Nixon is proposing a new Cabinet level Department of Energy and Resources. . . . There was a partial brownout of all Federal Buildings in Washington DC. . . . Nixon has established a price freeze for sixty days, and is considering a profit rollback. The markets didn't like it; the dollar was down and gold was up."

"There were seven inches of rain in Houston yesterday. . . . Dr. Kraft [the

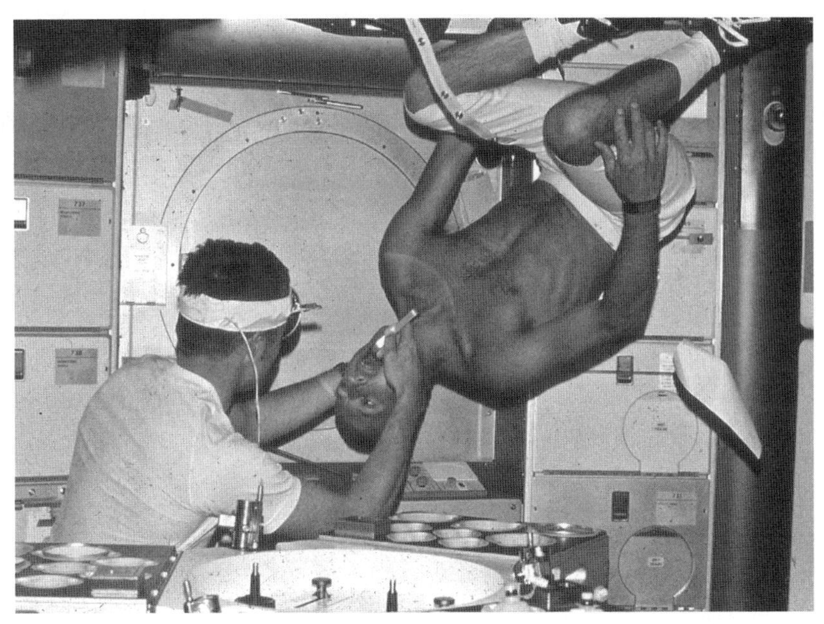

28. Kerwin performs a medical examination of Conrad.

center director, who lived a few miles west of the center in Friendswood] is spending the night at the Nassau Bay Motel.

"President Sadat visited Libya to discuss the planned merger of Egypt and Libya. . . . General Francisco Franco, now 80 and ruler of Spain for 35 years, is turning part of his duties over. . . . The war in Viet Nam may be nearing an end. . . ."

"A Soviet TU-144 crashed at the Paris Air Show yesterday. There were many deaths. . . ."

"In case you're going to South Padre Island, Texas, they have just elected a new sheriff who's a 27-year-old redheaded mother of two. She says, 'I'm a mean redhead, and if they ever call me 'pig' they had better be careful.'"

Paul: "I think we'll stay up here, Houston."

Houston: "Come to think of it, maybe you people are well off where you are."

Also on Day 22, there was "The Flare." For days, the sun had been tantalizing the crew with hints of increased activity. The crew got daily briefings on what was happening. They sounded like this: "Active Region 37 has rotated onto the disc . . . as a large spot group. And we had a subnormal flare there which began at 8:35. . . ."

The briefers hoped to alert the Apollo Telescope Mount operators (and all three crewmen shared that duty) to where a flare might occur. It was the Holy Grail of solar physics to capture a flare—especially the first crucial minutes of rise—with the variety of instruments onboard Skylab. It would be historic data. Each of the Skylab solar experiments had its own team of investigators. But since just one astronaut would operate all of them during a single fifty-minute sunrise-to-sunset "pass," the investigators had gotten together to plan a large number of Joint Observing Programs—JOPs—designed to handle all their various data needs during all levels of solar activity. And the granddaddy of JOPs was the infamous JOP 13, the routine for a solar flare. It required quickly and accurately pointing the Apollo Telescope Mount canister straight at the flare, then activating all the cameras in high speed mode with correct settings. It took LOTS of photographs; film was flying through the cameras. JOP 13 was not to be used lightly—scientists wanted desperately to get a flare, but nobody wanted to waste film on a false alarm.

How could there be a false alarm? The views and instrument readings that the crew had available had never been used this way before, so the "signature" of the beginning of a real flare wasn't known. As the mission progressed, it seemed that the best clue to a real flare was going to be an increase in X-ray intensity measured by one of the two X-ray telescopes. But another phenomenon also caused the X-ray count to increase—a trip through the South Atlantic Anomaly.

One of the first discoveries ever made by an orbiting spacecraft was made by Dr. James Van Allen, using data from a simple Geiger counter on America's first satellite, Explorer 1. He discovered two "belts" of solar radiation trapped above the Earth by its magnetic field. The inner Van Allen belt consisted of energetic (and dangerous) protons. Its center was about one thousand miles up, but at one point, just east of the southern end of South America, it dipped close to the atmosphere. That is the South Atlantic Anomaly. And Skylab passed right through it, not on every revolution, but a few times each day. For example:

Pete (on Day 18): "I have an in and out on the flare there, Houston; 650." [A reading of 650 on the proportional counter.]

Pete (a minute later): "Want us to go after the flare, Houston? It's 690, 700."

Houston (after checking): "Pete, you're in the Anomaly right now, and that's the reason you're getting the flare indication. So, we do not want you to press with a flare JOP."

Paul: "Hey, Houston, I think you guys have got to put those . . . anomaly passes, all of them, on our pads. If that ever happens out of station contact, we're going to come over the hill minus about 300 frames of film."

Finally, on Day 22, they got lucky. At eight minutes after nine, Houston advised the crew that a "subnormal flare" had started in Active Region 31. Paul was on the ATM console. Thirteen minutes later, Kerwin called back:

Joe: "Houston, Skylab. I'd like you to be the first to know that the PLT is the proud father of a genuine flare. . . . Just about the time you called, he got a high count. And this time it was confirmed by image intensity count over 300, by a bright spot in the x-ray image, and a very bright spot on the XUV monitor. He found the flare in Active Region 31, a factor of ten brighter than anything we've seen. In other words, it was unmistakable once it happened."

Paul got about two minutes of flare rise, surrounded by his crewmates, who had dropped everything when he called. Subsequent crews did much better; but they had the first one, and were "proud as new daddies," as Paul put it.

Pete and Paul, the operators of the Earth Resources Experiment Package, became increasingly skilled at finding and photographing Earth "targets," even through pretty extensive cloud cover.

Paul: "For information, it's hard looking out at 45 degrees forward. You look through a lot of atmosphere. It's hard to see detail. . . . and I got Fort Cobb, and the reservoir. . . . Okay, for special 01, all you're getting is clouds so far . . . very low sun angle clouds; It's like a scene from a biblical movie just before the heavens open up."

Joe: "It's going to break up in a minute over Lake Michigan."

Besides doing the scientific EREP passes, all the crewmen loved taking pictures of the world. They got pretty good at recognizing continents and islands—not perfect, but pretty good.

Houston: "Skylab, Houston with you for six minutes through Honeysuckle."

Pete: "We're just coming up on New Zealand. I think I'll get some pretty good pictures this pass."

Houston: "You're sure that's not Puerto Rico?"
Pete: "You said Honeysuckle before I said New Zealand."
Houston: "Okay."

Honeysuckle Creek was the tracking station in the beautiful hills south of Canberra, Australia, occupied by a few kangaroos. It's closed now and very peaceful. Pete knew he was nowhere near Puerto Rico.

Each man had his favorites. Pete loved to photograph Pacific atolls; Paul favored the Great Lakes, the Rockies, and Australia and New Zealand; and Joe specialized in the Rockies and Chicago, his hometown; he kept looking for Wrigley Field and the Brach candy factory, where his dad had worked. After the spacewalk on Mission Day 26, they asked permission to use one extra roll of film just for Earth snapshots. Houston approved.

Back on Day 16, the crew had heard that President Nixon had scheduled a Summit Conference with General Secretary Leonid Brezhnev of the Soviet Union at the "Western White House" in San Clemente, California, for 18–26 June. On Day 24 it got a little more personal. Pete received a call from Nixon inviting them to attend. They accepted, and the president wished them all a happy Father's Day.

On Day 25 Pete relayed Skylab's respects to the Russian cosmonauts; this crew had now broken their duration record of nearly twenty-four days, set on the ill-fated Soyuz 11 flight. (The Soyuz 11 crew had been lost due to a loss of spacecraft pressure during its return from Salyut 1 in June 1971.) The following day there was a reply from Vladimir Shatalov, "Congratulations and a safe return." The crew noted with satisfaction the last EREP pass, the last run of each medical experiment. There was one more major chore to do: a spacewalk to retrieve that precious Apollo Telescope Mount film.

Around this time, Kerwin had written a poem to his wife that tried to capture the sensations of living in space:

I'm getting used to knowing how to fly.
When I was young I used to fly in dream
Up ways so high and easy it would seem
As if Earth wheeled and slanted, and not I.

And now it's real. We move that way at will,
Like dust motes in a sunbeam. Push away,

Drift down your own trajectory, tumble, play
And who can say what moves and what is still?

In this high sunlit ship the laws of space,
Height without vertigo, mass without weight,
Entrain our nerveways to their easy pace
As if this rhythm were our native state.

What if Man were an exile from the sky?
Are we, perhaps, remembering how to fly?

Mission Day 26 was another EVA day, and the crew was up at 2:00 a.m. Houston time. Pete's jobs today would include brushing away a tiny piece of debris from the rim of the solar coronagraph (it was blurring the view) and attempting to free a stuck relay in one of the battery charger relay modules by hitting the airlock skin over it with a hammer (it was preventing the battery from charging). At 5:45 Pete took stock:

Pete: "All, right now, let me just stop one second. I got the brush, I got the hammer, I got two film trees and I got an EV-1 and an EV-2 [him and Paul] in the Airlock. Is that right?"

Brushing off the debris proved easy. Tapping the relay was a bit more complicated. Pete had Rusty Schweickart, who was acting as Capcom, describe twice exactly where to hit. Joe, inside, made sure the charger was turned off. Then Pete gave the relay housing several mighty bangs.

Paul: "There it goes. Yes. Boy, is he hitting it! Holy cats!"
Joe: "Houston, EV-3. He hit it with the hammer. I turned the charger on, and I'm getting a lot of amps on the battery. Do you want to have a look?
Houston: "Okay. It worked. Thank you very much, gentlemen, you've done it again."

Pete and Paul scrambled back into the airlock after just one hour and thirty-six minutes, with all the film. The crew pointed out that they had done their thing with a hammer and a feather, sort of like Galileo (or the Apollo 15 crew on the moon). And that evening, Pete read the following message from NASA: "To Captain Charles Conrad, Jr. On or about 22 June 1973, you and your crew will detach from Skylab One, leaving it in all respects ready for the arrival of the Skylab 3 crew on or about 27 July, 1973. You will then proceed by space and air to the *USS Ticonderoga* without delay, and report immediately to the Senior Officer Present Afloat for duty."

The next day was Day 27, Wednesday, 20 June. Skylab was cautioned that morning not to record anything requiring immediate attention on B Channel—they'd be home before it could be retrieved and acted on. The very last medical experiment was run—a final exercise tolerance test with Paul as the subject. Then there was a press conference.

The conference was relaxed and upbeat.

Joe gave his preliminary appraisal of the medical effects of a month in space: "Right now the score is 'Man, three; space, nothing.' . . . What's been such a pleasant surprise is how nice we feel. We're able to get up in the morning, eat breakfast and do a day's work. I'm tremendously encouraged about the future of long-duration flights for that reason."

Pete's appraisal of the most significant accomplishment was "that we have now a ninety percent up-and-operating space station to turn over to the SL-3 crew." He went along with Joe on the crew's condition. Neither Pete nor Paul thought they'd eat as much as they did. And Pete thought he was in better shape than at the end of his eight-day Gemini 5 flight.

Paul emphasized how important it had been to have very high fidelity trainers and simulators on the ground. "And the things that are easy to do in the trainer are easy to here, ninety-eight percent of the time. And vice versa." Their advice to the next crew: "Don't forget the learning curve, don't worry about your training, have fun."

With that over, they started packing and got so far ahead of the flight plan that they decided to go to bed another hour early. Tomorrow was going to be deactivation day.

First call on Day 28 was at 1:00 a.m., and for the first time, Houston woke the crew with music: "That's 'The Lonely Bull' for you, Pete." Pete said, "You should have started doing that on Day Two," and a tradition was born. Ever since Mission Control has specialized in playing wake-up music for Shuttle and International Space Station (ISS) crews, tailored to their personalities.

They raced around the workshop. In fact, Paul clocked one complete traverse from the Command Module to the Trash Airlock at "sixty seconds loaded with gear, twenty seconds at max speed"—just to help out the activity planners. They took front and side "mug shots" of one another for the doctors. Joe squeaked his rubber ducky, the one his brother Paul, the Marine pilot, had carried on missions over Viet Nam. Pete said, "It's like a day-before-Christmas party up here, Hank."

Houston (Hank Hartsfield, the Capcom): "You know, it's 5 in the morning down here."

Paul: "How about giving him something to do, Houston, will you please?"

Houston: "Can you stomp your foot up there in zero-G as easy as you can in one G?"

Pete: "You bet your sweet bippy, you can also go 'Ah—haaa!'"

Paul: "You can only stomp once."

So everything was sailing along. Then it happened. The Trash Airlock jammed.

Pete (ten minutes before eight): "Okay, Houston. We've got some bad news for you. We were jettisoning the charcoal canister through the trash airlock per procedures, and it has hung itself in the airlock. . . . We're working the problem, but—it'll be pure luck if we bounce it off that lip and get it out of there."

Pete, Paul, and Houston began to work the problem in an atmosphere of grim hurry. No place to dump trash would give the next crew a terrible problem. Story Musgrave, Joe's backup, went over to the mockup to try to reproduce the problem and solve it. Finally, at 9:15, Paul reported: "So having wound up there [at the end of a malfunction procedure that didn't work] we started working on it a little more. And by judicious application of muscle, we did manage to get it up and free. So the trash airlock is operative once more." In other words, they kept fooling with it till something worked—just like you fix things at home. Everybody sighed with relief and pledged never again to put something that big down without taping up all the edges. And they carried on. There would be only one more glitch before the mission ended.

To bed at 2:00 p.m. Houston time, the crew played "America" to the satisfaction of Mission Control. Up at 7:00 p.m., sleepy and in for a long day. They'd be tired and ready for bed about the time they hit the water. Joe told Karl Henize, the Capcom, "It's wonderful of you to pretend it's morning, just for us." Lots of last-minute questions, cross checking that they had the right procedures, messages, and times. There was another review on exactly how to mate the docking probe and drogue, which had nearly sabotaged the mission on Day 1, and how to proceed if they didn't mate. (They did.)

The last problem was that Skylab's refrigeration system now began warming up. Houston worked the problem for nearly four hours while the crew finished stowage and donned suits. Would their undocking be delayed, canceled? Finally Houston decided the system's radiator, positioned right aft at the end of the workshop, where the engine nozzle would have been, had frozen up. They maneuvered the cluster to point it at the sun. The crew closed the tunnel hatch and waited in their couches for a go to undock. At 3:30 a.m. it was delayed. At 3:54 it was given; the radiator was unblocked and the loop was cooling down.

Pete flew around Skylab for a farewell inspection and photos; it looked small and friendly as they backed away, with its lopsided solar panel and crumpled parasol against a cloud-flecked ocean background.

The first of two deorbit burns came at a little after five, followed by the last star sightings through the Command Module's telescope. Joe got drinks for everyone before strapping in for the final burn and decided to save his until after splashdown. At 7:30 Houston gave Skylab the weather in the recovery area.

Houston: "There'll be two recovery helos, with the call signs Recovery and Swim. And you're being awaited by the U.S.S. Ticonderoga. And we're waiting to see you back here in Houston, too."

Pete: "Alrighty. You can relay to the Tico, 'We've got your Fox Corpen and our hook is down.'" [Pete was playing the Naval aviator coming in for a landing on the carrier's deck. Fox Corpen is the ship's heading. It sounded great to the rest of his crew.]

The final deorbit burn was successful at twenty-one minutes after six (Pacific Time). Joe and Paul were surprised to note that they "grayed out" a little during the burn. Pilots knew that fighter plane maneuvers that produced high levels of acceleration—loops or very tight turns—could drain the brain of blood and produce a reduction (grayout) or complete loss (blackout) of vision, or even loss of consciousness. The Service Module engine only produced about one G worth of thrust. That was normally a trivial acceleration. But nobody'd been weightless for a month before.

Joe: "I went kind of gray and then I was coming back."
Paul: "I think what gets you on that is the spike [abrupt] onset."
Joe: "We'll see; there's no spike onset to entry."

Entry G force would build up to about four and one-half Gs but very gradually. No problem was really anticipated; but they did rehearse what switches had to be thrown to assure successful splashdown, and by whom.

Joe: "Remember, ELS Logic to AUTO if you're blacking out."
Pete: "Right."

Nobody blacked out. Pete got the Earth Landing System switch to AUTO right on time. And at 6:45 Skylab contacted Recovery, at 4,500 feet, with three good main chutes.

The USS *Ticonderoga,* CV-14, was a proud old ship at the end of its thirty years of service. Recovering Skylab I would be its last cruise. Everyone knew that, and it gave the ship a sense of celebration, regret, and tension.

Tico was commissioned on 8 May 1944, the sixth Essex-class carrier and the fourth ship to bear her name. She fought hard in the Pacific, surviving two kamikaze strikes in early 1945 to steam into Tokyo Bay on 6 September, four days after the formal surrender. She made several roundtrips stateside, bringing thousands of soldiers and sailors home in Operation Magic Carpet, then was placed in storage in 1947. She was reactivated and converted in the mid-1950s, adding steam catapults and an angled deck to take on modern jet fighters, the Skyhawk, Phantom, and Crusader among them. She served the Navy as an attack carrier for the next fifteen years. In 1970 Ticonderoga underwent her final conversion, configured this time for antisubmarine warfare, helicopters instead of jets. Among her missions were the recovery of the crews of Apollo 16 and 17 in 1972.

She steamed out of San Diego about a week before splashdown. Aboard was a team of recovery and medical experts from NASA. The medical team included physicians from the U.S. Navy, Air Force, NASA, and Britain's Royal Air Force. Also aboard were two women (unusual at that time)—the lead press pool reporter, Lydia Dotto, was science writer for the *Toronto Globe and Mail.* Doris Rodewig, an artist from New York City, was invited by the Navy to record the recovery. The two were given the admiral's quarters for the short cruise.

There was plenty of action aboard. With the recovery team in the lead, the ship's crew spent the time rehearsing the entire process, using a "boilerplate" Command Module that could be put in the water and hoisted. The medics were preparing their equipment in the six medical trailers deployed on the hangar deck to receive, examine, feed, and house the crew. The deck crew vacuumed and swept the twenty-two yards of red carpet between the port aircraft elevator and the trailers.

A lucky ridge of high pressure had kept bad weather away from the landing area, but as the morning of Friday the twenty-second dawned, multiple cloud layers threatened to give the helicopters a hard time sighting the capsule. Lydia Dotto wrote: "On the bridge, a dozen officers talk in muted tones, waiting for the fix on the spacecraft as it comes down. Navigator Commander Newton Youngblood and his men huddle over their charts, plotting the ship's course. Dials and gauges glow red in the darkened room. Now, as the cloud-shrouded sun brings a grey light to the sky, everyone waits for the sight of the three eighty-three-foot parachutes. . . ."

Just as the spacecraft reentered Earth's atmosphere, the clouds began to break up. A slash opened in the sky, the tops of the clouds glowing red from the rising sun.

The Command Module, its three orange-and-white parachutes gauzy in the morning light, dropped right through the break. As it splashed into the water, four recovery helicopters converged on the scene, dropping swimmers with rafts and a flotation collar to stabilize the spacecraft. Lt. (jg) Tim Kenney, commanding officer of the swimmers, gave a thumbs up to the choppers, meaning he'd established radio contact with the crew, and they were OK. With that, Capt. Norman Green took the con and steered *Ticonderoga* the final six and a half miles. Engines back one-third, and she steadied beside the spacecraft.

For the astronauts, water impact had not been very hard. Pete hit the chute release switch promptly, and the spacecraft bobbed to the upright position. Joe took his and his crewmates' pulses: lying on the couches, eighty-four for Pete and Joe, seventy-six for Paul; semistanding in the lower equipment bay, about ninety-six for everyone. They were fine, but those heart rates showed that they were fighting the unaccustomed gravity. Pete and Paul returned to their couches; Joe fetched the strawberry drink he'd prepared before reentry and chug-a-lugged it. His gut told him almost immediately that this was a mistake. He paid for it later aboard ship.

Hoisting up the Command Module and depositing it carefully on the elevator at hangar deck level was routine. For six anxious minutes those outside waited while yellow-overalled technicians prepared the module for opening. There was a moment's confusion because Pete had already unlocked the hatch from inside.

Pete was determined that this crew was going to egress unassisted. He knew the cameras would be on them. "There's no way we're coming out of here on litters," he told his crew. Mel Richmond, in charge of the NASA recovery team, said that Pete was right at the hatch when it was opened—"He was on his haunches, ready to jump out." According to Lydia Dotto, he looked "like a man from Mars peeking out of some outer-space vehicle for his first look at Earth." With a hand on each arm he was eased to the platform and immediately given a blue *Ticonderoga* baseball cap to replace the white fireproof model he was wearing. Kerwin was already feeling seasick after his strawberry drink; Pete said to Mel and Dr. Chuck Ross, the Skylab I Crew Surgeon, "We need to get Joe the hell out of here; he's not feeling that good."

Weitz appeared next, then Kerwin. Each tested his legs gingerly and waved to the cheering Navy crew. Then one by one with one hand on the railing, they descended the platform's steps, each accompanied by a NASA physician; Conrad with Chuck Ross, Kerwin with Jerry Hordinsky, and Weitz with Bob Johnson. With smiles and waves they walked slowly and with wide gait down the sixty-six-foot-long red carpet to the interior of the hangar deck and the Skylab Mobile Medical Laboratories. All three felt vertigo, the sensation that the world was spinning, when they moved their heads; and all felt abnormally heavy. Weitz likened it to riding a centrifuge at four times gravity.

A full day of medical testing was planned. All the researchers wanted to get that precious data on the crew's response to gravity on recovery day, "R+0," before any readaptation had taken place. But it was apparent that the men were fighting serious fatigue—they'd been up for seventeen hours on little sleep before arriving on *Tico*'s deck.

Conrad was in the best condition; his in-flight insistence on a lot of exercise had paid off. He got through all the testing, including a treadmill run and Lower Body Negative Pressure. Weitz tolerated the LBNP about as well as he had in flight. He undertook the ergometer run but was unable to finish. Allowed to lie down, he recovered rapidly and completed his other tests. Kerwin threw up the strawberry juice in the trailer. He felt a little better after that, but the LBNP run was only carried to the second of three steps, and the doctors decided not to ask him to run on the ergometer until the following morning.

After a couple of hours of medical examinations, the crewmen called their wives. Kerwin told his wife, Lee, that he was tired and seasick and had thrown up in the medical trailer but would be fine after a good night's sleep. Armed with that knowledge, Lee was able to respond later to a call from Dr. Chuck Berry. He told her that Joe was pretty sick, and they didn't know whether it was cardiovascular or vestibular. Lee said, "It's vestibular, Dr. Berry."

That afternoon Conrad and Weitz spent half an hour on the flight deck. They were surrounded by NASA people and a Marine escort, all wearing surgical face masks. They were greeted by Captain Green, who did not wear a mask but was careful to stand well downwind from the two astronauts. The reason for the face masks was that the astronauts still represented a valuable and unfinished medical experiment. NASA intended to collect detailed data for the next three weeks while they recovered and did not want illness to bias it. Green apologized for the strong wind blowing across the deck. "I haven't seen any wind or sunshine for twenty-eight days," responded Conrad, "so don't apologize. It really feels great."

By Saturday morning the astronauts were feeling human and hungry. All participated in a full schedule of medical tests. Kerwin especially seemed much better. The trio was allowed some time to walk around on the deck area for exercise and relaxation, while Navy people maintained the prescribed distance.

That afternoon Dr. Ross was summoned to the bridge by Captain Green. The skipper told him that a very important private phone call was being linked to the ship for him. The phone rang; Captain Green activated the speaker and a voice said "stand by for the President." Ross remembered, using the third person for himself:

"It was only a matter of a few seconds, with Ross 'reeling' from some disbelief, and not from the ships to and fro action, that the voice on the other end said "Hello, Dr. Ross, this is President Nixon." At this moment Ross still could hardly believe in the reality of the situation, but recognized quickly that this could not be one of the SL-1 crew or other playful astronauts in

Houston playing the 'supreme joke' to the discomfort of a NASA Flight Surgeon. Dr. Ross in a definitely higher pitched voice, while attempting to regain full self-control of himself, replied 'Yes sir, this is Chuck Ross.'

"With full appreciation for what the SL crew had just done for this country President Nixon requested a visit by the Skylab I crew the next day to the 'Western White House' in San Clemente. He did state that he hoped all the individuals were well enough to attend and it was his understanding already from NASA that if Dr. Ross gave the 'thumbs up medical clearance' the event could take place. Ross's mother had not raised an Einstein, but although he had not written the book 'Personal Presidential Communications for Dummies' (still does not exist) he thought that the event had most likely been blessed from NASA Headquarters."

Actually, Ross had some trouble convincing Dr. Hawkins back in Houston that all three were fit for the visit. Hawkins had attended to yesterday's reports that Kerwin was pretty sick and wanted him kept behind. Chuck "had to do a real medical sales job," he recalled but eventually prevailed. The only nonnegotiable condition was that the crew wear masks, since obviously the president wasn't going to. The word was passed: "You might like to know that the wearing of surgical masks by yourself and the crew is understood by the president and is thought to be the proper thing to do."

NASA flight surgeons never know quite what their duties are going to involve. That evening Ross discovered that due to predicted fog at San Clemente, the ship's helicopter would take him and the crew instead to the El Toro Marine Corps Air Station, whence they would drive to San Clemente with a military escort. And to preserve the crew from possible infection, he would be the limousine driver. And he was told what the crew already knew—the President was going to be accompanied by Leonid Brezhnev, the leader of the Soviet Union. Tomorrow was getting more complicated by the hour.

The crew wanted to bring gifts to present to Nixon and Brezhnev. Volunteers from *Tico*'s carpentry shop turned to and worked through the evening to produce three shadow boxes to hold flags and patches that had flown on Skylab. The astronauts complimented them on their work and gave them patches as well.

At 0800 hours (8:00 a.m. to you landlubbers), with the *Ticonderoga* five miles offshore and steaming toward its San Diego home port, the crew and

29. Conrad and Kerwin (sans masks) are greeted by Brezhnev and Nixon.

Dr. Ross launched from its deck in a helicopter piloted by Cdr. Arnold Fieser. It was the last operational flight to depart *Tico*'s deck. At El Toro the limousine turned out to be an old Rambler station wagon (to Chuck's relief.) The Marine escort in another vehicle was a bit speedy and lost them at a stoplight—but came back and found them again. They reached San Clemente on time and were escorted to the outdoor Protocol Area where the meeting was to take place.

Then came Ross's undoing. As the president and the Soviet leader approached, Conrad turned to his fellow crewmen and said, "This isn't right. We're not wearing masks in front of the president." He took his off and stuffed it into the pockets of his Navy whites, and Kerwin and Weitz followed suit. There was nothing Chuck could do.

The meeting went well. Gifts and compliments were exchanged, and the astronauts were invited by Brezhnev to visit the Soviet Union. They then flew directly from San Clemente to the *Tico,* landing about noon. The crew was relaxed, with just a few more medical tests before flying home to Houston. Ross was not; he still had to contend with Dr. Hawkins.

The crew took a short break upon arrival for a lunch of specially prepared Skylab food and fluids. Then the medical tests required for R+2 were accomplished efficiently and professionally with the crew looking very good. An abbreviated call was made from the docked *Tico* to Houston at 4:15 p.m. to present the most important clinical information while leaving some of the research data to be discussed in the following days back in Houston. Ross continues with his recollections:

"There was so much emphasis on making the 5:00 p.m. (PDT) takeoff time that the conference call was cut short. In a very short time this would be rectified from the flight deck of the returning aircraft. In fact some of the JSC hierarchy had a definite need to talk with Dr. Ross; have you ever wondered about the statement, 'Were your ears burning?'

"The SL-1 crew and members of the medical and recovery teams were transferred to a C-141 for the return flight to Houston. Shortly after take-off and when the aircraft had cleared the San Diego traffic area, a call came for Dr. Ross. The radio operator had provided him a good fitting earmuff headset with a boom mike. This snug but comfortable headset was necessary to prevent the next comments from loosening the headset from his head. There was no doubt that the made-up statement "hell hath no fury like a protocol ignored" was coming to fulfillment. Royce Hawkins was definitely upset as he started the communication with the statement: 'Chuck, we saw you on television with your mask on, but where were the crew's masks?' Ross managed to maintain his professional composure as he responded: 'Royce, this is a long story that is better managed face to face when I return to work at JSC on Monday.'"

Early in the week, when the crew and medical team were back in Houston, Captain Conrad took full responsibility for the decision of the crew not to wear the protective masks while visiting the president. At last the "heat" dissipated off the crew surgeon, and life started coming back to the routine of postmission follow-up medical work.

The crew members were welcomed that night at Ellington Air Force Base and reunited with their families. They only got to wave at their children from a distance—no kids, no chance for a school infection for the next eighteen days.

After the program ended, the accomplishments of Skylab 1 were summarized by the astronauts in talks given to the Congressional Committee on Space and Technology. Dr. Kerwin summed it up, after describing the medical findings: "You've all experienced teamwork in your lives, I hope. Real teamwork is memorable. And in space it's just the same. People perform up there the way they do down here. Their capabilities, individually and collectively, and their potential, and their weaknesses are the same.

"Hopefully, space stations will be a reality at some time during the next human generation. Five days before our crew was launched, we went outside in the evening to watch Skylab pass overhead. It moved pretty rapidly, but it shone as bright and steady as a star, and we knew it was going to be up there for a long time. To me, it was as though we were going up to homestead a new state—as though that vehicle were the fifty-first star on the flag. The territory is still open and there's a lot to be done up there. We'll be ready when you are."

8. "Marooned"

There are some things you just don't want to hear in space.
Among them: "There goes one of our thrusters floating by."

The launch of the second crew of Skylab was something of a rarity in the history of human spaceflight. While it's not uncommon for space launches to be delayed, scrubbed, and otherwise pushed back, the launch of the SL-3 Saturn IB was actually pushed forward. Though it had originally been scheduled for 17 August 1973, concerns over the condition of the parasol installed by the first crew and the station's "attitude-measuring" gyroscopes led to a decision to launch the second crew sooner so that the unmanned period could be shortened, and they could assume their role as Skylab's caretakers more promptly. On 2 July the crew was told that they would be leaving earlier than planned and had less than four weeks to prepare for being away from the planet for a couple of months. Launch would be 28 July.

For rookie astronauts Garriott and Lousma, the moment they had long awaited had finally arrived. After seven to eight years of training and simulations, the two, along with veteran Bean, were about to be on their way to space. Jack Lousma was struck by the way the Saturn IB looked as the crew arrived at the pad. "It was dark when we got out there," Lousma said. "I remember seeing it steaming away, and the oxygen venting, and the searchlights." He remembered thinking to himself, "It's just like 2001." ("Which was then almost thirty years away, but it's history now," he added.) It was at that point that he realized that he was finally doing this for real; after all those years, the simulations were over. "At least they looked serious about it."

There are a few special moments that somehow get placed into memory bank for the rest of one's life. Since the science pilot lies in the middle couch for launch, he was the last to board so that he wasn't in the way of the other two crewmembers as they got into their couches for launch. The ground crew would first assist the commander into the left couch and get him all strapped in and connected up—a tradition that can still be seen today on television in preparation for each Shuttle launch. Then, after the commander, it was done again for the pilot in the right couch. At these times the science pilot was left standing on the walkway for some five minutes all by himself with his private thoughts some 380 feet above the ground, and looking out over the entire launch complex. "There was a long training period leading up to this moment," Garriott recalled. "A fiery rocket would soon take our speed from zero to over five miles a second in less than ten minutes. Yes, it was probably the most dangerous ten minutes of the entire mission for us—and probably of our entire lives, for that matter—but we had planned for it for years, and we knew the options for escape if that should become necessary. Were we scared? I would say 'no,' but we knew the risks and had a healthy regard for the potential for disaster. Yet it was a very pleasant and introspective few minutes, which I have remembered for decades. Only more recently have I learned how other crewmen, and especially the other two science pilots, still recall and treasure these few moments waiting on the walkway." If Lousma was at all scared at that point, he handled the pressure well—"I fell asleep on the launch pad," he recalled.

Finally the countdown reached zero, and the wait was over. The SL-3 Saturn IB cleared the pad, and the crew was on their way into orbit. "One of the things I remember distinctly about launch was we had to get rid of the launch escape tower," Lousma said. After the spacecraft reached an altitude where the launch escape tower was no longer needed for an abort situation, its motor was fired to separate the tower and its shroud from the Command Module. "When we did, that uncovered all the windows. After climbing to a considerable attitude, the escape tower took off like a scalded eagle. You could see a lot more."

His first experience with staging, when one stage of the rocket burns out and separates and the next fires, is another memory that has stayed with him. "The engine shut down, and we had to coast for a little bit," he said. "The separation of the first stage was memorable for me. A shaped charge cut the [launch vehicle] cylinder all around like a cookie cutter, with a kind of bang, and all this debris was floating around out there in a circle. It was spectacular in that it was just, bang, and all of this stuff went in a sort of disc configuration out around us."

Garriott recalled the experience of reaching orbit as being exhilarating. "From pressed against our couch at several times our weight to floating in our harness in a fraction of a second. We were feeling great, literally 'on top of the world,' cruising along—well, coasting along—on our planned trajectory to reach a Skylab rendezvous in a few more hours. Long-duration weightlessness was new for Jack and myself, but it was not uncomfortable—at least not yet!—and we certainly were enjoying the view."

Unfortunately just as with the SL-1 launch of the Skylab's Saturn V less than three months earlier, the beautiful launch was marred by malfunction. "I was in the center couch and Jack was on my right with a small window near his seat," Garriott said. "He suddenly announced, 'Owen, there goes one of our thrusters floating by the window!'"

And indeed the object Lousma had seen float by was a dead ringer for a nozzle from one of the Service Module's quad thrusters. "I remember reporting it and thinking this was odd," Lousma said. "It was a conical shape just like a thruster, so it looked like a thruster bell, like a thruster nozzle. I don't think that I quite deduced the implications of that at the time because we were so busy with the rendezvous procedures. We were moving onward, noticed that, reported it, went on to the next thing."

While primary thrust for the Apollo spacecraft was provided by the one large service propulsion engine at the rear of the vehicle, directional control was the job of the four smaller quad units, positioned on the outside of the Service Module, near the Command Module. Each quad unit consisted of four engine nozzles arranged in the shape of a plus sign with one nozzle pointed toward the fore of the spacecraft, another toward the aft, and two more at right angles from those. The four quad units were positioned around the Service Module at ninety-degree angles from one another. From the crew's perspective, there was one on the left of the craft; one on the right; one at the top, one at the bottom.

"With a quick look out the window, we agreed it certainly looked like a thruster, but we hardly believed it to be literally true," Garriott said. Bean recalled the sighting being followed quickly by a thruster low-temperature master alarm. Added Garriott, "We promptly realized that there must have been a small propellant leak [oxidizer or fuel, meaning nitrogen tetroxide or hydrazine] which slowly crawled around the inside surface of the thruster and froze into ice in the shape of the metal thruster exhaust cone. Then when that thruster was fired the next time, even briefly, it must have shaken the ice loose and it slowly floated by Jack's window."

With that interpretation, Bean checked with the ground for confirmation and had to turn off the propellants to that quad of four thrusters. With that one quad shut down the spacecraft had three more quads still working fine. This had never happened before in spaceflight and was going to make the rendezvous difficult to pull off. There are no "time outs" in space to fix a problem.

Soon enough the crew began to close in on their target—first just a bright dot in the navigational telescope that grew brighter and began to take form as they got closer. Garriott's excitement increased at seeing his new home growing larger as the Apollo craft approached it. "Soon we were close enough to see the Skylab with the darkness of space as the background for viewing," he recalled. "Then the solar panels of the ATM and one wing of the workshop solar array could be resolved visually and even the orange parasol set by the first crew came into sight."

"Before the boarding, however, we had to complete a successful rendezvous and docking in a crippled spacecraft," he added. Rendezvous required the Apollo spacecraft to arrive in the near vicinity of the Skylab within 330 feet or so and match its velocity with that of the station. The commander had a schedule of quad thruster firings that had to be carefully executed to slowly match his speed to that of Skylab so that they would arrive on station with no relative motion. "Otherwise the crew might arrive at the Skylab rendezvous point with too much speed, or even worse, possibly collide with Skylab in a terrible catastrophe," Garriott recalled. "This actually happened during the manual rendezvous of a Progress vehicle at the Russian space station Mir some years later, with nearly catastrophic results. It should never happen in normal circumstances, but ours was not normal. We had lost one set of quad thrusters and that meant less than full force was available—technically, reduced 'authority'—from the control system. But equally troublesome, this failure also produced an asymmetric thrust since nothing compensated for the one lost quad on one side of the CSM. Any translation such as braking to slow down produced unwanted rotation, and then rotational correction to bring the spacecraft back to the desired pointing direction, or attitude, produced unwanted translation!"

Also every time Bean used the thrusters to slow down, he had to fire them

30. The Service Module thruster quads are visible in this picture of the SL-4 Apollo spacecraft docked with Skylab.

for a longer period than scheduled to compensate for the reduced authority. This sequence—slow down, correct pointing direction, slow some more—was repeated many times during the rendezvous phase, and it all had to be done with precision to complete a successful rendezvous.

As the Apollo craft zoomed along at almost five miles per second around the Earth, its velocity relative to Skylab was only a few feet per second and this had to be slowly reduced to zero at the rendezvous point. Alan Bean said, "Back in the simulator, Owen, Jack, and I were really good at rendezvous. We never missed a rendezvous in all our training time. They gave us failures by the zillions; we didn't blink—we'd rendezvous. During our training cycle they gave us all the failures they could think of. Because they knew we were hot and could do this stuff. We never missed one. So lo and behold, we get up in space, and I remember Jack saying a quad just floated by the window. We thought, 'That can't happen. A whole quad just can't let go.' About that time the master alarm came on for a low temperature of that quad. We quickly realized that it might have been a chunk of fuel or oxidizer ice shaped like the inside of the thruster and that's what Jack saw as we fired the thruster.

"We realized we were lucky we didn't have some sort of explosion and blow that leaky quad thruster right off and really have a problem. But it didn't. So then we had to isolate that quad and not use it again. We'd never done that in all our rendezvous training.

"We went through the failure mode checklist to isolate a quad. We went to the book; I had Jack read it to me. We had circuit breakers for each thruster; throw that one, not that one, and that one. It really incapacitated us a lot. The main effect we had was any time I did anything, we went off attitude in the other axes.

"Meanwhile we are coming up on burns [more thruster firings], tracking and all that other stuff needed to successfully complete the rendezvous. But still whenever I tried to brake, we went off in yaw. That was the big problem. And the amount of braking wasn't the same as with all four thrusters available; it was a lot less. That's where Owen and I got into a discussion that I often remember."

Garriott's job at this time was to help Bean make sure the spacecraft was on the defined trajectory to arrive on station at the rendezvous point with zero relative velocity. In other words he was keeping an eye on how quickly they were "slowing" as they approached Skylab. "My advice to Al on the necessary braking or deceleration required would have been greatly facilitated if we had only had a range-rate measuring device on board," he said. "But in 1973 these had not yet been developed. We had to estimate our 'range rate,' or the rate at which our distance from Skylab was decreasing, by taking two range measurements from our onboard radar transponder at two different times and then dividing the range difference by the time difference. Not the most accurate technique, but we had practiced as best we could. As we began to close, it became clear to me that the standard deceleration protocol, which Al was attempting to follow, was not slowing us down enough."

Lousma, who had been concurrently running the same calculations during the initial portion of the approach, was reduced almost to bystander status during the final phase. "I had to make his backup calculations on the closure rate," he said. "I was sitting there with this little HP calculator and punching all those numbers in, going through this formula and backing up what the ground saw and what we saw in the spacecraft. There had to be a third vote and that was me. I never enjoyed making that calculation. You had to get it right. If you missed one keystroke, you had to start all over again and it was a long one. But that kept me busy. It kept me from bothering everyone else and being worried."

Bean was doing the best he could to balance the competing concerns of attitude and velocity. "One of the worst things you could ever do was slow down too much," Bean said. "Because then you had to use fuel to get closing again, all the timing's off, you came into daylight too soon—all these things were going on in my mind at that time, really zipping. I remember thinking I'd braked enough. We didn't have range rate; we had range only. Owen could use the ranges and times and estimate range rate. He's a great 'back of the envelope' guy, and he would look at the ranges and make a recommendation. I remember braking and braking. When we did midcourse corrections, you only did them with the quad thrusters, we did not do it with the main engine. That's where the problem was.

"Anyway, I braked and braked, but I didn't know for sure what our range rate was. Owen was giving me recommendations, which was good, which we did in training. 'You need to brake a little more.' I remember Owen kept saying, 'We're closing too fast; you've got to brake some more.' Finally after braking for what I thought was at least twice as much as we had ever braked in training, I said, 'No, we've braked enough.' Owen studied the computer range and said again 'Alan, we are closing too fast; you've got to brake some more.' 'No, we've braked enough,' I replied. I was concerned that our closure rate might be too little at this distance to complete the rendezvous. As I looked out my window Skylab seemed very small and far away; at least that is what I thought.

"During training Owen always stayed in the middle seat next to me during the braking phase of the rendezvous, right in front of the computer. Now all of a sudden Owen released his restraints and floated out of his couch down into the lower equipment bay. To say this caught my attention would be an understatement. He'd never done that before when we had a difference of opinion. I'd better rethink my decision, because Owen makes a lot fewer mistakes than I do. And when he believes this strongly but doesn't want to argue with the commander, I'd be wise to listen up and so I did.

"Then I began to actually see that we were really closing. If Owen had not said that, we'd have zipped right by. I can remember Jack saying when we got closer, 'Don't hit it!' That was on my mind too, but I was keeping it where I could see it. You can't maneuver relative to an object unless you can see the object; I had to keep Skylab in the window and keep moving towards it. I had to keep moving along this 'line of sight.' It was not the precise maneuver we had planned and practiced but I knew we weren't going to hit Skylab, because I wasn't going to let us hit it.

"I was also concerned that if we went by Skylab, Mission Control would tell us to wait and re-rendezvous. And that uses more fuel. That would be real embarrassing, even though we did have this failure. I would say that I had the highest heart rate I ever had during my two spaceflights, no doubt about it, more than landing on the moon. So then as we get close, I could see we might be able to stop, maybe, but for sure we weren't going to hit it, and we actually stopped right underneath Skylab. Our best efforts and skills were tested. It was difficult, but it turned out okay; we did it.

"I've heard Kenny Kleinknecht [the project manager at JSC] and others congratulate us for doing it. The quad failure was a big one. They didn't even give us that in training, so we had never, ever practiced that. Looking back on it now, as a crew we did a really good job. But the hero was Owen. If he had not said what he did, I would have sped past Skylab, and we would have had to re-rendezvous."

After rendezvous the crew was to make one fly-around inspection of the whole Skylab at a relatively close distance, less than three hundred feet, to inspect the Skylab exterior. This too was complicated with one quad thruster inhibited. With considerable skill, Bean drove around their new home to be, being careful to not get too close, where the thruster jets might blow away the orange parasol deployed by the first crew, which was keeping the Skylab relatively cool.

Garriott kept a memento of that incident for years afterwards: "We had no general-purpose computers available in Apollo, only the special-purpose computers for navigation and other functions," he said. "So before flight I obtained a HP-35 hand-held calculator to assist me in tracking our motion around the Skylab. We still had to estimate our range and range rate by eye, but we measured angles with the Apollo 'attitude ball,' and I entered the numbers into the calculator. The HP-35 was quite helpful with a small program I had written manually and entered into the calculator on a small magnetic strip.

"When I resigned from NASA some thirteen years later, I still had this now ancient calculator in my possession. Technology was now leaps and bounds ahead of this old 'antique.' But I listed all the government property in my possession at that time, including the HP-35, with a request to pay for and retain it personally. Naturally, this was more than government bureaucracy could manage, so I had to turn it in, after which it was probably junked some years later and lost to posterity as a potentially interesting artifact."

The crewmembers in orbit were not the only ones having somewhat of a bad day. Lousma's wife, Gratia, had returned home on the launch day, 28 July. That same evening after she had seen Jack depart on his adventurous rendezvous with Skylab, she was back home mucking out her horse stalls, even as a heavy downpour of rain threatened to flood their home near a creek in Friendswood. With three small children at home, she had to worry about the possibility of having their car submerged, so she drove it to higher ground and then walked back home in the heavy rain. Finally she just stopped and sat down in the middle of the road and had to laugh at the contrasting situation, from celebrity to soaking stable hand, all within a few hours. (Coincidentally, Joe Kerwin's wife, Lee, had a similar experience with flooding in a thunderstorm not long after her husband's launch.)

The crew had managed to rendezvous with Skylab successfully and dock safely with their new home. By Mission Day 6, things were beginning to look up. The challenge of rendezvous was several days in the past, and after initial difficulties adjusting to life aboard the station, the crew was feeling better. Life on Skylab was beginning to fall into its routine for the second crew. But the problems with the Command Module's thrusters were not over yet.

"When we awoke that morning we were getting right to work," Garriott said. "I was checking my weight (body mass) in the slowly oscillating chair, the time period of the oscillation measuring the mass. Al might have been getting our EVA hardware ready, while Jack was getting out the prepackaged breakfasts for all three of us.

"Jack happened to look out the wardroom window where he saw a very unusual sight and called me over to look. It was the first of a good many beautiful auroras we would see, in this case near New Zealand. We admired the long folded sheets of green 'curtains,' whose slow motion was noticeable with careful observation. It was sometimes tinged with red at higher altitudes, caused by a different chemical reaction in the high atmosphere about ninety kilometers, or about fifty-five miles or more, above the Earth's surface but still more than three hundred kilometers beneath our Skylab perch in space—a most unique opportunity to view. I was just about to call the ground, half a world away, when a 'snow storm' came blowing by our wardroom window."

Since a real snowstorm never occurs in space, the crew immediately knew that something was leaking from Skylab somewhere. Judging that the leak was probably from the Apollo spacecraft docked to the far end of the station, Lousma and Bean zoomed off through the workshop, the airlock, and the Multiple Docking Adapter to the Command Module in a matter of seconds, where they confirmed that another of their spacecraft's quad thrusters had sprung a leak, even though all valves were turned off. With guidance from the ground, the systems were reconfigured so that all propellants to both of the leaking quads were completely cut off.

"I remember seeing that—shower spray was what it looked like—glistening in the sunlight," Lousma said. "Shortly thereafter, the low-pressure alarms went off. Al hustled for the Command Module and shut everything off.

"I think for me that was probably the low point of the mission because it threatened our ability to get our job done, and I wasn't willing to come home," Lousma said. "I've never been afraid of space, but that was a fear that I had—losing the mission—more than anything else."

Bean recalled that the crew got a call from Johnson center director Chris Kraft to discuss how to proceed. They told him that, despite the problems, they wanted to stay and complete their mission. "We were concerned that they were going to make us undock and come home, which we didn't want to do, naturally," he said.

Only two of the four quad thrusters were now usable and an extended debate was initiated, especially on the ground. There were two vital questions that had to be faced. Could the crew maneuver home safely in a Command Module with only half of its quad thrusters functioning? And more importantly was the problem isolated to only those two thrusters? With those were several related issues. The precise cause of the problems had to be identified. It had to be determined whether the two failures were connected. The likelihood of another failure had to be examined.

These in turn raised more questions: Could the crew successfully reenter with only one usable quad if there was another failure? Should they come home right away before there were any more failures? Was it possible to mount a rescue mission for the crew? Could a Command Module be reconfigured in time to allow one or two crewmen to come up to Skylab then return with three more passengers? Most of the answers had to be worked out on the ground with the large assembly of talented engineers and flight controllers. Of course the astronauts on orbit were very much interested in their thinking, and wanted to participate in the decision making as well.

"Basically, we felt secure," Garriott recalled. "Skylab was working well. There was plenty of food and water for many months. The only issue for us was a successful return to Earth. We had worked so long and hard to get here, we certainly didn't want to come home now."

But with so much uncertainty about the situation, work began on planning a rescue mission that if necessary could bring the Skylab II crew home safely.

To some, the situation no doubt seemed to eerily echo a movie that had come out only four years earlier. In 1969 Columbia Pictures had released the space thriller, *Marooned,* based on a novel by Martin Caidin and starring Gregory Peck, Richard Crenna, David Janssen, and Gene Hackman. While the original version of Caidin's novel was set at the end of the Mercury program, the story was updated for the movie version, which focused on a crew of three astronauts that had just completed a long-duration mission on an S-IVB–based orbital workshop. As they prepared for reentry, however, their thruster system malfunctioned, leaving them unable to come home. In hopes of bringing the crew home safely, a daring long-shot rescue mission was mounted. As in *Marooned* the thruster problems encountered by the crew on orbit sparked work on the ground to prepare a rescue mission. However, the real-life effort was not the daring desperation ploy of the fictional version. In fact planning for the possibility of a rescue mission had begun years earlier.

The first step toward the rescue mission was formalized with George Mueller's flipchart sketch of a rough version of what would eventually become

Skylab, which led to the creation of the Multiple Docking Adapter with its spare radial docking port. Unused during normal operations, the adapter provided means for two Command Modules to dock with the station simultaneously should there ever be such a need, among which was a rescue mission. If, for whatever reason, it appeared that a crew would be unable to return in the Apollo spacecraft they flew into orbit, a second Command Module would be able to dock with the station at the unused radial port. Plans then called for the disabled capsule to be jettisoned before the Skylab crew left on the rescue vehicle, freeing up the axial port to be used by the next crew. Until the rescue crew arrived, however, the disabled vehicle would be left attached to Skylab so that its communications equipment could still be used.

The next step in making a rescue mission possible was to modify a spacecraft to be able to carry more crewmen than the three in a standard Apollo Command Module. Without the technology for autonomous rendezvous and docking, the rescue craft would have to be launched manned, and each seat filled on the way up would be one less available for the ride back. Since there were three astronauts in the Skylab II crew, a standard Apollo capsule would not be able to bring them all home.

Ironically, Jack Lousma and Alan Bean, members of the very crew for whom the rescue mission was being planned, had played an important role in the design of the rescue-mission spacecraft. By late 1971 work on the rescue vehicle configuration was well underway, and testing had begun on some of the modifications. "Alan and I had worked on the configuration for the Command Module for five-man reentry," Lousma said, explaining that the two of them were picked to provide operator input on the design of the spacecraft not because they seemed like they might need to be rescued but rather because it was thought they could well be the first people that might have to fly it in the event that a rescue mission was needed to bring the first crew of Skylab home."

The pair, Lousma said, spent a considerable amount of time at Rockwell, going through the same sort of design reviews for the modified Apollo that would have been needed for any new spacecraft. "We configured it such that there would be two couches on the floor underneath the main couches, one on each side of the package between us, which was going to be the critical experimental data," Lousma said. "Three people would come

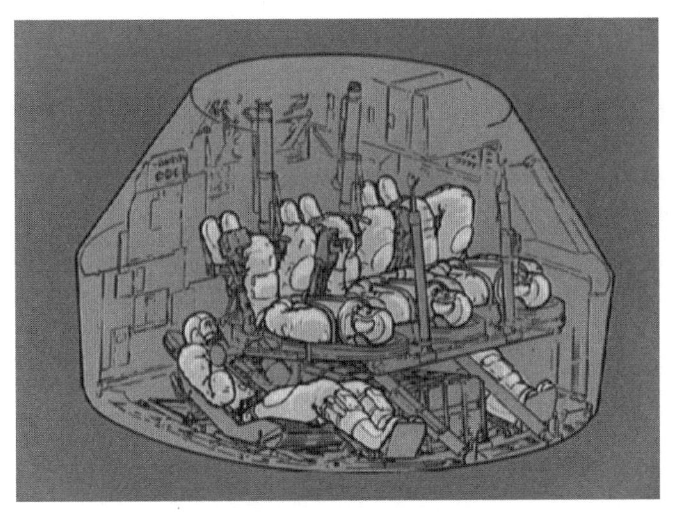

31. Modifications would have allowed the rescue Command Module to carry two additional astronauts behind the three standard couches.

down in the main couches, and two would be in the couches under the left or right seat.

"They had couches that fastened to the inside of the heat shield. It was like a molded seat you might lay in on the beach. It probably just had some tack-down, tie-down, or fasten-down points. So when Pete went up, that configuration was already confirmed."

The biggest concern, he said, involved the potential "stroking" of the upper deck of couches. Those couches, the three that were standard on an Apollo Command Module, were designed to stroke, or have their supports compress like an automobile shock absorber, in the event of a hard landing. While usually unnecessary for a water landing, the stroking was an additional safety feature included in the event that for some reason a crew had to make an unplanned landing on hard ground. If that happened, the supports would absorb some of the force, ideally preventing injury to the crew. For the rescue mission, the concern was that a couch that stroked would drop onto the astronaut in the couch below. However since no couch had ever stroked during the Apollo flight program, the risk was considered minimal.

The addition of the two additional couches came at the sacrifice of a substantial amount of stowage space in the lower equipment bay, so bringing the crew home from Skylab in the rescue vehicle would mean that they would have to leave behind much of what they would have otherwise brought back with them, including results of experiments conducted during their stay.

Between the two crewmembers was a stowage area that would be reserved for the highest-priority items to be returned to Earth, and any leftover space in the lower area would also be filled for the trip home. "There was a priority list of what we wanted to bring back because we couldn't bring it all back," Lousma said. "Otherwise, the whole bottom was filled with bring back. Whatever they thought was the most important would come back there."

"Ironically," Bean said, "the highest priority items in premission planning were the frozen urine samples and dried fecal samples. They would then be studied to ensure it was safe for the next crews to stay even longer in space."

With nothing they could do about the thruster situation for now, the crewmembers on orbit moved ahead with life aboard Skylab. Meanwhile, on the ground, two astronauts learned that they were being called up for prime crew duty for the rescue mission. Commander Vance Brand, science pilot Bill Lenoir, and pilot Don Lind were the backup crew for both the second and third manned Skylab missions. All three men were unflown rookies. The two pilot astronauts, Brand and Lind, had joined the corps as members of the fifth group of astronauts selected, while Lenoir was a member of the sixth group, the second class of scientist astronauts. In addition to their backup crew duties, Brand and Lind had also been assigned as crewmembers for the theoretical contingency Skylab rescue mission. Those duties consisted mainly of providing crew input on the planning. They were involved, for example, in testing procedures for use of the modified vehicle. In essence they were the prime crew for a flight that did not exist.

With the problems being experienced on orbit, however, that mission changed from theoretical to imminent. "I don't remember the exact time that I found out," Brand said. "Of course you know that the backup crew included three guys, and if you had a rescue, there's really only room for two crewmen going up so that five could come down. Fairly early on, without much delay Don Lind and I found out that we would be the rescue crew. We were pretty enthusiastic because we hadn't flown in a spaceship."

"I suspect that Bill was disappointed that it wasn't him, but Don was elated of course," Brand said, adding that all of the members of the crew had trained for each role and that any of them would have been qualified for any role. (Lind, in fact, went on to make the switch for his Shuttle mission from pilot to mission specialist.) "I was not in the discussion that selected the crew. We just found out. Both were capable of doing that job. Bill was a scientist but also an excellent engineer and pilot. Everybody cross-trained for everything."

Once they were assigned to the rescue crew, Brand and Lind hit the ground running preparing for the mission as did many engineers, flight controllers, and others throughout the agency and its contractors. "We had about a month to get ready," Brand said. "I know that we decided very quickly after they had the two thruster quad failures. Everybody felt really under the gun. The hardware was being prepared at the Cape in a typical fashion. The agency—but mostly JSC, really—was responding to have everything ready in a month. We were completely serious about this. If anybody was thinking about the alternative, which is what really happened later, that they were able to deorbit, we weren't thinking about that. We very much [believed] we were going up to rescue them."

Several tasks were occurring simultaneously involving several different groups. "You will recall the effort that was mounted when the first manned mission encountered a damaged Skylab and the parasol and all that," Brand said. "Well this was, while not quite that big, on the same order. It was very significant. Everybody was pulling together."

Engineers were preparing the modifications that would allow Apollo Command and Service Module CSM-119 to be used to carry its two pilots and the three Skylab astronauts safely home from orbit and rapidly readying the Saturn IB to launch it. "The Cape had accelerated their preparation of the SL-4 vehicle, and all of the stuff that was to configure it for a rescue was in place," flight director Phil Shaffer noted. "So I think we could have gone fairly quickly." In addition engineers on the ground were also working to figure out exactly what had caused the thruster problems in orbit. Relatively quickly they came to the conclusion that the two leaks were isolated incidents with little chance of the other two quads failing.

Brand and Lind spent long hours in simulators not only training for the specific requirements of this unusual mission but also making dry runs on the ground to make sure that everything would work as planned. In addition, they were providing crew input to the other groups as they worked on different aspects of the mission. "We were involved in not only training but the planning, certification and verification, and stowage and that the couch [redesign] would work. We were just involved in a lot of the general

32. Vance Brand (*left*) and Don Lind in the official rescue crew portrait.

planning on how you would do this, which made it especially interesting," Brand said, adding that the numerous obligations kept him and Lind quite busy during that time. "Those were very long days."

Any fear that the crew in orbit had that their mission might be brought to an abrupt end after the second thruster failure was allayed fairly quickly. A few days after the failure, they were told that the rescue flight could not come and get them for at least a month, meaning that there was little point in not letting them finish out the full duration of their mission.

"Probably the long pole in the tent was getting the vehicle ready to go at the Cape, the Saturn IB and the integrated stack," Brand said. "I recall seeing a launch preparations schedule. I think we would have been lucky to be off thirty days after that. But we were talking about that, aiming for that."

Though the purpose of the mission was unique in NASA's history, its actual flight profile was not that unusual. The launch, rendezvous, and docking portions would be very much like the last two flights to Skylab. (Hopefully much more by the book than the last one.) "It was pretty much a standard rendezvous," Brand said. "They had two docking ports, and we would have just used the unused one."

The time spent on orbit would have been relatively straightforward. "Not much more than required," Brand said. "We would have to make sure certain things were brought back. The primary thing was just getting the people back." Likewise the return to Earth would have been fairly standard despite weight that normally would have been cargo instead being extra crew. "Because of all the similarities with rendezvous, etc., there wasn't so much risk," Brand said. "I guess you would have to say that looking at the overall thing, the main risk is just in chartering another mission. There's always a risk with any mission because you could lose an engine or something.

"Of course, the other risk is anytime you do things in a hurry, there's always a chance you might have overlooked something, though we didn't think we did. And we probably both would have had a little more to do in flight because there were two crewmen instead of three." About a month after work began on the rescue mission, the agency was adequately confident that it could be flown successfully. "They got a long way," Brand said. "We had hardware."

There were a few interesting points about the reconfigured spacecraft, though, according to Don Lind: "One of the funniest things was when they had to reconfigure a Command Module with five seats, and we had to run all the tests and so forth. Well, a Command Module has two stable configurations [when floating in the water], one in the normal point-up position, which they call Stable 1. But it would also float in good stability with the cone pointing straight down. That puts the seats not exactly strapped to the ceiling, but in a very strange position very high up on the wall as it starts to curve into the ceiling.

"So we had to test this. We took a test crew that was going to be rescued. Vance and I had some experience with this thing in Stable 2 with just the two of us. I realized out in the ocean with the waves pitching and rocking back and forth, it was incredibly difficult to tell which direction was down. So when we did this with five crewmembers, I was briefing the other three, and I said, 'You won't be able to tell which way is down, so when I tell you to unstrap, be sure you're hanging on to something because you may feel like you're falling straight up.' Everybody looked at me like, 'Oh, come on, Lind, how dumb do you think we are?'

"Well, it turned out when we got in Stable 2 that Bill Lenoir was the first one to unstrap. And as he did so, he just opened the seat buckle and fell up and slammed against the bulkhead. He looked at me like 'Lind, if you say anything, I'll get you.' Of course, the other two were hanging on when they unbuckled. So there were interesting little light notes even as we were getting ready to fly."

For Brand and Lind, however, helping to successfully plan the rescue mission did not mean that it was time to relax. Instead they were given a new task and had to shift gears and start again. More long hours in the simulator awaited.

Having proven that a rescue mission could be flown, the agency began looking into whether it could be avoided. Brand and Lind worked with a team analyzing how well a Command and Service Module could maneuver without the two thrusters that had failed to see whether it could make a safe return, thereby avoiding the rescue mission.

"Near the end of our preparation period, management said, 'Well, we believe we can do this, now let's set about to see how we can get them down without expending the resources for a rescue mission,'" Brand said. "So just overnight we changed goals.

"We got the simulator adapted to the changed situation," he said. "I spent a lot of time in the simulator on that. I must say in all of my work on the ground in the space program that was probably the most interesting time that I can remember. That whole exercise was very satisfying."

However, the short-deadline nature of the work definitely could be a challenge to proper coordination. "I found out one piece of information that I thought was critical just when I was walking down the hall at work," Brand said. "I spoke to the Draper representative, and he said, 'Oh, by the way . . .'

"He said, 'You know that when the crew up there gets ready to deorbit and they have to use plus-x, if you don't hold full left THC [translation hand controller], that might surprise them. They might go out of control and mess up the flight.' So it was built into the procedure. I mentioned that at some point to Alan Bean, and got the information up to him. And I thought, 'Gosh, why didn't we know that?' Maybe it was before we had an opportunity to simulate that, because I'm sure we would have found it out in simulation."

Despite the significant amount of fuel that had been lost during the leaks, running out of fuel was not something they needed to worry about. The Command and Service Module, after all, had been designed for going to the moon, and flights in low Earth orbit used only a fraction of its capability. The powerful primary Service Propulsion System main engine had to be capable of making the trans-Earth injection burn that pushed a spacecraft out of lunar orbit and back toward its home planet, and it stocked plenty of fuel for making that burn. The Service Module's Reaction Control System may have lost a lot of its fuel, but the main engine had plenty to spare.

Two reentry procedures were developed. The first assumed that the two remaining good Service Module thrusters would be usable. It involved piloting the spacecraft more or less as had been done during the rendezvous and docking, compensating for the missing quads.

The second procedure was even more creative and would not have used the propulsion systems in the Service Module at all. Instead, the entire reentry would have been handled with the smaller Reaction Control System (RCS) thrusters on the Command Module. Combined, those RCS thrusters could have generated enough thrust for the retroburn that would slow the spacecraft down and bring it out of orbit—but just barely enough.

"We had a procedure to do it," Brand said. "These thrusters were only designed to give you attitude control, so you had to figure out a way to beat the system to get translation out of it. I think it involved having two Command Module hand controllers going in opposite directions at the same time, to actually get translation."

The Command Module would have had just barely enough fuel. "I don't think there was much room to waste any, but there would have been enough left after that to control the attitude of the spacecraft [during reentry]. Somewhere I still have those handwritten procedures, copies of them, and they were rather bizarre."

Flight director Phil Shaffer explained: "The solution to the attitude control problem turned out to be putting the CG [center of gravity] of the CSM in the right place. When you translated fore and aft, it would rotate the spacecraft around where the real CG was. Once we figured out we had enough stuff on board to place the CG where we wanted it, then it became just a procedure, which Vance did a wonderful job of working out in the simulator." In particular, Shaffer said, Brand and Lind had to put in a good bit of time figuring out how much burn it was going to take to get the desired reaction. "So it really worked," he said. "Vance was the hero of the rescue team. Later Alan told us that the heads-up on the THC duty cycle requirements had been extremely helpful."

While the rescue crew was hard at work on the ground putting the procedures together, the crew in orbit was becoming anxious about when exactly they would see those procedures. While they had hoped that it would not be necessary to send up a rescue mission that would end their stay on Skylab prematurely, they were now eager to see that the ground had in fact figured out a way for them to come home safely.

"Alan was understandably impatient," Brand said. "It was, 'When are you going to get those up here?' And, just as in the case of Apollo 13, the people who were simulating these were just wanting to be 99.9 percent sure that everything was OK. So we put Alan off a little bit."

Lind said that while the rescue crew task of figuring out how to retrieve the on-orbit astronauts and the backup crew task of figuring out if the wounded spacecraft could make it safely home were both very challenging, they were very different experiences for him. The latter he described as "purely a technical question—do you have the capability to control the Service Module during reentry in all the modes and all the reasonable failure modes? So it's just a mechanical question about can this vehicle survive under any reasonable circumstances in the configuration it has."

The issues involved in the rescue crew mission, he said, were more varied. There were the technical questions of configuring and operating the rescue spacecraft, but there were other logistical concerns not involved in the backup crew work. One of the biggest of those questions that he was involved in, he said, was figuring out what exactly besides the crew would be brought back. "When you put five guys in that Command Module, it's rather intimate to start with," even before the process of loading scientific cargo begins, he explained.

For Brand and Lind the work was accompanied by mixed emotions. After years of waiting, they had finally been assigned a spaceflight. They had done the training and simulations to prove that the mission could be carried out and that they were fully prepared to fly it. Next, though, they were given a task that could cost them their spaceflight. If they succeeded in proving that the crippled Command and Service Module docked at Skylab could carry its crew home safely, then they would also prove that there was no need for the two of them to fly to rescue them.

"It was kind of a two-edged sword," Brand said. "In a way, we had so focused on [the rescue mission] that it was a little disappointing that we wouldn't get to do it. But on the other hand, we understood completely, and we set about working as hard as we could in traditional backup crew mode to help do the flight plan and preprocedures and everything so we could get them down on their own." He said the disappointment of losing the flight was tempered by the knowledge that NASA was making the right decision by not flying the rescue mission. But it was still a bittersweet experience.

"We would jump at any chance to fly. You know, being an astronaut is a lot like being on a roller coaster. You have these high highs and low lows, disappointing events coinciding with the low lows and maybe getting assigned to something and just being top of the world, and so it cycles."

"It's hard to describe our feelings," Don Lind said, "We were the backup crew, and we needed to work out the procedures with the quads so that they could safely come home. You're really dedicated; you really feel not just a professional obligation but also a personal obligation to the fellows on the crew that you know so well to do that job very well. So we did the very best job we could and were able to convince management that we had enough redundancy to safely bring the guys home with the quad problems.

"But we were also the rescue crew. And if we hadn't been so efficient as the backup crew, we would have flown on a mission. After the whole thing's done, you say, 'You know, we're good guys, but boy, are we stupid guys.'

"When you're in that kind of situation, and many of us in the space program had been in the military, so when things really count, you simply knuckle down and work very efficiently. Sure I had to get home and see my family occasionally, and yeah, you require sleep and that sort of thing. Your main emphasis is we've got to get this job done in a very limited time; we've got to work very efficiently. You obviously don't take a day off to go play golf; that's just not in the priorities. You have to relax a little bit, but you have to get the job done, so if you have to get up early in the morning to get in the simulator, you get up early in the morning and get in the simulator.

"You never really hope that anybody has any problems; you just don't allow yourself those thoughts. But it was a long time before I flew. I was there for nineteen years before I flew. I had been in a group that was being trained to go to the moon, and I thoroughly expected to be the second scientist on the moon. Jack Schmitt was obviously going to fly the first because he had the whole geological community behind him, but I was obviously going to be the second one. And by darn they lowered the budget and canceled the last three flights, and only twelve guys walked on the moon, so big disappointment. [Official flight rosters were not made for these missions, and only Deke Slayton knew whom he would have assigned.]

"Then it happened again in Skylab because Vance and I had been the rescue crew, and Vance and I and Lenoir had been the backup crew on the last two missions. It was completely obvious to everybody that when they flew the second Skylab [workshop], which was already built and paid for, that we were going to be the prime crew. Of course, then they lowered the budget, and they cut the second Skylab in half with a welding torch. And it's now in the Smithsonian museum as the most expensive museum display in the world. Those things are professionally frustrating, but hey, that's part of life. After a while, you quit whimpering and press on."

Despite working themselves out of a spaceflight on the Skylab rescue mission, both men would go on to eventually make their way into space. For Brand the wait was relatively short compared to the other astronauts still unflown at the time—a "mere" two years. Along with Tom Stafford and Deke Slayton, Brand flew the Apollo-Soyuz Test Project, the first joint U.S.–Soviet space mission. Stafford, the flight's commander, was a veteran astronaut who had flown most recently on the 1969 Apollo 10 mission that had tested the Lunar Module in orbit around the moon. Slayton was both the astronaut corps' senior member and like Brand an unflown rookie, having been selected as one of the original seven Mercury astronauts but disqualified from flight status due to a heart condition. The ASTP crew launched on the final flight of the Saturn rocket and the Apollo Command Module and docked with a Soviet Soyuz crew in orbit. (Interestingly, the story of *Marooned,* with its depiction of an international cooperation rescue mission, has been cited as a factor that helped inspire ASTP.)

When asked which mission he would have preferred to fly given a choice Brand said, "For the sake of [the Skylab II crew], I guess I would have picked ASTP, but if needed, I would have been very enthusiastic about a rescue mission.

It'd be something that, the rest of your life, would really stand out."

Lind, on the other hand, would not fly for twelve years after the rescue mission he missed out on, a total of nineteen years after he joined the astronaut corps. Though originally brought into the corps as a pilot astronaut, Lind, who had worked as a NASA space physicist prior to his selection, flew as the lead mission specialist on the 51-B mission of Challenger in 1985, the second Spacelab flight.

Looking back, he said the wait was well worth it. "Oh, yes, absolutely. Because the nineteen years was not just standing in line waiting," he said. "For example, I had a [position] in the Apollo program that was very, very satisfying." Lind explained that he was involved in the development of the lunar laser ranging experiment, which involved reflecting lasers off mirrors placed on the lunar surface to make precise distance measurements between the Earth and the moon. He said that his contributions helped make the ranging mirrors the only Apollo experiment still used over thirty-five years after the "corner reflectors" were left behind. "There were some very interesting, satisfying experiences going along, even when I spent six and a half years training for two missions that didn't ever fly."

9. High Performance

What is it like living in space, not just visiting for a little while but actually setting up a home and living and working there for two months? Beyond the novel and unique circumstances encountered immediately, what is day-to-day life like as a "resident in orbit"? In other words what is it like to "homestead space"? And, when you return to Earth, how do you hang on to an accurate memory of the unique experiences you've lived through?

For two members of the Skylab II crew, the best way to remember the details of their homesteading adventure was to maintain an in-flight diary. It would have to be done in the minimal time available after all the science and other work was accomplished. Yet both commander Alan Bean and science pilot Owen Garriott maintained a journal during their time on Skylab, preserving not only a chronology of mission events but also a personal record of their thoughts and impressions during their stay in space.

"We launched and arrived at Skylab on July 28, 1973, called Mission Day 1," Garriott explained. "As you may imagine, we were pretty busy at first and even though I hoped to make entries in my in-flight diary every day, some days were just too full. Still, as I reread the entries today, now over three decades later, the mission flow and a sense of continuity remain. It was actually Mission Day 4, or July 31, before I had a chance to make my first entry."

Alan Bean wrote in his journal after going to bed at night, as a way to wind down his day. Neither of his fellow crewmembers was even aware of the existence of this diary at the time—nor, for that matter, until more than thirty-two years later when he contributed it for this book.

(The excerpts from the Bean diary in this chapter have been modified with direction from Bean, for the sake of clarity. The entire diary is reproduced in unabridged form as an appendix.)

Alan's first writing was done on Day 10 (6 August 1973), although he starts by referring back to events prior to launch:

Bean, MD-1:

Launch Day. I am writing this in the morning of day 10. Could not sleep, EVA today, so thought I might catch up. Slept well early tonight [the night before launch], took Seconal and hit the bed about 7 pm, so did Jack and Owen. Awakened on time by Al Shepard. He and Deke [Slayton] kept track of us the last few weeks more than usual. This has mixed blessings. . . . First there were the microbiological samples. Then physical. Then eat. . . . . Al Shepard rides with us in van as far as the Launch Control Center. I watched him because he held the RDZ [rendezvous] book—when he got up to get off, he forgot [to leave the book] and I had to ask. On the way he told us he was our last minute back up—he then mentioned John Glenn having his suit at the suit room prior to Al's first flight [ready to take Al's place].

Despite the thruster problems during their approach and rendezvous phase, the SL-3 crew was able to dock with Skylab with no further problems. The hatch was opened, and Skylab became the first spacecraft to be lived in by two different crews.

Lousma described his first moments in his new space home: "I remember being in the Multiple Docking Adapter, in which everything was oriented around the circumference. And I never did figure that out for two months." Most of the architecture in the workshop, including the lower deck used for experiment and living areas, had a normal Earth-like configuration, where there was an "up" and "down" as on the ground. But when an astronaut floated through the Airlock Module or the Multiple Docking Adapter, he was never sure what orientation to expect. It always required examination of the experiments mounted around the circumference to get in the proper position to operate the hardware.

Very shortly after the crew entered the Skylab, though, a new problem arose. As they began settling into the station, the symptoms of space sickness began to be felt.

Garriott, MD-4 (9:30 p.m. on 31 July):

Writing in "mid-air"—difficult!—First day, thru rendezvous, no noticeable or unexpected symptoms, altho didn't want much to eat. After rendezvous I began working in the MD/OWS, did notice symptoms of "stomach awareness". Jack did become sick, but problem with adaptation not fully realized.

By MD-2, wanted almost nothing to eat & intermittently became very queasy.

Believe I had one Scop-Dex that evening & an hour later began considerable improvement in feeling & outlook. Still not hungry. Jack sick several times early, then got on Scop-Dex with some improvement. Al usually pretty active but indicates problems with too much head movement. Message arrives saying take Scop-Dex tomorrow & if start, take another one every 4 hours. On this day (MD-2), I do ~30 min of head movements after pill. Jack & Al don't [do head movements], altho they took drug.

NASA was already well aware of the possibility of "space sickness," but the fact that the first Skylab crew did not encounter much if any space sickness may have led people to think that it might not be encountered on later flights as well. "My diary concentrates on the 'stomach awareness' or 'space sickness' for several reasons," Garriott explained. "It was a major objective of the flight experimentation to find out the degree of discomfort and hopefully how to minimize or avoid its occurrence entirely. We were equipped with the best medication available at the time, pills of scopolamine/Dexedrine, which we had all tried before flight in situations challenging to one's vestibular system. For example, in aircraft 'zero-G flights,' which make many or even most passengers (including experienced pilots) nauseous, a 'Scop-Dex' capsule will usually eliminate any tendency to become sick to one's stomach. We had a rotating chair on board Skylab, tested preflight many times. Anyone with a normal vestibular system is essentially guaranteed to vomit when exaggerated head movement coupled with rapid chair rotation is continued for ten or fifteen minutes!"

Lousma had the most immediate problem with nausea, followed by Garriott, and then Bean. The crew was supposed to begin promptly the process of reactivating Skylab after its period of dormancy following the first crew's departure, but the symptoms they were experiencing made getting much done quickly a daunting prospect.

"I started not feeling good when I took my suit off in the Command Module," Lousma said. "We didn't take any medications before we left because we didn't want to slow our reactions, and I didn't expect to feel bad. But when I took off my suit, I started not to feel so good. When I got into the space station, the Skylab, I didn't feel any worse until I really started moving around. But when I started to work, and getting things unstowed and set up, then I started feeling like my 'gyros' were going around and around. I thought, 'If it's going to be like this for two months, it's going to be a long two months!' "

The problem, Lousma added, got worse the more he moved around but would abate somewhat when he rested. "I think probably the best cure was to recover to the point where you don't feel so bad and then strike out again until you did. Every time you could last a bit longer before you had a problem with it. For about two days, I felt just a little bit of vertigo. But I continually improved. I think the thing that was most debilitating was, because I didn't feel good, I stopped eating for a while. I just didn't feel like eating. So I think I got behind the power curve in terms of energy—just didn't have enough nourishment—and it took a little time to build that back up."

The mission commander's experience was much the same. "We all got kind of upset stomachs to different degrees," Bean recalled. "If I would be still, then it would gradually go away. But then I wouldn't be doing any work. So my feeling was, if I would stay still, then I felt okay, but you couldn't activate the workshop that way. I would work as much as I could, and when I'd zoom around and unpack, pretty soon I'd start feeling an upset stomach. So I'd just have to slow down. It just took a lot longer. I never did vomit, but if I'd kept going, I would have."

Bean also recalled that not only did the symptoms slow work down but they made the thought of eating unappealing. Since the crew was going to have to eat, doctors on the ground recommended that they try eating four or five smaller meals a day to see if it would be easier to get through their daily menu with it divided up into smaller portions. Bean was disappointed at the prospect of a work pace already slowed by nausea being further reduced by having to stop frequently for meals. "I kept wondering if the nausea was going to be like this for the whole fifty-six days," he said. "I kept thinking, we can do it but it's sure going to slow down what we want to do. We had all these plans of activating the workshop real quick and getting right to productive experimental work. We wanted to be the best we could be as a team, and so this was distressing, and yet, that's the best we could do."

Lousma recalled, "We all kind of helped the other guy out, and I was probably the guy that needed more help than anybody. But we worked together to do the things we needed to do to get situated. And sure enough, after about five days we got well, and we could just zip around and do our jobs and everything."

The problem kept the crew from getting the start they would have liked on their work, but it did not affect their relationship with Mission Control.

"We were also honest with the ground," Bean noted, "even though it was a little embarrassing; we thought of ourselves as a 'right stuff' crew, and 'right stuff' astronauts don't get sick! Later, when our flight was over, we proved that 'right stuff' crews can get sick, provided they find a way to overcome it and perform well before they come back to Earth."

In addition to their vestibular concerns, the crewmen noticed other physiological changes, including some involving bodily functions of considerable interest to elementary school children, judging by their inevitable questions to astronaut speakers about these matters. "Of course, in reality, everyone has these questions in mind, but it is only the uninhibited children who bring up the issue immediately," Garriott noted.

Bean, MD-3:

We are farting a lot but not belching much—Joe Kerwin said we would have to learn to handle lots of gas.

Got to stop responding to ground so fast and just dropping what I am doing—causes us to run behind on the time line. Do not know just what to do about this. . . .

Still losing a lot of things, too big a hurry. Wish the flight planners would let up. The time taken to trouble shoot the condensate system shoots this whole timeline. Got to stay on schedule.

The intestinal gas issue was not a direct effect of spaceflight itself as it was also encountered by the SMEAT crew on Earth. The cause is more likely that the human body generates a certain mass of gas depending on one's diet, and in a low pressure environment (5 psi in Skylab and SMEAT versus 14.7 psi on Earth at sea level), the gas expands to about triple its normal volume.

Garriott, MD-3:

On MD-3, everyone improving but still slow & inefficient. Incidentally, we have all been working from 0800 till 2200–2400, with almost no breaks. Only a few minutes devoted to looking out window (Fantastic—Gulf of Mex, Hou ⟷ Yucatan; Pacific Coast, Hawaiian Isl., Med[iterranean], etc!) Believe we each had 2 Scop-Dex. Seemed comfortable, except at meals. Meals are bad for everyone. No one sick, Jack worked all day, with difficulty. EVA day keeps slipping!

While the crew was able to go about most of its assigned tasks, albeit more slowly than planned during those first couple of days, changes had to be made regarding another major task—an EVA that had been scheduled for Mission Day 4. "The ground controllers were most sympathetic to our problems, and we all agreed that we should slip our first EVA day until we were feeling better," Garriott said. "It could be a disaster if either of the two EVA crewmen, Jack or I at first, were to vomit while outside in our pressure suits."

Though their trip outside was delayed, the crew was gradually becoming more efficient at the work to be done inside the station and to get caught up had already started working extended hours (something that was to be a theme throughout their stay on Skylab). Still they took rare opportunities to appreciate the unique vista their accommodations accorded.

"An early surprise for me came on MD-3 when I was not yet used to the great distance to the horizon in all directions," Garriott recalled. "We had just passed over Houston, where our homes and families were located, and I was watching out our wardroom window to see all the familiar terrain pass beneath us—the large white buildings at the Manned Spacecraft Center (appearing as small dots), the freeway to Galveston, Clear Lake where the whole family enjoyed (sometimes!) small boat sailing races. It all was past my view in only a few minutes, when I looked toward the top of the wardroom window and there was an island in the middle of the Gulf of Mexico! But there are no large islands in the Gulf! I immediately realized that my field of view extended all the way from Houston to Yucatan, the 'island' I was now viewing."

Garriott, MD-4:

Supposed to be a day off, altho MD-3 was too. No one took Scop-Dex. Al & I very near normal, Jack much improved. We all went thru M-171 protocol [experiment on metabolic activity] on bike, w/o instrumentation & no bike mods. I did 50 head movements of 131 type [vestibular function experiment], w/ no effects. Believe my vestibular system nearly adapted after ~72 hrs, certainly almost so. Appetite improved, but still not good. Had to force down a filet tonight. Whew! Paul B. [flight physician Dr. Paul Buchanan] had good news for later meals—eat what/when we want.

As the ground-based PIs (principal investigators) wanted all the vomit and fecal material to be returned for postflight analysis, it was all placed in sterile bags and inserted into a specially designed pressure-tight enclosure that could be both warmed and vented to a vacuum. In this way all water was evaporated in a few hours, leaving a dry and easily managed residue for return to the ground.

Bean, MD-4:

Jack was taking a cooked fecal bag out of the dryer—laughing—here is a real nice ripe one. I said, bet you are a good pizza cook. No, said Jack, pancakes. We had too many fecals and vomitus bags to cook—

MD-5:

We're in extremely high spirits today, first day we all feel good. Owen said that today we ought to ask for a reduction in our insurance rates because we were no longer running the risk of drowning or auto accident.

The crew had run the LBNP (lower body negative pressure) experiment on Day 5, which subjected their bodies to stress similar to what would be expected when standing erect back on Earth, and all did pretty well. Now that everyone was feeling better, it was time to reschedule the spacewalk. Day 6 seemed about right, and plans were set in motion. Those plans, however, fell through when the second thruster leak was discovered. As the crew and the ground worked to determine the cause and implications of the second thruster failure, the EVA was once again delayed, this time until Day 10. In the meantime, life and work on Skylab continued.

Garriott, MD-8:

Yesterday we all felt perfectly fine. Fully adapted & enjoying 0g. Appetites improving but not up to normal, and "weights" (or, more accurately, body mass) stable for several days.

Their vestibular problems were far enough in the past that the crewmembers were enjoying the experience of weightlessness. They were also working to get caught up after having fallen behind schedule during their days of malaise, undertaking all of the tasks they were able to do. The crew could accomplish all the medical experiments and Earth resources protocols, but the solar physics agenda was greatly constrained because most of the cameras required film that could only be replaced by EVA.

"On MD-9, we accomplished two of the Earth resources operations," Garriott said, "which was kind of a big deal because it required a major change of Skylab's attitude in space. The whole station was reoriented with use of the large control moment gyros and pointed toward the Earth instead of

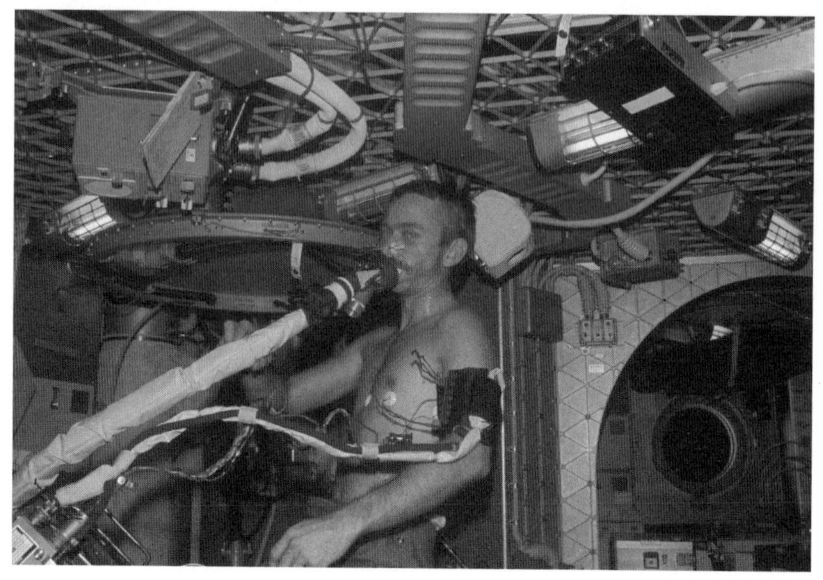

33. Garriott exercises with the cycle ergometer.

the sun. We worked these experiments early, since we still had no film in the solar cameras."

Bean, MD-9:

Left SAL [Scientific Airlock] vent open last night after water dump. Thought I was so good at it, did not use check list—fooled because this was first night without experiment in SAL. [Skylab featured two Scientific Airlocks, which allowed small experiments to be exposed to vacuum outside the spacecraft. One, on the "sun-side" of the station, was used for the sun-shield parasol, and couldn't be used for experiments the rest of the mission. The other, on the opposite side of Skylab, could still be used.]

Owen let Arabella out of the vial. She had been in there since days prior to launch. She had not come out so Owen got the vial off the cage, opened the door, shook her out where she immediately bounced back and forth, front to back, four or five times, then locked onto screen panels at the box edge provided for visualization-there she sits clutching the screen. Owen and I talked of giving spider food because she has not moved one half day. Owen said "no" because when she gets hungry is when she spins her web. She can live two-three weeks without if she has to.

First back-to-back EREP. Jack [looked for] Lake Michigan. But got Baltimore instead. Or Washington, his prime site.

Saw what we thought was a salt flat but turned out to be a glacier in Chile. We could see Cape Horn—Cape Horn and Good Hope all in one day, fantastic.

Owen wanted to know if we had tried to urinate upside-down in the head [the waste management compartment]. He said it is psychologically tough. Jack said he tried it and he peed right in his eye.

Diving thru workshop different than in water—here the speed that you move (translate) is controlled entirely by your push off so for some spins or flips, you can have a ½, 1½, 2½, 3½, etc. body rotations. Difficult to push off straight and to get spins you want. You must watch your progress as you spin—it's tough to learn but to keep from hitting objects, it's a must.

It was a great day—first back to back EREP *and it came off perfect. Jack and Owen good spirits for* EVA *tomorrow—we worked all afternoon and evening on prep, much more fun than on Earth in 1g.*

Owen worked 22 hours today because he counted his sleep cap time. Every day is filled with memorable experiences—sights, sounds, emotions, hope, fear, courage, friendship. I just wish we could go home to our wives at night.

My urine volume lower than Owen and Jack. Been drinking a lot but must do better. Been concentrating on eating too much. Owen said meals were the high point of a day on Earth and here too. Only difference is there it's the start, up here it's when you finish.

I cut a hole in the bottom of my sleeping bag near the feet—too hot, had to tie a knot to keep from freezing in the early morning.

Heard about leak in AM *[Airlock Module] primary and secondary cooling loop. Pri should last 17 days and secondary 60 days. Wondering what ingenious fix they will come up with [on the ground].*

No CSM *master alarm today. Almost a "no mistake" day.*

Arabella and Anita became well-known names in 1973. The public was enthralled by the two "cross spiders" prepared for spaceflight by a high-school student in Massachusetts, Judith Miles. "Fortunately, 'cross' refers to a large mark on their backs, and not to their disposition," Garriott remarked. Miles proposed an experiment to study their method of web formation in weightlessness, which is a clue to their mental activity as they adapted to the microgravity environment. Arabella was released into her fully enclosed box from the small metal container about the size of one's thumb. Her initial webs were very scruffy-looking, but every day they improved after she consumed the last one and spun a new one. The webs finally ended up looking every bit as good as they would in an Earthly garden. Despite Bean's concerns, the spiders remained healthy, and after about three weeks Arabella was returned to her container and Anita released. She proceeded to exhibit the same behavior as Arabella, even after being cooped up in that small container for about a month.

"I remember when we got Arabella out," Lousma said. "This was Owen's job; he's the scientist. We hooked up this box with this open door. He said, 'Hey, Jack, how about helping me get this spider.' So we got the spider out. And it didn't know where it was, the poor spider. Finally, she figures out she can stick herself on something and somehow fasten herself."

Arabella and Anita captured the public's fancy, and Lousma, who gave the public a look at Skylab with his televised tours, admits feeling a bit of jealousy at the time over the spiders' status as spaceflight superstars. "It really disappointed me a little bit that on the ground the general public got more insight into what was happening with Arabella than what we were doing."

The two spiders were the subject of a "gotcha" the second crew considered leaving for their Skylab successors. They had a large (fake) spider and web to place over the docking adapter hatch when they left. Unfortunately they mistakenly thought it had been left on Earth and didn't set it up.

Around this time Garriott began to get calls from the solar science team on the ground about one somewhat obscure item he had not yet completed. The first mission had found that the XUV monitor, which enabled them to see the sun in extreme ultraviolet (XUV) light, had extremely low sensitivity, and the TV display was so faint as to be useless. So the ground developed a small conical light shield that the operator could place over the TV display and peek through the small hole at the apex of the cone. "Even this was still too dim," Garriott said, "so we were provided with a recent technological marvel called a Polaroid camera! When the camera shutter is opened, it remains open until enough light has been passed through to properly expose the film. So this was ideal for us—we mounted the camera at the cone apex looking down on the TV image. When enough light had been accumulated, the shutter closed automatically and the camera developed the print. As you probably remember, it then went 'bzzz' and delivered the print out the bottom of the camera.

"Every day they asked if the XUV monitor camera had yet been installed and operated. 'No, not yet, but I'll get to it soon,' I replied. After three or four days, it became clear they were really interested in how it was going to work, so I took time to set up the conical cover and the camera. When complete,

34. While their early efforts indicated trouble adapting, the spiders were quickly able to spin webs that appeared as they would have on Earth.

including the preinstalled first film pack, I thought I should check out the camera operation before trying it on the sun. As Jack floated into view, I snapped a quick photo, followed by the 'bzzz' and out came the developed print—of a recent Playboy centerfold! All ten sheets of that first package were similarly pre-exposed and we all had a great laugh. But we never said anything to the ground about it until they made their next inquiry about the camera. I then reported that 'Yes, the camera operation was normal and providing quite interesting photos.' That was all that was ever mentioned in-flight, and only on the ground two months later did we congratulate Paul Patterson on the Naval Research Laboratory solar physics team for his creative 'gotcha' and amusing surprise."

Bean, MD-10:

EVA day. I had a tough time sleeping. OK for first 6 hours or so then off and on—finally writing at normal wake up time, 1100Z (0600 Houston) because they let us sleep late. Bed is great. I am going to patent it when I get home. The bungee straps and netting for the head and the pillows were my idea. Might come in use someday because no other simple way to make 0G feel like Earth.

Jack sleeps next to me then Owen at end—the reason, his sleep cap equipment fits better.

Funny how good we feel now, I think [at the beginning of the mission] we all would have said "to hell with this, let's go home". No one ever said it in words but that was the way we all looked at each other around day 2 and 3.

Sleeping is different here because the "bed clothes" do not tend to restrain or touch your body. This causes large air space about your body, that your body heat doesn't hold. It's difficult to snuggle down. Have to put undershirts (long) and t-shirt on during the night. I cut feet out of the long handles then use them for pajamas. Also I mod'ed [modified] my bed by cutting a hole in the netting near the feet, too cold at night so close it up with a knot.

Little worried, Funny—Owen's PCU *[Pressure Control Unit] is #013 and his umbilical is #13. I'm not superstitious, but . . .*

Started taking food pill supplements today. Kit is junkie's paradise.

Jack discovered new way to shake urine collection bags to minimize bubbles. I called ground and said, "we even have our professionals—Owen ATM *check-out, me condensate dump, Jack urine shaking."*

After being delayed for nearly a week, the time for the crew's first spacewalk finally arrived on Day 10. Jack and Owen were assigned to go outside on this first one, leaving Al inside to tend the store and assure everything went well. When the missions were originally planned, before the launch of Skylab, film replacement was to be about the only thing to be done on this spacewalk. But now the work needed to be almost doubled. The parasol that had been extended through the Scientific Airlock during the first mission to shield the workshop from the direct sunlight had been in place for over three months. Its ability to shade the workshop was beginning to deteriorate. Still aboard the station, however, was the second thermal mitigation system that had been launched with the first crew back in May, the Marshall Sail twin-pole sunshade. Installing it should again cool temperatures that were very gradually beginning to rise again as the parasol's efficacy diminished in the sunlight.

"We had the twin-pole sunshade to deploy over the top of the parasol in addition to the film replacement," Garriott said. "So after the film installation was first completed, I had to connect eleven five-foot sections of aluminum poles, twice, forming two long poles. These were then extended to Jack some forty or fifty feet away, where the poles were mounted in a v, and a large 'sail' pulled across them with nylon lines. This may have been the only 'sail' this Marine has ever rigged, and without a bit of wind to fill it out!"

As had been the case with the solar-array wing deployment conducted by the first crew, Skylab had not been intended to support spacewalks like this. No provisions had been made for spacewalkers to get around, save for the limited path installed to access the ATM film canisters. For the work he was to be doing, Lousma had no translation aids provided to help him reach the area from which he would be installing the sunshade. Garriott remained near the Airlock Module hatch to remove the segments of the poles from their packaging, mate and lock each piece together and then extend the long poles to Lousma who had positioned himself far out on the truss structure. (Even without translation aids on Skylab's exterior to help him reach his destination, Lousma was in no danger of floating off into space, since he was connected to Skylab by an umbilical running back to the Airlock Module.) Once he was in place where he would be doing most of the work, Lousma had a set of foot restraints designed to attach to the structure at that location and secure him in place. "You just kind of clamped them on, and you could stand there and enjoy the views," he recalled.

After getting into position, Lousma next had to mount an adapter to the truss that featured two slots into which the long fifty-five-foot poles would be inserted. Garriott began putting the pole segments together with a standard bayonet-type connector. He fitted each segment into the next, depressed a spring, rotated the segment about twenty degrees, and latched it into place. Then a rubber ring was rolled over the fitting, securing the connection. On a later spacewalk, the crew found that this rubber locking ring had rolled back away from its connection, but the bayonet connection had been adequate to hold the segments together. When the two long poles were assembled, Garriott passed them on to Lousma, who fixed them into their slots, so that they stretched all the way to the far end of the workshop. Lousma then had to deploy the sunshade onto the poles, stretching it across the poles with long ropes or "lines," eventually covering almost the whole workshop exposure and the old "parasol" deployed by the first crew.

While assembling the poles, Garriott encountered an unexpected problem. During the preflight testing of the sunshade equipment at the neutral buoyancy tank at Marshall, a difficult decision had been faced: whether to take the flight hardware underwater and then into space and risk corrosion and malfunction or only test it on the dry floor out of the water without the added realism that practice in neutral buoyancy would provide.

"We finally decided that for the twenty-two pole segments, a floor test without pressure-suited operation, would be adequate," Garriott said. "This was about the only compromise made in testing under the most realistic conditions possible. Naturally, this returned to bite me in space. When I had to remove each individual rod segment from their aluminum transporting frame on which they were all mounted—manually, in a pressure suit—my 'fat' fingers in their thick gloves could not get under the rods to lift them against the elastic straps that held them tightly against the transporting frame! I ended up having to squat down in the pressure suit, holding the frame beneath my foot, use one hand to lift each rod upward against the surprisingly tough elastic, and then use my other hand across my body to wrench each rod from under the elastic strap. It may sound simple, but it turned out to be the most difficult physical task of the whole EVA, which we might have been able to modify had we tried it all in a pressure suit on the ground. And I had to repeat all this about twenty-two times! (Send this to the 'lessons learned' department!)"

Lousma recalled that the neutral-buoyancy training had served him quite well. "We learned how long it took us to do each task, and I think it took us twice as long in space. That wasn't because we weren't prepared. It was simply because we had the time and wanted to do it right. And we worked slowly and double-checked and rechecked everything as we were doing it."

There were all kinds of concerns that the twin poles were going to be too "whippy" because of their relatively thin diameter compared with their fifty-five-foot length, which Lousma said made them not unlike a giant fishing pole. "I wasn't worried about that too much," but he could tell a difference as they got longer. Lousma also encountered one unanticipated problem during the spacewalk. "The twin-pole sunshade worked very well, except for one little episode," he remembers. "When you look at the Skylab photos the sunshade is kind of brown, but has a white streak in there."

When the sunshade was packed for launch, it was "folded like an accordion" into a bag. However, because of the rush to get ready to fly during the ten-day period prior to the launch of the first crew, the adhesive used to attach the pieces together had not had time to cure fully before the sunshade was folded up and packed. As a result, when Lousma unpacked the sunshade in orbit and began to deploy it, the adhesive prevented it from unfolding as well as it was intended to do.

"So I had to bring that whole thing back toward myself," he said. "It was all out of the bag and billowing up all over, and by hand I had to unfasten all of those folds. Then I had to attach the two corners that were nearest me with a long lanyard, and drift out to two places on either side of the MDA to attach the lanyards. When the large sail was deployed, the twin poles were flopped down on top of the parasol and against the Skylab workshop, and the lanyards tightened. It nearly covered the workshop and worked quite well. So that was done, and I thought, end of story.

"But it turns out that I had missed one of those folds, and so it was out there like that for a long time, and getting browner and browner. Then the sun did the rest of the job and unstuck that one little piece. And so you see that white streak in there, that was the one that had remained folded for the longest time."

Lousma estimated that the sunshade deployment took up about three hours of a six-and-a-half-hour spacewalk. In addition to the routine ATM tasks and the sunshade, he and Garriott also explored the exterior of Skylab to try to gain clues as to the location of a coolant leak. The source of the mysterious leak would plague the second crew throughout their tenure on the facility. "During our inside time we also had to do quite a bit of exploration, taking some panels off," he said. Removing the wall panels allowed access to the station's "plumbing" but proved to be a difficult task since it was another maintenance activity that had not been anticipated preflight. Assuming that there would be no reason to detach the panels, engineers had designed them to remain firmly in place with no simple mechanism for removal. Despite their efforts, the second crew was unsuccessful in finding the source of the coolant leak. Ultimately a method was devised for the third crew to recharge the coolant supply, yet another unanticipated procedure.

The ATM film exchange provided Garriott with the opportunity to do something he had been looking forward to. "One of the first things I did for fun was something I had planned before flight," he said. "Is there anyone who has not looked over the edge of a high cliff or a tall building and felt an extra surge of emotion and adrenalin at the view? So here I stood at the front end of the ATM solar telescopes to replace film, but could also look straight down a 435-kilometer (270-mile) 'elevator shaft' to the ground! It is a different perspective when in a pressure suit with nothing between you and a hard vacuum other than a thin, Plexiglas faceplate, as compared to looking out the window of a jet aircraft or even the wardroom window of Skylab."

Bean made a special addendum to his diary about the spacewalk:

Jack said, being out on the sun end, was a little like Peter Pan—or that you were riding a big white horse—feet spread wide across the whole world—the Earth

35. The Marshall Sail deployed on Skylab.

is visible on both sides, at the same times and you can see 360 degrees—riding backwards.

Watching out the window as Jack worked in the dark; I could not look at him in the light as he was too close to the sun, it was fantastic to see the sunrise. It began as a light blue band which grew with a fine yellow rim near the limb—the blue gets larger then.

Just before sun up you could see flashes of light toward the horizon where thunderstorms were playing. This pinpointed the coming horizon which was not yet discernable against the dark of the Earth from within the lighted cabin.

Gold color grows in last 15 sec to change much of dark blue into bright orange. As the sun rises the Earth's horizon slowly moves from head to toe on Jack as he is silhouetted against the blue line. It gives the feeling of going around a big planet, a big ball rather than just a disk moving from in front of the eye. The science fiction movie effect was fantastic.

And Garriott's diary summed this all up with only five words:

EVA day—went very well.

Garriott, MD-11, 12, 13:

Full ATM ops. On 11, got a flare right off in AR85! Had been working that AR [active region] all orbit. Very fortunate.

MD-12, dble EREP + more good ATM.

MD-13, M-3 X-ray flare, well covered & then a C-2 or-3 [a classification of intensity], all from AR85. The last covered only by XUV Mon on VTR [video tape recorder]. Also a good S-063 [ozone photo] w/ EREP in AM, and S-055 CalRoc successful! Vy good day, indeed.

Everyone in excellent spirits. Tomorrow is more or less "off day," but we'll stay busy.

Can become disoriented w/ rapid spins. We all still feel some sense of up & down, related to orientation of 1-g trainer & eqpt installation.

Fish orient "down" twd [toward] wall, usually and fairly quiet. But if "stirred up" a little & held in middle of room, still do outside loops, pitching down.

Fed both spiders today. Not sure if they will eat.

With the second crew's Apollo Telescope Mount operations now well underway, the collection of instruments was producing groundbreaking results. The flares, for example, were very exciting for the crew to witness. These energetic outbursts on the sun showed up particularly strongly at ultraviolet and X-ray wavelengths visible to the observers on Skylab with their TV screens—because they were above essentially all of the Earth's atmosphere—but not to ground observers. The ground saw the active regions (ARs) in visible light and could direct the crew's attention to promising locations on the solar disk, but unlike Earthbound astronomers the crew could see the first indications of an outburst from Skylab with ultraviolet and X-ray displays. The "CalRoc" Garriott mentioned on Day 13 was a coordinated observation of the same active region on the sun by Skylab and a rocket flight to high altitudes by the Harvard College Observatory experimenters.

The fish were part of an extra, small experiment that Garriott had asked to do well before launching, for the crew's own interest. Arrangements were made by a veterinarian on the staff of the Houston space center, Dr. Richard C. Simmonds. The experiment included two small mummichug minnows and fifty unhatched eggs in a small plastic bag that the crew taped to a wall or bulkhead. The minnows had the strange, and quite unexpected, response to weightlessness of swimming forward but looping or pitching down. Watching the transparent eggs develop and the fry after hatching also proved interesting. Even the fry born in this weightless environment exhibited some of the same looping behavior. These observations eventually led to one scientific paper and later several more space experiments—and more scientific papers—on later Shuttle Spacelab flights.

Bean, MD-11:

Passed the LBNP today for the first time. Think I was too far in it and squeezed around stomach, cut off blood, will move saddle from 9 to 6.

Did a lot of flying about the workshop just before sleep tonight. Skill needed, but great relaxer.

Wish Owen would move Arabella. Arabella finished her web perfectly. When Owen told Jack at breakfast, Jack said "well that's good, I like to see a spider do something at least once in a while".

MD-12:

My green copy of Childhood's End *floated by. If you wait long enough, everything lost will float by. A dynamic environment no one can be stranded in center of a space because small air currents have an effect.*

Tried to fly (like swimming) last night. But air currents much more dominant.

Fire and rapid Delta P drill today. Owen needs this the most but hates them the worst. I tried to stick with him and do this together, Jack goes alone—when I am distracted, Owen will be doing other things not drill related and I must get him back.

Slept better last night (upside down) because it was cooler from the twin boom sunshade.

Arabella ate her web last night and spun another perfect one.

Garriott, MD-14:

Another good day! Houston reported filament lifting, got to (ATM) panel as large loop was ~R=2.5 [extended out to a radial distance of about 2.5 solar radii]! Followed all the way out beyond R=6. Excellent, I thought. Also made hemoglobin check (~16-16.3, all three)... TV of Arabella, etc. Supposedly a "day off", but we made 4 ATM passes, 3 S-017 OPS, etc. Back at it tomorrow! Also talked w/HM [Helen Mary, his wife] and family. Said flare was big news locally, w/scientists.

More than just "another good day," Mission Day 14 was a history-making one for the Apollo Telescope Mount, which was used to capture an unprecedented image of the solar corona. One of the instruments on the ATM, the White Light Coronagraph, would hide the bright face of the sun behind an "occulting" disk and image a superimposition of all of the visible light wavelengths in the corona. The sun's very bright upper region, which is what is visible to the eye on Earth, is about one million times brighter than the faint corona, which can only be seen from the ground at the infrequent times of a total solar eclipse. On Day 14 the ground saw what appeared to be the start of a solar eruption at visible wavelengths and brought it to the attention of the crew, even though they were not manning the ATM panel at the time.

"I got there in time to see what is now called a 'coronal mass ejection,' or CME, in progress, where the ejected material in the form of an enormous magnetic loop was moving out through the corona," Garriott said. When he first saw the loop, its height had already reached about the width of the sun, and by its peak a few hours later, it was more than three times the sun's diameter. "The radial extent of this giant magnetic loop could be measured on our TV screen. Then on the next orbit about ninety-three minutes later it was obviously stretched out much farther and it could be measured again. A simple calculation allowed the minimum speed of the ejection to be estimated, which turned to be about 500 kilometers per second! At that speed, it would reach the Earth in about three days. As far as I know this is the first visual observation of this phenomenon ever made." Since coronal mass ejections can have a noticeable impact on Earth when they reach our planet, the groundwork laid on Day 14 toward better understanding them has had lasting benefits.

Garriott recalled getting immediate feedback on the day's events during a phone call with his wife. "I had a telephone visit with the family at our home in Nassau Bay," he said. "The wives brought us up to date on local news; for example, they told us that the TV reports and the solar scientists comments seem quite enthused about the 'flare' observations. It was our pipeline to the 'real world!' "

Measured hemoglobin levels on all three crewmen had reached the upper end of normal, which Garriott suggested might have been due to the loss of water in weightlessness and a reduced total blood volume in circulation. Bean said that during his time on Skylab, he had to make a conscious effort to avoid becoming dehydrated. "The thing that I noticed for myself is, I had to make myself drink water," he said, "because I wasn't thirsty then. And the next day I would have less energy. My urine volume would be low, and it finally dawned on me that I was getting dehydrated because I just wasn't thirsty. So it got to where every time I came near the table, I'd take a drink even when I didn't want it. And that helped. But I would fall back. After about four or five days of drinking water, maybe the fifth day I would not do it so much; I'd get complacent. Then I'd notice the sixth day that I got tired early, and then I would remember my low urine volume that morning. So I remember that as being a continual problem for me."

In fact, Bean said during the entire mission, he had to make a conscious effort to stay in good shape and not allow his desire for productivity push him past the point of exhaustion. "Every day I remember trying to do as much as we could that day without hurting the next day," he said. "I'd say to myself sometimes, 'Uh oh, I worked too long,' I was on the edge of fatigue each day at the end of the day. And if I didn't get the sleep and food and water I needed, then I'd be fatigued the next day. I always felt like I was right on the edge, and I had to be really careful to keep myself healthy in order to do the next day the best I could, and feel really good all the next day, and be in a good mood. People get in a bad mood, I think, if they get tired and fall behind. We had good relations with Mission Control. In fact, our relations with Mission Control were great except when we wanted more work and couldn't get them to schedule it for us."

Bean MD-14:

Day off—we had mixed emotions. We were tired and needed rest yet our chance to do good work was almost one-fourth over. When each flight hour represents 13-14 Earth training hours then you can make (up for) a lot of pre-flight effort with a little extra in flight effort. We did however do some ATM and some S019. We ask for extra. Plus housekeeping. Wipe, dry biocide wipe, the place is immaculate and not a predatory germ within miles, much less traveling at 18,000 mph.

Got a thrill today. Tried to put out a urine bag [through the Trash Airlock] with the end filter for the head in it in addition to three urine bags. It would not eject. I tried to close the doors and breathed a real sigh of relief as it came closed. I removed the filter from the bag and tried again, this time it moved 1" or so then stuck. I tried to close the door but this time it would not. My heart was beating fast. Could this be happening to us. Could we not have a way to get rid of our garbage? I tried the ejection handle again, and no luck, the door was stuck. Finally the only way was to force it. I tapped it again and again at first no success, but finally a little at a time she broke free. The heart still beat fast but maybe a lesson was learned. Why did they not build the lock as an inverted cone so whatever was in there could always be moved down the ever expanding diameter.

Owen did the spider TV three times. Once because he recorded it on channel A, once because the TV switch was in the ATM and not OWS position, the last time it was okay. He got behind and I did some of his housekeeping as he was still up when Jack and I were headed for bed. Jack said "Owen, do you have anything left I can help you with". Owen said "no". But that's the way Jack is.

Notice we do not seem to reflex to catch something when we drop it as we did the first few days. It's enjoyable to just let a heavy object float nearby.

Garriott, MD-15–16:

Things beginning to ease up just a little. We're considerably more efficient and flight plan may be a little less tight. Al now asking for more work(!) . . . All feeling excellent. Al doing lots of acrobatics (he's good). Jack is walking around on the "ceiling."

Garriott was not the only one to feel that the crew was beginning to hit its stride around this time. Being as productive as possible had been one of Bean's foremost goals for his crew from the outset, and the limitations they had faced early on had been a disappointment to him. He and the others had been working all during the mission to become more efficient, and around this time, they could tell they were getting close to the mark.

"We were in there working as best we could; and we were following the flight plan accurately; we were following our checklist, and as a result we were getting a lot of things done," Bean said. "I felt like it took us until around Day 16 to really be as efficient as we ever could be. That was my feeling, and also looking at the data later on. We began to be pretty good at it."

"So we sat down and had a crew meeting and decided that we needed to have an inventory from the ground as to where we were and what we had to do to catch up," added Lousma.

Bean recalled: "Maybe at a third of the way through, or a fourth of the way through, we called the ground and asked how we were doing. I knew we'd fallen behind because of being sick, and I thought maybe they'd tell us we'd done 90 percent so far of what we should be doing. And they told us we'd done 50 percent, 60 percent, something shocking. Well we knew we weren't going to go back to Earth doing 50 percent. They will have to shoot us down because we aren't going back till we've done the best we can do. We were going to find a way, and that's kind of when we decided we were going to have to do things differently because we had to catch up, at least we all thought so. So we began to try to be more efficient. You know, we thought we were being efficient, but this motivated us to become more so."

Every possible step was taken to increase efficiency. The crew stopped eating all of their meals together, so that two crewmembers could be working at all times. As soon as the crew woke up, someone would begin manning the ATM station while the others went about their morning routine. Bean said that not long afterwards he realized that Garriott was the best of the three at manning the ATM console, so he and Lousma began swapping duties with Garriott so that he could man the ATM more.

The crew, and Bean in particular, began working to move items to and from storage during the day to reduce the amount of time that had to be spent on housekeeping. "We were working as much as we could," Bean said. "We were really hustling around." Finally, the crew reached a level of efficiency such that they were getting all of the work done that they had scheduled on a given day. But, having gotten behind at the outset, simply reaching 100 percent efficiency was not enough for Bean. "We began to try to get housekeeping done before it was scheduled, so we could say to them, 'We've already got the trash thrown for tomorrow, we've already got the food moved, go ahead and put us on the ATM,'" he said. "As I remember, we had to convince them to give us more work. We were ahead, and then they would call up and they wouldn't have anything new the next day, and we would be twiddling thumbs. We were ready to go, but they hadn't geared up for us yet. I remember us talking with them for about two or three days before Mission Control finally said, 'OK, let's give them a lot more work.'

"We then got going, and so we were just zipping around there as good as we could from wake up till rest before sleep. Because you can't just stop working and go to sleep, we knew that you had to kind of take thirty minutes or an hour," Bean said. "We were working all the time, except Sundays. Then we began to work Sundays after a while, because there wasn't a damn thing to do, at least I felt that way. What are we going to do, sit around and just look? Not likely! We had trained hard for two and a half years, and we are going to make the most of our limited days, only fifty-six, in space. At least as much as we could. So we got going!"

Before long, the ground had to work to keep up with the crew. As Lousma recalled: "We got so good at what we were doing that it took so much less time than they had anticipated that we asked for more work, and that's where they devised the Earth observations experiments: 'Can you see this; can you see that; what can you see physically or visually from space?' We would photograph those places and report on them. Every mission after that—I don't know if they do it anymore, but Shuttle missions had Earth observations briefings and some special things to look for. So that was all derived as a result of our mission. They also jury-rigged some additional experiments using hardware that we had on board. They had some kind of experiment that had to do with transfer of fluids; it was not one we had planned to do. The point is they gave us extra work to do and things that we hadn't planned on doing, so we actually ended up with more experiments than we started with."

A major thing the crew had going for them, Lousma believes, was how well they got along. "I think our crew was somewhat remarkable in that we were such good friends," he said. "We trained for two and a half years, and I don't ever remember a cross word. I don't remember one during the mission, or since."

Even as the crew was becoming more efficient at their work, they were also becoming more efficient at their play. After over two weeks in weightlessness, the astronauts had become acclimatized to the unique acrobatics that microgravity allowed. "For an unusual experience, one could walk around upside-down on the ceiling of the laboratory area," Garriott said. "It was fun to play 'Spider-Man' and walk around on the ceiling or elsewhere."

While the entire crew had gotten their "space legs" by this point, it was Bean whose microgravity maneuvering was the most impressive. "I was amazed at how proficiently Al performed flips, twists, and other acrobatics while jogging around the ring of lockers in the OWS," Garriott said. "While Jack and I looked every bit the novices we were, only after inquiring did I find out that Al had been a gymnast in college! If only we could submit video instead of personal appearances, we might have had a shot at the next Olympics."

The long straight layout of the pressurized volume of Skylab was the basis of another amusement for the crew. "Another challenge," Garriott said, "was to launch oneself at modest speed all the way from the bottom of the living and experiment deck and try to pass through the OWS, the Airlock Module, the MDA, and reach the CSM without touching anything—a floating distance of some fifty feet with narrow hatches between each module. With practice we could all do it—sometimes."

Bean, MD-16:

Had a thriller, was writing in my book when caution tone then warning tone came on—Jack in the toilet—Owen and I soared up and found cluster ATT [attitude] warning lt [light] and ACS [attitude control system light] on. We looked at the atm panel and found much TACS firing and x gyro single, Y gyro okay, Z gyro single. A quick look at the atm panel showed multiple TACS firings. Both Owen and I were excited, it had been some time since we practiced these failures, plus we are in a complicated rate gyro configuration—we both really were looking at all things at once—DAS [data acquisition system] commands, status words, RT [rate] gyro talk backs, momentum and CMG wheel position readouts. We elected to go ATT HOLD but TACS kept firing, so we then turned off the TACS, looked at each rate gyro and set the best one back on the line. We would have gone to the CSM but with our quad problems that would be a true last resort. No, we had to solve it right then. We put the rate gyros back into configuration then enabled TACS, then did a nominal momentum cage—this seemed to make the system happy—namely TACS quit firing. Owen and I had settled down by then and were solving the problem again and again to insure we have not forgotten any step. We came into daylight—were only two degrees or so off in X and Y so went to S.I. [solar inertial]—maneuvered too slow so we set in a five sec maneuver time and selected S.I. again—Houston came up and I gave them a brief rundown—Owen, never giving up time, started my ATM run for me while I went down for dessert of peaches and ice cream.

EREP passed today, Jack got four targets, we then had an EREP CAL [calibration] pass taking specific data on the full moon—all three of us working well together, we have trained a long time for this chance and we want to make the most of it.

Jack made a suggestion to walk on the ceiling as the floor for a few minutes—we did and in less than a minute it seemed like the floor although covered with lights, wiring runs and trays. Our home seemed like a new place—cluttered but nice—the bicycle hung overhead and was different as was the wardroom table but many lockers and stowage spaces were much easier to see and reach—I might use this technique to advantage when hunting a missing item or looking in a locker drawer.

Had to ask Capcom, Story Musgrave, to give us more work today and also tomorrow—we are getting in the swing—when you're hot, you're hot. We will have about 45 more days to do all the things we were trained to do for the last 2½–3 years—time is going fast and we must make the most use of it. Most of what we learned will have no application after Skylab—such as how to operate specific experiments, systems, where things are stored, experiment protocol, how to operate the ATM, EREP, etc.

The gyroscopes that allowed Skylab to maintain its attitude proved to be an occasional hassle. One failure of the gyros was particularly memorable for Bean, who committed a rare violation of procedure in the heat of the moment: "I remember the time we lost attitude hold. The alarm went off, maybe even in the night; I don't even remember when it was. We had a procedure if it did, and I can remember not following that procedure. It's one of those deals where you make [someone else] follow the procedure, but when you're there, you don't have to do it."

Rather than trying to regain attitude control with the control moment gyros, Bean opted for the more immediate method of using the TACS thrusters, which had a limited, and unreplenishable, supply of cold gas, a large amount of which had been expended in the barbecue rolls before the arrival of the first crew. "I can remember not following the procedure and wasting some of the gas, wasting some of that to 'zero out' the rate gyros, instead of doing other things," he said. "I can remember the ground didn't say anything. Then later, about a day later, they came up with a new procedure, 'just in case,' which really was the same procedure, except, 'Why would you guys do what you did?'

"At the time I threw that switch, I knew it was the wrong thing to do. It was too late then. It didn't even seem right then, it just seemed like the expedient thing. We solved the problem quickly that way. But it wasn't a good thing to do. I can remember me throwing that switch and thinking at the time it was a bad idea."

Bean, MD-17:

Had bad experience today, sneezed while urinating—bad on Earth—disaster up here.

Did 10-15 minutes on dome lockers. Handsprings, dives, twists, can do things that no one on Earth can do—fantastic fun and I guess good limbering up exercises for riding the bike.

I went up and looked out of the MDA *windows that faced the sun, but at night. What an incredible sight, a full moon, Paris, Luxemburg, Prague, Bern, Milan, Turin all visible and beautiful wheels of light and sweeping under the white crossed solar panel of the atm. Normally you cannot look out these windows because of the sun's glare, I could not watch Jack and Owen on their* EVA. *Now we are over the Bay of Bengal. In just 16 minutes we swept over Europe and Eastern Asia, Afghanistan, Pakistan, and finally over India. Too cloudy to see Ceylon [Sri Lanka]. Sumatra and Java will be here soon. We repeat our ground track every 5 days but 5 days from now as we go over the same point of ground the local time there changes so that in 60 days we will have seen all points between 50*N *and 50*S *at 12 different times of the day and night. At least once we can watch Parisians [Paris residents] getting up, having breakfast—*

Owen and I spent his first night in 17 days just looking out the window during a night pass. We came over places that aren't our EREP *targets, the Dardanelles were visible, then he pointed out the Dead Sea, the Sea of Galilee—I said I had been as high as anyone on Earth and had visited the lowest point on Earth, the Dead Sea, last year. Owen talked of the night air glow—the fine white layer about a pencils width above the surface of the Earth.*

We had looked last night for Perseid meteor shower with them burning up below us. Did not see any during S019*—to hit the atmosphere, to make a shooting star, they all fly past us—with no meteoroid shield, hope we do not contact any one of them.*

MD-18:

Fixed my sleeping bag today, safety pinned on two top blankets and took up slack in blankets—too much volume of air to warm at night. Have been waking either around 1 to 2 hours prior to 6 o'clock [normal wake-up time]. Houston time and having difficulty going back to sleep. Maybe this will help, sleeping upside down has helped, the cooler OWS *as a result of the twin pole sunshade deployment is perhaps the greatest contributor.*

Normal morning sequence is wake up call from Houston, I get up fast, take down water gun reading, then put on shirt and shorts for weighing. Take book up and weigh while Jack gets teleprinter pads and Owen reads plan. I weigh, Owen weighs, then Jack. I fix breakfast after dressing, with Owen a little behind.

Jack cleans up, shaves, does urine and fixes bag and sample for three of us, I finish eating as Jack comes in and I then clean, shave and sample urine, I'm off to work at first job as Owen goes to the waste compartment. Jack is eating and about 30 minutes later we all are at work.

A sudden realization hit me this afternoon—there is no more work for us to do—ATM is about it. Except for more medical or more student experiments what a sad state of affairs with this space station up here and not enough work to do.

We could think up some good TV productions getting 5000 watt-min of exercise per day and that should be enough.

Boy oh boy have I been farting today. You must learn to handle more gas up here and I wondered if we would forget when we went home. Owen said can't you just see Jack in his living room with all his family and friends around and he forgets.

I am so glad that Owen and Jack and I are on the same crew. Our personalities fit one another well—Jack always working, always positive, always happy—Owen always serious, well maybe not always.

Owen looks funny lately as he has not trimmed his mustache hair nor shaved under his neck too well—our little windup shaver and the poor bathroom light being the problem. I don't look too great either, my hair getting long, wonder if "O" or Jack will cut it on our day off.

Owen got his ego bent last night. He had been conscientious about weight loss, wanting more food, and salt—peanuts are a favorite, Dr. Paul Buchanan called on his weekly conference and told Owen, [that] Jack and I were doing okay but he needed to have a chat with him [Owen]. Paul said, Owen, we have been looking at your exercise data over the last two days and don't think you are doing enough, maybe your heart isn't in it—Owen about flipped because he takes great pride in his physical program pound for pound he does more than Jack and I. He could hardly hold back, afterward he worked out till sweat was all over his body, then called on the recorder to tell Paul and those other doctors the facts of the matter. Maddest I've seen him in months. [Garriott explained that it turned out the ground had not yet read the data off the recorder, and the issue was smoothed out later.]

As Bean noted, the sleeping bag modification referenced at the beginning of this entry was the second major mod he made to his "bed." The sleeping quarters were designed in such a way that an air vent would cause air to flow from the feet to the head when a crewmember was sleeping in the bed.

Bean found it difficult to sleep in that configuration and unstrapped the cot from the "vertical" bulkhead where it was mounted and inverted it so that he was sleeping "upside-down" compared to the other two astronauts. Garriott noted that while Bean's modification to invert his cot worked fine on Skylab, where each crewmember had his own "bedroom," it could have been more problematic if the station had been designed with the three sharing one larger area since it could be disconcerting to carry on a conversation for any length of time with someone in a different body orientation.

Near his upside-down bed, Bean kept a sign posted on the inside of a locker door, which he made a point to read at the start and end of each day, and which thirty years later he still says was an important part of his life on Skylab.

A man is what he thinks about all day. "The only time I live, the only time I can do anything, the only time I can be anyone is right now.

Each hour we have in flight is the culmination of approximately 12 to 13 preflight hours (1½ days). These hours well spent are our only tangible product for literally years of work and preparation.

Our doubts are traitors and make us lose the good we oft might win by fearing to attempt.

Did we enjoy today.

Ask for questions.

Importance of the individual.

Write in my crew log.

Garriott, MD-18:

Not enough to do today! Al doing most of HK [housekeeping tasks]. Mentioned to Al—he agreed—that often when "sitting" still with eyes closed, there is an apparent sense of motion. Sort of slow vibration (2 or 3 second oscillation), back and forth . . . Maybe body is actually perturbed slightly by air draft, but I think not. Does seem to be a vestibular "false motion."

Bright flashes occasionally. Always dark adapted. Believe have seen with eyes open. Usually spots, not necessarily pin-points. Occasionally a longer streak. Only one eye at a time.

The odd body oscillation Garriott noted he later determined was probably real motion caused by each stroke of the heart pushing arterial blood out through the body. The crew's vestibular systems were probably unusually

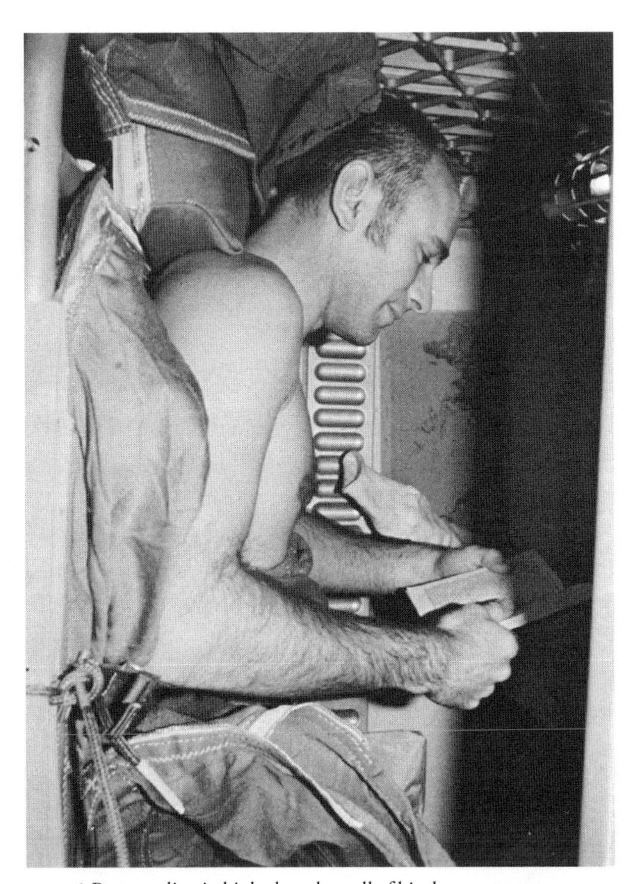

36. Bean reading in his bed on the wall of his sleep compartment.

sensitive to any body motion as the large gravitational acceleration could not be sensed in free fall.

The bright flashes of light were explained later as passage of an energetic particle through the retina creating a flash that the crew could see. It almost always seemed to occur when Skylab was near the South Atlantic Anomaly where the Earth's magnetic field is a bit weaker than at most locations, and trapped energetic particles can dip down to lower altitudes like that at which Skylab orbited. The phenomenon was not isolated to Skylab—other astronauts since have also reported seeing bright flashes while crossing through the anomaly.

Bean, MD-19:

We have been trying to get the FLT [flight] planning changed. I especially have had a lot of free time, Owen and Jack to a lesser degree. Jack keeps on the move all the time, Owen has a long list of useful work that he brought along, things that other scientists have suggested, worthwhile. How do I accomplish this feat of us producing our maximum without infringing on Owen's time. He deserves some amount per day to do with as he chooses.

In a way space flight is rewarding but on a day to day it is awfully frustrating. Jack today spent whole night pass taking star/moon and star/horizon sightings on his own time to satisfy an experiment. When the pass was over, 20 marks made, he was debriefing and as he was talking he said, well, I did those sightings with the clear window protector still on. He had not noticed it in the dark. The data would be off by some small amount and that just didn't suit Jack. He told the experimenter on record that he would repeat them later.

> Teleprinter message: To Bean, Garriott and Lousma
>
> We have been watching and listening as the three of you live and work in space. Your performance has been outstanding and the observations that you are making are of tremendous importance. Through your efforts Skylab 3 is a great mission.
>
> Keep up the good work.
>
> Signed,
>
> Jim Fletcher [NASA Administrator]
>
> George Low [NASA Deputy Administrator]

Received this today. Why do they not send something similar when we are not doing too well, like days 2-4. We appreciated this but just wondering not only about them but about myself.

Went to bed on time, do not feel as energetic as usual so feel something was coming on. Sleep is the best thing to repair me, it always works on Earth.

Bean, MD-22:

Our first real day off. Best news was in the morning science report where it said we would catch up with all our ATM science as well as the corollary experiments except for medical which was reduced by 25 hours the first half of the mission, we would do the rest—I called and discussed the additional blood work, histology and urine analysis [specific gravity] that Owen had been doing and wanting them to count that.

We did housekeeping a bunch and had to plan two TV spectaculars. Since we have ATM all day we had to schedule it in the 30 min night time crew rest. Hair cut next, then acrobatics, then shower. Lots of planning for 3 ten min shows but think the folks in the old USA will enjoy.

The shower was cooler than I like it—the biggest surprise was how the water clung to my body—a little like jello in that it doesn't want to shake off. It built up around the eyes, in the nose and mouth (the crevices) and it gave a slight feeling of trying to breathe underwater—would shake the head violently and the water would drop away (not down but in all directions) some to cling to other parts of my body, some to the shower curtain, some sort of distended the water where they were and snapped back. The soap on the face stayed and diluted with rinse water tasted sour when I opened my mouth. The little vacuum has sufficient pull but is rigid and will not conform to the body—so does not do too well there, but is okay on the inside walls, floor and ceiling. Jack had said it was better to slide my hands over my body and to scrape the water off and over to the shower wall. This worked for hair, arms, legs, but difficult for my body especially back—two towels were required to dry off because the water did not drain.

MD-23:

Flew T 020 for the first time. Jack as usual had the dirty work but was trying harder because of his error yesterday. The work was slow and tedious because it was the first time around and because the strap design was poor.

Jack said 'I've done some pretty dumb things in my life but I never got killed doing it—in this business that is saying a lot'—

Owen said "now the dumbest thing I can remember was flying out to the (solar) observatory near Holloman, NM—short hop so I decided to do it at 18,000 ft—as I neared there I started letting down, called approach control—we talked and as I descended their communications faded out—I kept thinking why should they fade out—it suddenly dawned, shielded behind mountains—full power and a rapid climb in the dark saved my ass—I think of the incident several times every month over the last three years."

Garriott recalled, "I thought I had never mentioned this to anyone, anywhere, since it was such a dumb thing to do. I had forgotten about this one time in Skylab. I had worked all day on ATM things in Boulder, Colorado, that day, then drove to Buckley Field in Denver, to fly solo by T-38 to Holloman AFB and work the next day at the solar observatory in Cloudcroft. A beautiful clear night, stars but no moon. When I heard approach control at the airport, I started down. DUMB! When their voices started breaking up and then faded out, I asked myself why. When realization came quickly—mountains!—it was maximum power (burner) and steep climb until I heard their radio transmissions again with no further problems. I have continued to think about this incident frequently for the next thirty-five years, but it is so embarrassing that I have never admitted it to anyone—except on this one Skylab occasion!"

Bean's diary for that day continues:

Jack was saying that when we got back he and Owen might be considered regular astronauts—Owen laughed—it was beyond his wildest dreams to be classed as a real astronaut.

Been wishing Owen and I had taken pictures of the Israel area the first time we stayed awake to see it—I want to give pictures of this region to some of my religious friends.

Jack's having his ice cream and strawberries. Jack's food shelves when we transfer a 6 day food supply are almost full of big cans plus a few small—Owen and I have half full shelves with more or less equal amounts of small and large cans—Jack really puts down the chow.

All are in a good mood, morale is high in spite of all the hard work, we are getting the job done.

MD-24 A tough, tough day. Worked almost all day on trying to find the leak in the condensate vacuum system—hundreds of high torque screws, stethoscope, soap bubbles, 35 psi nitrogen, reconfiguring several pieces of H2O equipment—we never found the leak—that effort must have cost $2.5 million in flight time.

Owen got this word that the citizens of Enid [Oklahoma, his home town] would be putting their lights on for him to see—I went up with him—it was the clearest, prettiest night we've had—we could see Ft. Worth-Dallas particularly—a twin city, one of few—then Oklahoma City then Enid then St. Louis then Chicago—Owen made a nice narration. He said started to say he saw Tulsa up ahead and realized it was Chicago. Paul Weitz said that was the one thing he never became accustomed to on his flight—the speed which you cover the world, especially the U.S.

This was not quite the end of the story, however, as Owen heard more about the incident following his next conversation on the family private communication loop. It turns out his wife, Helen Mary, also raised in Enid, felt she had to call the radio and TV stations in Enid and try to explain how Owen could be so thoughtless as to not even mention their major citywide effort to be seen directly by him.

Perhaps his predicament was best explained by Alan's comments on Mission Day 35 after most of the fuss was over: "That night we both went up to see the lights of Enid—he talked of Mexico, Ft. Worth, Dallas, here comes Tulsa, look at St. Louis, Chicago—everything but Enid—Helen Mary called up there and tried to soothe the people—she gave Owen hell—I kept telling him to say something about Enid; they had a direct TV hookup, radio hook to us and all lights including the football field."

"That's been an embarrassment to me ever since," Garriott said more than three decades later. "In fact, I undoubtedly saw Enid, but because there were so many lights all across the area, I wasn't certain just which ones were from Enid, and by the time I thought I had it figured out, we were past Chicago, less than two minutes later."

Bean, MD-26:

Owen reported an arch on the UV monitor in the corona yesterday. We called it Garriott waves to the ground—he was in the LBNP and was embarrassed and told us to knock it off—we were happy for him. Today he heard the ground could not see it in their taped TV display—he went back and checked and found it to be a sort of phantom or mirror image of the bright features of the sun except reflected in the camera by the instrument. He'll get over it (maybe that's why he was distant).

Crippen woke us this morning with Julie London singing "The Party's Over." Jack wanted to make this Julie London Day, so did Crippen so he could call her but Owen won out with Gene Cagle Day [who played a major role in the ATM development at MSFC].

MD-27:

Would you believe it we get better H-alpha pictures at sunset than we do at sunrise because our velocity relative to the sun is less and that effectively changes the freq[uency] of the filter in each H-alpha camera and telescope—not a small item either.

Owen's humor—I said "watch your head" as I pulled out the film drawer. Owen replied "I'll try but my eyeballs don't usually move that far up."

We were laughing about this malfunction ("mal") we had after we discovered the water glycol leak—I wanted to call Houston and say "Jack is working on the CBRM *[charger battery regulator module] mal, Owen on the camera mal—tomorrow after we fix the door mals, the 9 rate/gyro mals and the nylon swatch mal, I'll start work again on the coolant loop or the water glycol leak mal."*

Everyone feels better about EVA*—I worry too much and Jack will pull it off. Funny how easy it looks now that we are going to do it—did it get easier as we understood the plan or did we just want it to be doable? Morale is high—did perfect on my* MO *92/171 (medical experiments).*

Mission Day 28 brought the crew's second spacewalk. Like the first spacewalk of the mission, the second would include an extra task to repair a problem with the station. In addition to the routine task of changing out the ATM film canisters, the EVA crewmembers would also install a cable for the "six-pack gyros."

The Skylab's attitude, or orientation, control system relied on two sets of gyroscopes. The large Control Moment Gyroscopes were used to torque the whole Skylab to a new attitude or hold it in position. A set of smaller attitude control gyros was used to monitor the attitude of the station. These smaller gyros had proved erratic since the station's launch, and while they continued to function, the decision was made to activate a new six-pack of gyros on the second EVA in hopes of providing improved attitude control.

During the four-and-a-half-hour spacewalk, the two astronauts changed out the ATM film cassettes and left two samples of the material used for the parasol outside to be recovered on a later EVA so that the effect of exposure could be monitored.

In the days leading up to the spacewalk, Bean found himself having to make a difficult decision—who would go outside and who would stay inside? The original mission plan had called for each of the three astronauts to get two turns at an EVA, and the second was to be performed by Bean and Lousma. Bean's diary captured the decision-making process for who would go on the EVA:

Bean, MD-24:

Heard tonight we may put in the rate gyro 6 pack—I told Owen [that] Jack & I would do it because they did the twin pole and because that sort of work fits my skills better than Owen's—hope it did not hurt his feelings but that is the way I see it and that's my job—Owen even brought it up by saying "I think you want to put out the 6 pack and that's okay with me—I'm glad to do it but know you want to"—I said you're right we don't need this job but if it comes up we will pull it off.

MD-25:

Today was a special day—found out we were going to put in the rate gyro pack—who to do it—Owen still wants to do it and so do I. Made up my mind that it would be Owen and I but after reading the procedures realized that I should stay in because of my CSM experience—Owen and Jack are just not up on it and it is the best decision—Jack will do the 6 pack as he is the most mechanical-Owen does not do those things as well as Jack, it will be taxing to tell him tomorrow—I was awake about two hours trying to put the pieces together and think Owen and Jack outside—me inside is the best way.

MD-26:

Told Owen and Jack about EVA crewman, they both seemed happy, told them what factors were involved and who I felt most qualified for each position. Called Houston and told them later, they seemed happy. I started looking at the equipment for the job—all in good shape.

"I felt that was my job," Bean said. "I wanted to go EVA too. But I felt it was my job. I was mostly concerned with the Command Module and attitude system. Even more important, Jack was the strongest guy. If anyone could twist those connectors that had never been designed to be loosened in flight, Jack was the man. He needed to be out there. And Owen could support him out there. We didn't need me."

Though Lousma was also trained as a Command Module pilot, Bean felt it made more sense for Lousma to perform the EVA than to man the Apollo spacecraft. "To suddenly move Jack from the right to the left seat in the CSM, not a good idea. It's a better idea to let me do what I've been doing all this time. Let Jack do the twisting. He was the strongest; he's also good at repairing things. He was the right guy.

"So we told the ground, 'You know, we've got an EVA coming up in just about three days. We've been thinking about our crew assignments for a couple of days. And we think that this would be good. What do you think?' And sometime later on the next day they said something, like, just in the update, 'We think that's right.' So we were always ahead of them in these kinds of things, or we tried to be anyway, so that we had the right people doing the right job."

The six-pack gyros were so called rather logically because they consisted of a set of six gyroscopes. The six-pack itself was actually installed by Garriott inside the Skylab, but turning over attitude control to the new system required going outside and connecting up a set of cables to circumvent the station's original attitude control rate gyros in favor of the new ones. Using a special tool designed for the task, Lousma had to twist the old connections from their sockets and then twist the new cables into place—a task notoriously much more difficult in microgravity than on Earth. If you twist something in microgravity without gravity holding you in place, you also twist yourself unless you're secured in place. The work that was to be done had not been anticipated during the Skylab's design, and as a result there was nothing on the structure for the purpose of keeping an astronaut from spinning around an object that he was trying to twist.

"So I ended up wedging myself somehow so that when I turned on these [connections] that were hard to get off I didn't rotate myself out of the picture," Lousma said. "It took a fair amount of sweat and so forth to figure out how that was going to be done. It was one of those things that the water tank misleads you on. It's not perfect in neutral buoyancy."

Bean, MD-28:

EVA day. I was talking to myself during EVA and Jack wondered what I was saying—I told him I was just shooting the shit. Jack quipped, "get any?—what's the limit on those?"—Owen was saying " come on, . . . hustle . . . give us some of that positive mental attitude"—PMA (he doesn't believe in it. But knows I use it on me and them also.) "Go Earl [motivational author Earl Nightingale]," Owen said. I said "you need it, it works on you whether you like it or not".

Jack had a difficult time with a connector or two—it was difficult for me to keep from asking questions of Jack as I wondered if that would be the end of the show but he said don't talk for awhile and just let me work on it. He did for a very long 5 minutes and then reported connected.

Owen was elated with the view over the Andes—the 270 degree panorama with 3 solar panels in the field of view to form a perspective or frame work. They were flying over all the world outside of the vehicle going 17,000 MPH. Lost three shims and one nut taking off the first ramp.

We have only 1 to 2 min of TV because of recorder time left so have to hold it for Owen's return to the FAS. Owen had to come out of the foot restraints to remove the ramps from the SO 56 and 82 A doors. Sun end BA lights worked this time—Jack said he can see many orange lights, we were over mid Russia—not many cities—Orion came up, a beautiful constellation, Owen still working on bolts at Sun end.

*Got a master alarm-*CMG *gas-*S/C *going out of attitude—I put it in Att Hold—*TACS X *was at 16 degrees.*

Sometimes, like on a tall building, get a controllable urge similar to jumping off which is to open a hatch to vacuum—or take off a glove or pop a helmet—fortunately these are passing impulses that you can control but it is interesting to know they take place.

Great EVA *today—all happy tonight.*

Owen bitched about the medical types that take care of our food because they told [crew physician] Paul Buchanan our food cue cards were wrong for optimal salt and they had not bothered to update it and had been making them up with supplements.—Owen flew off the handle because he has been wanting salt.

It is comforting to know someone (many someones) on the ground are working our space craft problems faster and much better than we. We generally perform a holding action if we can. Till help and advice comes, then take the info or suggestions and do them. This is the only way we can free our minds to do the day to day task, the production tasks where someone is trouble shooting our problems.

Rearranged my bunk room—put a portable light on the floor near the head of my bed and turned my bed bag upside down so that I could grab the items inside easily. I used the door next to the bottom locker (pulled out about 30 degrees) as a writing desk. Stole the power cable for the light from the spiders' cage—hope Owen doesn't get upset. He has been getting messages to feed them both filet and keep them watered. Will we bring them to the post flight press conference?

The teleprinter is a device about which you can have mixed feelings—it would be hell to get the information any other way so it must be cared for as an expenditure of effort. But at the same time every time you hear it printing you know it is more work for you to do. Wish we would get a non work related message sometime.

Garriott recalled the view during the spacewalk being amazing. "As I was sending film canisters back to Jack with a telescoping rod, I had a few moments to just enjoy the scenery. At that moment we were moving eastward across the South Pacific approaching Chile. To my right I could see the high Andes Mountains, topped with snow and even high lakes and salt deposits, extending all the way to Tierra del Fuego. Looking to my left the Andes extended all the way to Peru. Large cumulonimbus clouds [thunderclouds] reached upward to quite high altitudes near the equatorial tropics, their vertical extent noticeable even from 435 kilometers high, and long shadows were cast in sunset colors, over 100 kilometers down-sun. Then looking straight ahead of our ground track, I could see over the Andes, across Argentina, to the Atlantic Ocean! *Magnifico!*"

"The EVAs were really the most memorable part of being up there," Jack Lousma said. "The launch and reentry obviously get your attention, but every other day kind of fades into the day before and the day after except for the times we did the EVAs. Those were just spectacular. Of course at that time we didn't have the continuous communication [provided by the communication relay satellite]. We could just talk when we were over a ground station, and if we were lucky, we could miss every ground station for a full orbit. You're out there all by yourself. You just kind of felt self-reliant, more self-reliant than you might otherwise feel.

"But the EVAs were spectacular. I remember going out to the Apollo Telescope one time and having Alan turn off the running lights on the Skylab. We were in the darkness over Siberia somewhere, and there's no light down there. I almost couldn't see my hand in front of my face; I'm whirling around the world at 17,500 miles an hour, hanging on by one foot; I can hardly see anything. And I thought, who in the heck has ever done this before? Nobody—or at least, it was a rather unique opportunity. It was those kind of things that I relished, that made the whole trip memorable."

Bean, MD-29:

Felt good to have ATM film again. Operating the ATM telescopes and cameras is one of the most enjoyable tasks here. It is challenging, you can directly contribute to improved data acquisition—Owen has effectively changed the method of operating it in just ⅔ of a month. The Polaroid camera and the persistent image scope have made a significant difference.

I had Houston give all the ATM passes tomorrow, our day off, to Jack and I, so Owen could finish some things he is behind on and do some additional items that he has planned prior to flight—the flight is ½ over and he has had little spare time—he needs some to be happy.

Using the head [waste-management compartment] for sponge baths because sponges squirt water out when pushed on the skin. Bathing has become more pleasant as I have been less careful about sprinkling water about. I tend to now splash it somewhat. And after the bath is complete, wipe up the droplets on the walls. None on the floor like on Earth if you do the same.

We passed Pete, Joe and Paul's old space flight mark, in fact we now hold the world record for space flight—it feels good to be breaking new ground said Jack today. We will be ½ into our mission tomorrow night.

Our TV got too hot during EVA and quit working—we will do the rest of the mission on one TV I guess. Funny, they did not insulate it sufficiently. We had a plan to put it at the solar air lock for EVA but can't do it now.

Jack has a small sty on his left eye, he wanted some "yellow mercury" but settled for Neosporin. Jack treated himself but Owen will examine it tomorrow—Paul Buchanan said for us to be extremely careful because that could be contagious. Perhaps a streptococcus of some type.

Reviews of the shower were somewhat mixed. While some of the crewmembers found that it was an agreeable luxury for occasional use, it could also be rather time-consuming. "'Bathing' in the shower facility provided meant floating in an erectable water-tight cylinder, preparing warm water, spraying yourself to get wet, soaping up, rinsing off, collecting waste water, and then reversing the whole process. It was an enormous waste of time," Garriott said. "It typically took more than an hour to complete. Especially true when a wet washcloth or sponge, soapy if desired, can do just as well in only five or ten minutes and one feels, and actually is, just as clean as in the shower. As a result, Alan took a total of two showers, Jack one, and I took zero for the whole mission. Yet we all remained quite well cleansed, especially after working out each day for exercise."

Garriott, MD-30:

First relaxed day! We stayed busy, ATM all day (Al & Jack), but not too much hurry.

I sent down three TV bits (Arabella—>Anita; rocket stability; water droplet). Good science debrief, even Bob MacQueen got on mike! A new precedent.

"The standard procedure had always been for the Capcom, another astronaut, to do all the talking with the flight crew," Garriott explained. "They are each well known personally to the crew and can possibly appreciate the crew's situation better from their own experiences. Whatever the purported reasoning may be, there is likely a bit of 'turf protection' involved as well. Communication is a crew task and nobody better interfere! Even the flight directors, who have the official responsibility for all mission decisions never get on the 'air-to-ground' loop to talk to the crew. (OK, a few instances of center director or others excepted. Maybe even the president of the United States.)"

"But with Science (capitalized) recognized as the main purpose of the entire program, why not let one or more PIs—principal investigators—discuss how things are going, especially when requested by the crew? Common sense did prevail, and Dr. Robert MacQueen became the first PI (of the White Light Coronagraph experiment and representing the entire solar physics team) to discuss some science issues directly with us. We discussed his coronagraph observations, flares and precursors, and several other items, most or all of which could have been handled by our proficient Capcoms, including (late) astronomer Karl Henize. But it did set a useful precedent, and it was repeated later in the mission with other disciplines."

Bean, MD-30:

We are going to sleep just under one hour to the mission midpoint. Our science briefing today showed that we had made up the ATM observing time we missed early in the mission and predictions are for us to exceed even the 260 hr ATM sun viewing goal. We are ahead in corollary experiments.

Took my second shower, noticed that I could not hear most of the time unless I shook my head because large amounts of water go into my ear openings. I was the only one showering today.

Exercised today although that is not my plan for day off—not doing the exercise would be a nice reward but did not have time EVA day.

Jack made an excellent observation when he said NASA should play down the spider after the initial release because it tended to detract from the more meaningful experiments we are doing up here. Will the tax payer say, now that I know what they are doing up there, I don't like my money going for that sort of thing.

Owen tried to do a science bubble experiment with cherry drink but [it] didn't look too promising to me. He kept losing the drop of drink from the straw.

I have noticed if I do not force myself to drink then I will drink much less than on Earth and will dehydrate—I do not seem to automatically desire the proper quantity of water. I suggested that when we get back we may not naturally readapt to one G and become dehydrated there. Owen does not agree at all.

MD-31:

Spent part of the morning composing a message to the dedication of the Lyndon Baines Johnson Space Center. Thanks to Owen's and Jack's suggestions, it turned out acceptable I think. It must have because at the dedication Dr. Fletcher read only President Nixon's and ours.

When I used to float from compartment to compartment I would be a little disoriented when I got there—now I look ahead as I enter and do a quick roll to the 'heads up' attitude for the space I'm entering.

Owen said his only regret was that he would never adapt to zero-G again—he thinks Pete [and his crew] is the only one besides ourselves that has ever done so.

MD-32:

About every other night I get up because of unusual noises—mostly they are all thermal noises. The most unusual view occurred once as I was in my bunk and peered up to the forward compartment. The ½ light from the airlock revealed three white suited figures, arms outstretched, leaning several awkward ways—silent, large with white helmet straps-one drying, the others waiting to dry. I was shook a little by the eerie sight so I went to the wardroom and looked out. The dark exterior with white airglow layer and white clouds filled the lower right portion of the window like it was one foot away. It startled me even more.

MD-33:

Morale is high—work level is high.—Last night after dinner Owen asked Jack if he (Jack) would like him (Owen) to take his last ATM pass. Jack said no he was looking forward to it—he wanted to find some more Ellerman bombs—[bright points in penumbra near sun spots] as he got some earlier—I interrupted and mentioned that the flight planners had voice uplinked a change in the morning, assigning me to the pass. Owen laughed—here we all are fighting for the last ATM pass [of the day].

Kidded Owen about wearing his M133 cap—I said Jack and I better watch our ps & Qs tomorrow, Owen will be in a bad, criticizing mood—he took the kidding well, hope it will have effect.

I have been decreasing my number of mistakes significantly by only doing one job at a time. Invariably if I do [more than one task], I do not get back in time or do not catch simple error in the first set up.

Bean found it difficult to avoid multitasking—starting up one experiment and then moving on to another one while that one was running. The

ATM provided particular temptation in that respect. After he started a task, there would be nothing else he could do for a little while, and he frequently found himself working on something else to fill the time. However, he found that while that period of time was long enough to make him want to do something else, it was short enough that he was gone from the ATM too long when he did.

On Day 37, Garriott wrote in his diary: "Almost everything on my personal list of extra items has been worked in. TV Science Demos are not too good. Still may get some worked in." For a generation of school children, these "science demos" were one of Skylab's most familiar legacies.

"Before flight, I prepared a list of (hopefully) interesting demonstrations that I might videotape or record on film that could be turned into instructional films for students, probably high-school level, but possibly older and younger," Garriott remembered. "They would be unique to the weightless environment and also challenge their thinking about physics in this exotic environment. I obtained a few one-quarter-inch by two-inch rod magnets before flight (a few dollars from Edmund Scientific), stowed them in my personal gear, and made use of other on-board hardware items for these demonstrations."

Indeed, the magnets were quite effective in demonstrating for students the unique environment of microgravity. When released from Garriott's fingers, they oscillated back and forth like any terrestrial magnet, but now in three dimensions instead of one or two like an ordinary compass. When two magnets were put end-to-end, their oscillation rate was much reduced. When placed side by side, they hardly oscillated at all because their two magnetic fields canceled each other out.

Another experiment mounting frame was used to spin extra large, flat metal nuts off the bolts on the frame. It is well known that nuts do not stay on bolts well in weightlessness, because the lack of gravitational forces reduces friction and makes them easy to "spin off." A very stable spin was produced in this way as compared to spinning them by hand alone when they always have a considerable wobble. When a magnet was taped to their face, the spinning nut was found to precess very nicely in space.

After the crew returned home, these films and videos were edited and a script prepared to show how the experiments all function in weightlessness. Each is about fifteen minutes long and was prepared with help from astronaut Joe Allen and a local contractor. They were distributed by most NASA centers and have been viewed by many millions of students in their classroom settings. They are titled *Zero-G, Conservation Laws in Zero-G, Gyroscopes in Space, Fluids in Weightlessness, Magnetism in Space,* and *Magnetic Effects in Space.* Bean recalled being impressed both with the experiments that Garriott did and with his dedication in using his Sunday free time to carry them out. "Owen did some experiments on Sunday, which I see even today on TV," he said. "Those were good experiments. Good stuff."

Though much of the central footage for the videos was shot live in orbit, Garriott had a few additional scenes to add upon his return to Earth, including one featuring a somewhat unwilling accomplice. "In the *Conservation Laws* film and video, I decided to try to film and explain how a cat always lands on its feet when dropped from a modest height," Garriott said. "I had demonstrated a similar result while in space. So I used our own house cat, Calico. But only once, as he learned very quickly what I had in mind. He was to be dropped from only about three feet onto a pillow just out of the camera view just in case it didn't work as planned. When dropped with feet upward, in a small fraction of a second he had rotated around with exaggerated tail and body motion to place feet downward AND had extended his claws, which raked across my nearby hand! ('Serves you right,' I can almost hear now from all the cat lovers reading this.) I carefully hid the blood appearing on the back of my hand from the still-rolling camera. But it does make for a fascinating explanation and it is none too obvious, to explain how a cat or a high diver or an astronaut can start with no body rotation whatsoever (no angular momentum) and then reorient themselves to face in any direction desired, before coming to a complete stop again. The cat does it instinctively and very fast, with high frame speed required to see it."

Bean, MD-38:

Got my first flare today—A C6 in active region 12. I noticed it while doing some sun center work as an especially bright semicircular ring around a spot. There were 8 or so similar bright rings but this one became exceptionally bright both in hydrogen-alpha and in the XUV. I debated with myself about stopping the scheduled ATM work and going over and concentrate on the possible flare. As of this time it had not reached full flare intensity and it is not possible to know whether it will just keep increasing in intensity or will level off then drop. As I elected to stop the experiments in progress and repoint I noticed about 5800 counts on our Be (beryllium) counter.—A true good flare would be 4150 counts.

It never got much higher. Owen hustled up at once to help—He noted [experiment number] 55 was not at sun center so we repointed it (It was 80 arc sec too low) We took pictures in all except 82A which is extremely tight on film at this point. Owen stayed up late doing ATM because of the activity.

Paul [Buchanan] had Joe Kerwin talk with us the other night. Joe had heard we had asked to stay longer and he indicated they had discussed it on Day 22 or 24 and decided against it. He seemed to think they had made the proper decision. Joe indicated Pete, Paul [Weitz] & he were in preflight condition—the only real funny was the fact their red blood cell mass was down 15% or so and their bodies did not start making it up till about Day 17. Why it waited that long they do not know. I wondered if it were possible to affect the mechanism so that it stopped forever. Seems far fetched, but a thought.

As an astronaut you become very health conscious—if we were not so healthy we could become hypochondriacs. I've worried about a rupture in the LBNP. My legs losing circumference, back strain on the exerciser, gaining too much weight, not having a good appetite in zero G, heart attacks, you name it—I worried about it—As Owen said—from a health viewpoint these may be the most important 2 months in our lives. He could be right with the changes going on and the remoteness of medical aid.

Garriott, MD-39:

Writing at ATM panel, first time I've had enough time to write up here! We've had fantastic solar activity the last 3 or 4 days. SSN (sunspot number) greater than 150 (178 once, I think). Subflares more or less routine. We don't respond in flare mode to save film.

Sunspots have been observed from the ground for centuries. At one point in history even acknowledging the possibility of sunspots was a dangerous belief since it seemed to indicate that the sun was not "perfect." By the time of Skylab, however, a lot more was known about them, such as that the spots come and go over about an eleven-year cycle (or twenty-two years, for those who watch their magnetic polarity). At times of least activity, they may all be gone; and when Skylab actually reached orbit, it was near the time of sunspot minimum. But much to the crew's surprise, amazement, and pleasure, the sun decided to "act up," and generated more than one hundred of these spots and regions across its face at times. A very "measle-y" appearance but great news for the solar physicists. During the second crew's two-month stay on orbit, the sun made two full rotations and also changed its sunspot activity from the low teens to over 150. It provided a marvelous opportunity to study the sun in all its suits of clothes.

Bean recalled that "along about halfway through or so, we began to realize that when Owen manned the ATM, which was our most continuously operating experiment and the primary experiment, that things went better—the coordination with the ground, the knowledge. So very soon, I said, 'I'm not going to do the ATM anymore, let's let Owen and Jack do it.' I tried to put Owen on there as much as I could, as much as he could take. Because we felt the data was better when he was there. As I remember, when flares came up, he was generally there.

"We were journeymen there I felt, Jack maybe was better than me, but Owen was much superior. And we could do the other stuff like heating up the furnace so it'll melt metal. You just turn on the switch and do the checklist, and we didn't have to make much in the way of decisions about what we were seeing there. It's just a fact. Owen was just superior at it. And it fit him. He enjoyed it, and knew more about it, and loved it. It was his kind of stuff."

Bean, MD-39:

I mentioned to Owen that our attitude varies like the sun's activity—now it's way up because the sun's activity is up. Owen avowed that ours is up or way up. Always positive. I mentioned that the first few days it didn't seem way up. He allowed that. It wasn't down—sort of like rain on a camping trip—You just have to be patient, good times are ahead. He also allowed that that would be a quotable quote when we got back.

Forgot to mention Jack saying that if they extended us we could always do BMMD calibrations all day—Right, 1 on the BMMD, 1 on each SMMD, then rotate every hour.

In the middle of last night I heard a loud thump—It actually shook the vehicle. I got out of bed and looked at all the tank pressures in the cluster and even in the CSM. Nothing seen—I recorded the time 0725 and told Houston this morning. They called back and said they broke the data down at minute intervals and found nothing—they are now breaking it into ½ sec increments. Something happened, but what I don't know. Perhaps it was a sharper than usual thermal deformation.

Lousma's crack about the calibrations was a bit of astronaut humor, since the crew felt that they wasted far too much time simply calibrating the body and small mass measuring devices, when in reality the calibration numbers never changed. Garriott had made a reference in his diary just a few days earlier to how much time was spent on the calibration. "By golly, I would make note of it in this diary and remember to tell the PI, astronaut Bill Thornton, what a pain in the butt and waste of time all this was!"

Bean, MD-40:

Jack is pedaling the bike with his arms—good for shoulder and arms, he can do 125 watt/min for 5 minutes.

Owen just flew by with the evening teleprinter messages—We try to find a new record, our old record is from the dome hatch to the ceiling of the experiment compartment.

Owen was on the ATM almost all day doing JOP [Joint Operation Plan] 12—Calibration rocket work to compare with a more recent sun sensor instrument to insure our instrument calibrations have not drifted.

EREP tape recorder easy to load, tape has not set and does not float off the reel & make a tangle.

Story Musgrave said there was a sound in the background like a roaring dragon—It was the sound of Jack pulling the MKII exerciser.

Jack's triangle shoes are wearing out—Hard work on the bike & MKII mostly. He is going to recommend SL4 bring up an extra pair.

Took a soap (Neutrogena) & rag bath today after work out—Do a soap one every other day—And a water one the other day. We are all clean—Body odor just is not present nor is a sticky feeling after exercise.

Several things I have learned up here but the most valuable for ATM operation is "Do not try to do anything else while you operate ATM—You invariably make ATM mistakes." Another is "2 to 3 minutes is too short a time to let your mind wander on another subject when you are within that time from a job that must be done then—such as a switch throw, photo exposure, etc."

Our condensate system vacuum leak has fixed itself—somehow when I connected it all back up after the dump probe changeout, it did not leak. The ground thinks it's a fitting on the small condensate tank.

The second paragraph of that entry refers to length of the teleprinter tape for one day's messages from the ground, some twenty-five feet, all of which had to be read, divided up among crewmen, and then executed to accomplish that day's activities.

The JOP 12 was another of the crew's calibration activities, in which they compared Skylab measurements with similar measurements made from a rocket launched from the ground reaching very high altitudes, to assure that the Skylab instruments had not drifted in sensitivity.

Garriott, MD-40:

3 ATM passes today. Got a good flare in AR12 [active region 12], an M-class (reported by the ground, a rather large one) and another probably high class C flare in AR9. Really prepared for it—"text book" situation. Everyone should have good data. Then good post flare on 3rd orbit.

Later saw aurora Australis (southern hemisphere), w/ photos, then several hours later, 0303Z, large extensive aurora borealis (northern hemisphere). Good photos, 1 and 4 second exposures.

Paul B. said we had a 7 day extension! We all thought beyond 60 days, but he only meant for next week: days 40–47! Oh well—

Note difficulty in finding "dropped objects." Eyes not accustomed to focusing at intermediate distances. Seem to always look at "bottom" surfaces.

"This was an unexpected phenomenon," Garriott explained. "When we dropped or lost a small item, we usually could not find it again promptly. We always seemed to look on hard surfaces where we would normally have left it. But three-dimensional space was just too difficult to search visually. Soon we found a solution, however. Air circulation was always from our living areas (should say volumes, since we can use all three dimensions now) and then collected at a single intake filter high in the OWS dome area. Every morning we could visit the intake duct, probably find a little lint from clothes and so forth, and also all the little items, pencils, notes, that we had lost the day before."

Lousma discovered another trick that helped him deal with the same problem. When he found it difficult to locate something lost in the three-dimensional space in front of him, he tricked his mind into looking at the situation differently by literally turning the problem upside-down. "When we were looking for something that we lost, the best way to find it was to turn upside down," he said. "Because you normally look at the top of everything, you don't think about looking under there. But when you're upside down and look for something, you look at those places that you don't normally see, or that your eye doesn't get drawn to, because you tend to expect things to be sitting on something."

37. Lousma demonstrates basic grooming on Skylab.

The auroras were a particularly beautiful sight from orbit. Bean recalled being surprised at how impressive they were viewed from above. "Strangely enough, because I wasn't that interested in auroras, I remember seeing both auroras," he said. "Owen became the first person, I think, in history, to see both auroras the same day. He saw them during the same orbit, about forty-five minutes apart.

"They looked different, and they looked strange. They were bright, and streaming. They were just easier to see than from Earth. I can remember just being amazed at the size of them, and the nice colors of them."

Bean, MD-41:

As I was waiting to start the EREP pass & we had a 3 min maneuver time to Z-LV. I did two chips [small segments of the operations plans] of the ATM contingency plan for no EREP then powered it down for EREP. Bet that's a space first.

Grand Rapids blinked its lights for Jack Lousma tonight. He said some good words over the headset.

MD-42:

Owen got a X class flare first time manning the ATM panel this morning, we all hustled up there to help. It was well done. The big daddy flare we have been waiting for. All of us were laughing and cutting up. Owen had said yesterday he had used all his luck up. Guess he didn't or he's running on Jack's or mine.

Took apart the video tape recorder and removed 4 circuit boards, 63 screws did the job. No sign of circuit problems, burns, loose wires, etc.

Owen & Paul had it out on the exercise, as Paul said last night Owen was slacking off. Jack was up at the ATM and was laughing and hollering as was I. We have been calling Owen "slacker" this evening. [Bean and Garriott said postflight that they consider this one of the funniest episodes of the mission.]

Owen and I got 10 EREP film cassettes, 1 EREP tape, 3 Earth terrain camera mags and a S019 mag out of a . . . bag where the SL-2 crew had left it. Wonder if we could use it on our mission or on a mission extension.

I am very happy with the way our crew is performing—We are doing the job without problems & without giving problems. In my view, it's a professional performance.

Garriott, MD-41:

Paul B. [flight physician Buchanan] complains about slacking off on [exercise] work. Probably data error. More tomorrow . . .

MD-42:

ATM discussion .. .. White Light transient, big flare . . . Sort of "chewed out" Paul B. on his data interpretation. Apparently [they] had lost several days of data. Al says I was "too hard". Jack thinks okay, just "business like". However, don't want a reputation [for] "bad disposition". Hmm, have to work on that.

[Today is daughter] Linda's [seventh] birthday. Sent greeting via Capcom.

MD-43:

Talk of reentry, etc., beginning. Good 3WLC transient [a "WLC transient" is a white light coronagraph transient, now usually called a coronal mass ejection or

CME]. Almost passed it up. Very good one, I think. Would have missed it, probably, except that I had a lot of "observing time"—free.

Al decided that I could go EVA on the 3rd one! Glad to get all three!

. . . health and spirits are higher than ever. We'd all like an extra week extension. Hard for Al to stay busy—about right for me.

. . . Al sent down the wrong chest girth for Jack. Something like 142 CM inspiration and 96 CM expiration. Ground medical report said too much, even for a Marine!

. . . May try taped message to Crippen tomorrow.

MD-44:

Lots of good southern aurora. Greenish at lower altitude—reddish above. Lots of structure, kinks, vertical striations, changes by the minute. Some of it almost directly beneath Skylab just before sunrise.

Easy day. Did TV show with magnetic demo[nstration].

MD-45:

Pretty good day, nothing special. Up an hour early, plus bed ~2 hours late for ATM, though. Not too tired. Still in running for [a mission] extension.

"On the evening of MD-46, I finally played the trick that had been in work for over two months," said Garriott. "It even had the flight controllers puzzled for twenty-five years! My objective was to pretend that my wife, Helen, had come up to Skylab to bring us a hot meal, even though this was an obvious impossibility. Here is how the scheme worked. I recorded her voice on my small hand-held tape recorder before flight, pretending to have a brief conversation with a Capcom, with time gaps for his replies. The Capcom would be my only 'accomplice,' but his role would be carefully disguised. It was also necessary to have some recent event mentioned to validate the currency of the dialogue, so it would seem it could not have been recorded before flight. The short dialogue is printed below in its entirety. I knew that both Bob Crippen and Karl Henize were going to be Capcoms for Skylab, so they were brought into the planning, given the script and rehearsed on their timing. They kept the short script on a piece of paper in their billfolds, awaiting the right moment.

"For our flight in August-September, there would be many occasions of natural disasters involving forest fires or hurricanes, which would be widely known throughout the United States. So a few comments about one or the other were made on the tape. This led to four different scripts being recorded, one for each of the two Capcoms and one each for the two natural events. I would play the tape on the normal air-to-ground voice link with my wife's recorded voice and the Capcom would respond as if totally surprised by the female interloper."

Near the end of one period of voice contact Garriott said to the ground, "I'll have something for you on the next pass, Bob." Crippen replied, "Roger that, Owen." Then quietly and surreptitiously, he reviewed the brief script that had been in his pocket for all these weeks. Soon after coming into voice range, the ground heard this voice on the standard air-to-ground link:

Skylab (a female voice): "Gad, I don't see how the boys manage to get rid of the feedback between these speakers. . . . Hello Houston, how are you reading me down there? (5 sec. pause) Hello Houston, are you reading Skylab?"

Capcom: "Skylab, this is Houston. We heard you alright, but had difficulty recognizing your voice. Who do we have on the line up there?"

Skylab: "Hello Houston. Roger. Well I haven't talked with you for a while. Isn't that you down there, Bob? This is Helen, here in Skylab. The boys hadn't had a good home cooked meal in so long, I thought I'd bring one up. Over"

Capcom: "Roger, Skylab. Someone's gotta be pulling my leg, Helen. Where are you?"

Skylab: "Right here in Skylab, Bob. Just a few orbits ago we were looking down on those forest fires in California. The smoke sure covers a lot of territory, and, oh boy, the sunrises are just beautiful! Oh oh. . . . See you later, Bob. I hear the boys coming up here and I'm not supposed to be on the radio."

"Then quiet returned to the voice link, but we were told later, Bob Crippen had lots of questions coming his way in the Control Center," Garriott said. "What was going on? Where was this voice coming from? Bob must have been a very good actor, because he claimed complete ignorance and innocence of how it happened. Everyone heard it coming down on the air-to-ground loop. The whole two-way conversation sounded like a perfectly normal dialogue. No breaks or gaps, and they all heard Bob respond in real time. Could I have recorded Helen's voice on a 'family conversation' from our home? Yes, but there was no recent one. How would she have known about the fires, or who was to be on Capcom duty and how could she respond to Bob's comments in real time, as everyone could hear?

"No one ever worked out how this was accomplished. Finally, at our twenty-fifth reunion celebration in Houston in 1998, and with many of the flight directors and controllers present and still with no clue as to how it was done, I described it all as above. My prejudiced opinion is that this was the best 'gotcha' ever perpetrated on our friendly flight controllers!"

Crippen recalled: "That was kind of a fun trick. There was head rubbing. Everybody in the MOCR, or the control room, was looking like, 'What the hell is going on?' We did a good job. It was fun. Working those missions got to be tough. We did all kinds of things to try to come up with levity. That was a nice one that the crew got that the ground control didn't know about."

Bean, MD-46:

This was a good day right up to the end. I had a M092/171 scheduled after dinner and at about 2 min from completion I had to punch out. I had a very warm tingly feeling in my arms and shoulders. Don't know whether it was too much hard driving today or just what—my urine output 2 days ago was larger by 100% my normal—It even beat Owen & Jack. It probably means something.

Flight Director Don Puddy said, "Crip's birthday is today and we have a surprise for him. Maybe you could sing Happy Birthday from orbit. (Incidentally, our wives and kids were at MCC tonight) We rounded up Owen's sound effects tape, found the party sounds and when he came up the next site we played the tape, told him we were having a party in his honor and sang Happy Birthday. Jack stood back and hesitated to sing for some reason. Crip was moved I could tell—they brought out a cake for him—He is one swell guy, and efficient too.

This was the last comm. pass tonight so he told us that he hated to be the bearer of bad news but our request to stay longer had been considered but that it was decided to hold to the present entry schedule of Day 60. We answered with a simple, "OK, thanks."

We talked of it the rest of the evening—I ran around saying how great that was—now we could get home—Now we could get off the food—our Command Module would never last more than 60 days—Owen said, "He never thought we would be extended because there was no positive reason for doing so, ATM film used up, more EREP sites than ever thought possible, we're all healthy, all corollary experiments overkilled—to sum up—more risk with little to gain—we could not think of any directorate but our own who would support us. ATM wants us back for data to look at prior to SL4, EREP wants its data, medical wants our bodies. Jack was disappointed.

Got call from the ground wanting to know who had been riding the ergometer during Jack's M092/171—I said me. I knew Paul would ask about it later (by the way, this occurred yesterday) tonight Paul wondered if I thought I could monitor M092 from the bike—I said yes, but that I knew the medical directorate would not like it. I asked if he could ride a bicycle and carry on a conversation at the same time. He said he went over to the simulator and tried it & it seemed OK to him.

In Apollo you go for just a visit or trip to zero G. In Skylab you live it.

"In earlier manned spaceflight programs and missions 'launch to landing' flight plans were prepared in detail and then executed with updates as required," flight director Phil Shaffer said. "Basically it was held intact to satisfy mission requirements established before the flight design process began. In Skylab, sections of the flight plan such as launch and rendezvous or deorbit and entry were similar, and a complete nominal plan was generated for the on-orbit operations for such activities as determination of consumables usage budgets, but the actual daily on-orbit plans were generated in real time to recognize situations and conditions as they were in the present time frame.

"As a result, the folks at NASA headquarters thought they should be directly involved in planning the activities to be planned in the near-real-time flight planning processes. This preference was not known and was not prepared for by the flight operations people until late in the premission time frame.

"I believed there was an inherent conflict when upper-level management people stop limiting themselves to setting objectives, requirements, and guidelines and begin trying to control implementation and execution, especially when control was down to the level of specific procedures. I did not believe they were trained for this and were not required to be sufficiently familiar with the specific configuration of systems and hardware. I believed the selection, scheduling, detailed planning, implementation, and execution responsibilities rested with the flight control and flight crew people who were both trained and familiar. In any case the result was the establishment of the Mission Management Team (MMT) that met outside Mission Control and provided inputs to the planning teams that were sometimes inappropriate.

"On Alan Bean's flight, this conflict surfaced when the MMT sent direction that the crewmember serving as the LBNP experiment monitor was to discontinue the practice of riding the ergometer [stationary bike] during the performance of this experiment. Their concern apparently was that since the experiment subject could lose consciousness when the pressure was reduced on the lower part of his body, the monitor could not respond quickly enough in terminating the depressurization. However, the practice of riding the ergometer during LBNP activity provided a free exercise opportunity and in fact the monitor did not have to get off the ergometer to reach the control for repressurization. He could reach it from the seat.

"An MMT representative came into Mission Control to deliver the input for the day, and I had the good fortune to be the flight director that day. He directed me to tell the crew that henceforth and forever more the LBNP monitor would not ride the ergometer during the experiment. So I asked 'And the rationale for this is . . . ?' and he told me how dangerous the MMT thought the practice was. I started to ask him where they thought the monitor ought to be but didn't as it probably would have started a nonconstructive debate. Instead, I told him 'I need to talk to Bill Schneider, now.' Bill was the program director for Skylab from NASA headquarters. The MMT messenger looked at me for about a heartbeat and left.

"In short order Bill showed up at my console, and I told him, 'Bill, you guys are making a big mistake with this direction to not ride the ergometer during LBNP operations.' I described to him the proximity of the ergometer to the LBNP and its controls and then told him, 'I want you here on the console with me when I tell Alan that the LBNP monitor can no longer ride the ergometer during the LBNP experiment because he can not adequately monitor the subject. Further, Bill, I want you to respond to Alan directly when he comes on the downlink and tells us how little he thinks of that idea.' Bill looked at me for an instant and said, 'Don't tell him . . . we are not going to do it that way.'

"And we didn't; we continued to take advantage of the free exercise period during LBNP operations for the rest of Skylab."

The decision not to grant the extension marked the beginning of the end of the Skylab II mission. The crew began preparations for their return home in earnest. The next day Bean noted in his diary that he had received several changes to the entry checklist (reflecting the new procedures needed due to the thruster malfunction) and had spent an hour or so reading through the revised version.

"The other evening I spent an hour or so in the CSM touching each switch as I went thru the entry check lists," he wrote in his diary. "Nice to find out one does not forget too rapidly."

The approaching end of the mission meant that the crewmembers also had to begin a staged shift of their circadian rhythm—the body's sense of when it should be asleep or awake—to prepare for the return to Earth. The schedule for the final day of the mission had already been planned out to assure that the crew had the opportunity to get as much rest as possible before beginning reentry. For the remainder of the mission, they would gradually change their scheduled sleep and wake periods to transition their circadian rhythms so that they would be ready.

Garriott, MD-47:

Two busy days. More aurora, ATM sees a more quiet sun now, several EREP passes; the bad news last night was no further mission extension was possible. 59 ½ days would have to be it. We would be eating into the third crew's food to do that, which we ended up slightly infringing anyway—mostly the sugar cookies, I think. And I'm sure the ATM film will be exhausted before then, as we are already having to ration ourselves. Just too many fascinating things to record!

Bean, MD-51:

Day off. We did our usual 2 EREP & ATM plus not much else. We go to bed 2 hours early tonight to shift our circadian rhythm around— We did not want this but can live with it. I went to the CSM to get a Seconal to sleep on time.—Owen couldn't find the OWS Seconal—it was in some other drug cans that the ground had him move. Later he inventoried some drugs— This sort of thing always puts him in a bad mood.

Pedaled the ergometer for 95 straight minutes, to establish a new world's record for pedaling non-stop around the world—and as Jack said, I did it without wheels too. Owen was interested and thought he might do it later in the week when our orbit had decayed and then beat my time by a second or less. Bruce McCandless [Capcom] pointed out that he must exceed by at least 5% to establish his claim.

Owen did some good TV of how the TV close up lens could be used medically—He looked at Jack's eye, ear, nose, throat & teeth and discussed how the TV might be used by doctors to aid us in diagnosis and in treatment of problems we might have, say, an eye injury, a tooth extraction, suturing a wound or any number of things from a broken bone to skin rash. Owen has a mind that dwells on the scientific aspect of all that he does. He knows much about much—he is interested in all branches of sciences. He is a great back of the envelope calculator—able to reduce most problems to their simplest elements. He has done great school room tv demonstrations of zero g water, magnets, his spiders.

MD-52:

Our circadian rhythm is in good shape today after the shift. Found out today that we had 6 hrs from tunnel closeout to undock—then 1 hr 45 min from there to deorbit burn then 24 min to 400,000 feet. A nice slow timeline that will allow us to get set up, double/triple checked for our entry.—Maybe we [can] stand up 2 hours later—well, we'll see.

ATM operations have become much simplified the last week—with all the solar activity the film is gone.—It was a freak on the sun and we were lucky to see it.

Garriott, MD-52:

"Day off" yesterday [but did] several TV shows, magnetic effects demo, medical demo. Jerry Hordinsky [the next crew's physician, who was filling in since Paul Buchanan was en route to the recovery ship], mentioned a "limited test" in which subjects were given Scop-Dex to see if it affected their medical tests. He said in one subject there was a minor effect. Jokingly, I asked how the other test subject did. Jerry replied that "he had no effect".

[Remarked Garriott: "Some study! I intended my question as a joke, but there really were only two subjects!"]

MD-53:

Some free time still. Every one still feeling tops. We're winding down now, getting ready for final EVA and reentry. Doing a few 2-hour time shift adjustments to get ready for reentry [west of San Diego]. Finally pulled out a library book a few minutes ago and read for 10 minutes. Jonathan Livingston Seagull. *Space is too fascinating a place to experience, to waste time doing what can be done just as easily at home, that is, read books! I'll philosophize when I get back home.*

Bean, MD-53:

Owen & Jack were doing TV of paper airplane construction and flying. The trick is not to cause them to have lift or they will pull up into a loop—with more space they would continue in loop after loop. The designs were different than we've all made as kids—more folds in the nose in the inside edge of the wings.

We interrupted our work to do some special TV*—I took 2 of the* M509/T-20 *pressure bottles put a twin boom sunshield pole between them and taped that up. I then put some red tape and marked 500 on each. We now had a "1000-lb barbell". We showed Jack with Owen and I lifting the barbell up to Jack. He grimaced till he was red as he lifted it up. . . He lifted it again and as he came to full up he released his triangle shoe locks and kept going off the top of the camera field of view!*

We then did the Bean push up—both hands first, then with Jack on my back, then also Owen on his back—then a one arm push up then the finale a no arm push up with all 3. The piece-de-resistance was a 3 man high with Owen at the bottom, me in the middle and Jack on the top. Owen was great. He wobbled around like we were toppling. We now must put it all on movies to use after we get back.—Funny, you never know what movies people will find funny—It gives a welcome relief from the science we do.

MD-55:

Out the wardroom window we saw a bright red light with a bright/dim period of 10 sec. It got brighter and drifted along with us for 20 min. or more. I said it was Mars but Jack & Owen said a satellite—it was because it also was moving relative to the stars. It may have been very near, it was the brightest object we've seen.

Also saw a laser beam from Goddard. It looked like a long green rod perhaps as long as your fingernail held at arm's length when viewed end on, that is 20 times longer than in length (which was parallel to the horizon) than in width. Tomorrow the ground will tell us that Goddard did not have our trajectory right and did not point at us—we may have seen the side view somehow. Owen said at the time a laser should appear only as a bright point of light and not a bar.

Entry −5 day. CSM *checks went well—somehow I knew they would. We only look at the* G&N *[guidance and navigation] and the real problem might be the* RCS *[Reaction Control System, with the failed thrusters]. Well, we'll know soon enough. There's no reason to believe anything's wrong with the two remaining quads. The days can't pass fast enough. We have done our job and are ready to get back. At least I am, I don't know about Jack, but Owen would like to stay.*

Garriott, MD-55:

Another full day. [Lasted until] An hour after scheduled bedtime . . . Al and

Jack saw laser on one pass. I missed it, twice. Tomorrow again . . . "Ice Cream" party tonite . . . [Jack's wife] Gratia said our "stunts" were on national TV*! Oh well.*

ATM *about [shut] down . . . I'll miss old Skylab. Really hate to leave for a variety of reasons. Mostly all the unique things to do and see. A geographer's paradise. Jack and I would both like to spend days at the window w/ camera. Next time!*

While the crew had early on abandoned always eating meals as a group in favor of increased productivity, the "ice cream parties" were one social occasion that remained a part of the routine throughout the mission. The crew had arranged their menus such that they all ate ice cream on the same nights.

"On one of these occasions we all gathered around the wardroom window to eat ice cream and strawberries and watch our ground passage all across Spain, Italy, the Mediterranean, Greece to the Near East," Garriott said. "Another memorable experience, keeping in mind the history of Western Civilization!"

That experience was one that stayed with Bean also. "We could look out the window and eat," he said, recalling that the area around the Mediterranean Sea looked "just like an atlas, except it seems like there was a volcano making smoke. I remember those as really nice times."

While the view of Earth from an orbiting vehicle is universally hailed by anyone who has seen it as an unforgettable experience, Bean said he was also awed looking out at the spacecraft he called home. "I can remember being amazed looking out the windows at the structure of the Skylab," he said. "How heavy and big. These beams were big; the things that rotated the ATM were just huge. And here it was up in orbit, and going about 17,000 miles an hour, and you think it's a fragile spaceship, but really it's more like a bridge. It's more like one of those old bridges that you cross that have all those trusses. It reminded me of that. In fact, there were trusses all over this thing. That was always amazing to me, how much heavy weight there was."

The crew, and Lousma in particular, had made national television earlier in the mission with the video tours of their home, featuring a glimpse of life in space, complete with such mundane, yet out-of-this-world tasks as a weightless haircut. "It was fun to do them, because you could be humorous, and show everybody what it was like," Lousma said.

Garriott, MD-56:

On to the "overage food". Lots of meals left—tuna and bread for lunch. Pork loin and asparagus for dinner.

When the Skylab workshop was launched, it carried with it provisions for all three crews. They were divided up according to the nominal mission lengths—one twenty-eight-day increment and two fifty-six-day increments. In addition, however, additional provisions (the "overage" Garriott mentioned) were included in anticipation of the possibility that one or more of the missions might exceed the nominal length. Despite the overage provisions, however, members of the third crew have reported that some of their provisions seemed to be missing by the time they reached Skylab; most notably strawberry drinks and butter cookies.

Food was not the only item affected by the mission duration limitations, either. Jack Lousma explained, with tongue planted firmly in cheek when discussing his crew's virtuosity respecting their successors' supplies: "One of the things, of course, on the Skylab was that most all of our equipment and gear and food and clothing and whatever didn't go up on the [separate crew launches] to get there, but they went on the original launch of the Skylab.

"And when we got up there, we were all scheduled to have a certain amount of everything. There was a group of stuff for the first crew, and they pretty much kept to their stuff; they didn't get into ours. And there was a certain amount for the second crew—that was us. And we pretty well confined ourselves to our stuff. We didn't get into the third crew's stuff at all.

"Actually, what we did was, we knew we were supposed to be up there fifty-six days, or whatever multiple would get us over the landing site and that these guys were going to be up there fifty-six days too. We wanted to stay longer than them.

"So at Day 40 or so, we asked if we could stay ten more days. It went in multiples of five; every fifth day you were over the right landing site. And Mission Control deliberated on that for about a week. And they finally came back about Day 50 and said, 'You guys have used up your food—or you will—and you've used up your film. And we don't want you getting into the supplies for the third crew.'

"We wouldn't do that anyway; we were very careful about that. But on the other hand, we were having somewhat of a problem because we were limited in our supplies of underwear. The plan was we would all have a change of underwear every two days for [the planned mission lengths]—twenty-eight days, fifty-six days, fifty-six days.

"Since there were no laundry facilities on the Skylab space station, soiled clothing was jettisoned into the evacuated LOX tank via the Trash Airlock and was replaced with new clothing. The allocation was for one change of outer garments every two weeks and one change of underwear every two days. So the ground had the delicate dilemma of deciding how to provide enough sets of skivvies for both crews from a carefully calculated, limited supply without compromising the duration of the present and next missions, the doctors' hygiene restrictions, and especially the crews' most personal expectations with respect to living and working in space with the same comfort to which they had become accustomed with regular changes to clean skivvies.

"On the morning of the last appointed day of the last set of skivvies, it became clear the ground had solved this problem, at least to their satisfaction. The answer was uplinked on the teleprinter while the crew slept.

"The solution to this problem was printed in a common humor form of the era known as a 'Good News, Bad News' joke. The message was: 'With respect to today's regular change of underwear, we have Good News and Bad News for you.

'The Good News is: You will get to change your underwear today!

'The Bad News is: Al, you change with Owen; Owen, you change with Jack; and Jack, you change with Al!'

"All of this was in keeping with a motto the Skylab II crew shared among themselves: 'Never lose your sense of humor!' "

The final EVA of the mission was the shortest of the three, with a duration of less than three hours. Garriott again ventured outside, this time accompanied by Bean, on his only EVA not taken on the surface of another world. The pair retrieved their second and final set of film canisters out of the ATM for the return to Earth, just days away. They also picked up one of the two parasol material samples that had been put out on the previous EVA.

As with the second EVA, Bean found it a difficult decision to choose who would go on the final spacewalk. According to the original plan, at this point Garriott would have gone on both of the first two spacewalks, and Lousma and Bean would have had one EVA each. Instead, both of the other two crewmembers had two spacewalks, so Bean had to decide which of them would get a third.

On Mission Day 43, Bean wrote in his diary: "Made a decision for Owen & I to do the EVA. Talked it over with Jack before I asked Owen—Reason was that he would probably get another chance to fly & to EVA, but Owen would not. In my opinion Owen has made this spaceflight much more interesting than it could have been with three operational types."

Ironically not only would Garriott fly again, but he would end up with a longer total spaceflight duration than Lousma, who also only made one Shuttle flight. Neither, however, would ever go on another EVA.

For Bean the spacewalk was an unforgettable experience, unlike anything else he encountered during spaceflight. The highlight was a darkened half-orbit with no responsibilities. While he was working on the instrument doors on the ATM, the ground radioed up that they would need to test the doors in the light and thus told him they just needed him to wait out the roughly thirty-five-minute night pass. "They said, 'We want you to stay out there overnight, and then when the morning comes, then we'll test the doors.'

"So I had nothing to do then for the night pass, and I remember we weren't in night yet, we were going into it across the Mediterranean, looking down at Italy and Sicily, with the volcano [Mt. Etna]. Next, looking down at Egypt, the Nile Delta was very obvious.

"Off in the distance was Israel and Saudi Arabia, and it was dark there. I could see the flares from all these oil rigs, and they were just all over the place. Most of them were in the water, in the Persian Gulf, though I couldn't tell it then, but when we got closer I could, 'cause it was still sort of dark on the ground, light where we were. I remember thinking that was an amazing sight."

"And then, I'd been a gymnast in college, so I kicked out of the foot restraints and did a handstand on the handholds there, and I felt like I'd set the world record handstand for height and speed. I remember that as fun. Then we came back into the daylight."

Another memory that stands out for Bean from his Skylab EVA experience was, after the Airlock Module was depressurized, first egressing through the open hatch into open space. "I can remember that being more scary than the hatch on the moon," he said. "Because the hatch on the moon was smaller and you went out backwards. And also when you went out, you were looking at the door and the frame and then you looked over here, and there's the dirt. It wasn't like you were going to fly away [The moon even provides about one-sixth of the Earth's gravitational force]. When that hatch opened on Skylab, and we were sitting there looking out, it just seemed like we could fall out! I mean, there was nothing there.

"As I tell people if they ask, it was much more science fiction to go EVA in Skylab than it was to go EVA on the moon. The EVA on the moon was much like training; you were in light, the sky was black, but everything else was the same. You were standing there, like we trained over and over. But when you go EVA in space, it's like crawling out the window on an airliner and just going along the wing, and looking in the engine. I mean, something that would be impossible to do. But I think it's the nearest analogue to what we actually do on EVA. We crawl out on the vehicle, and go along the side, and there's nothing you can do on Earth like that."

Finally, after nearly two months in space, the time had come to return to Earth. Of course, leaving Skylab meant that at least two more adventures still remained for the crew. The first, more immediate and dramatic, was reentry. Back on the planet below, the rescue-mission crew had proved in the simulator that it would be safe for the three astronauts to return home in their crippled Apollo spacecraft. Now, however, it was time to move those procedures out of the simulator and into real life, maneuvering home with the two thruster quads still available. Once that adventure was complete, the second would begin. Though perhaps less exciting than the former, the second adventure would last longer and prove to be a bit more challenging—readjusting to life on Earth after living for two months in weightlessness.

Backup crewmember Vance Brand was among those waiting in the control room during reentry. He, Don Lind, and others had spent a lot of time developing and testing the procedures the crew would use to fly their spacecraft home. Now, it was time to put those procedures to the ultimate test. The atmosphere in Mission Control, Brand recalled, was a mixture of confidence and concern as the astronauts began their return to Earth. "We were confident, but you know any little thing could mess it up, so nobody was overconfident," he said. "We expected success."

By and large the actual reentry flight was not too much more stressful than it ever is to fly a superheated metal box down from hundreds of miles high at speeds many times faster than the speed of sound. After procedural adjustments made to compensate for the locked-out thrusters, the crew

38. A long pole was used to extend the film canisters to Garriott at the sun end of the ATM.

managed to return to Earth without serious problems, other than some difficulty in reading the deorbit checklist.

The checklist of course had been revised in the wake of the thruster failures, and Bean had made extensive notes above, below, and in the margins on almost every line all the way through the book. When Garriott began to read the checklist, he found it extremely difficult to make out Bean's distinctive handwriting during the dynamic reentry phase. Further, he had not participated with Bean on any of the rehearsals of these procedures. So he was almost lost in trying to read the sequence of these very critical steps.

Garriott said he was considerably embarrassed by not being able to help Bean more by reading the extensively modified deorbit procedures to him, allowing Alan to focus on just "doing the right thing." Bean recalled: "The thing I remember about reentry was not positioning some RCS switches correctly. We got behind and Owen could not read my notes in the checklist because of the limited space (and my 'unique penmanship'). I said, 'Give me the book, and I'll reconfigure the switches.' So he gave me the book; then I reconfigured a few. I had a lot of other things going on, and I didn't reconfigure them all. About ten minutes later, we began to drift out of attitude and we got a master alarm, and I then reconfigured the rest. I switched to 'direct' and returned to the proper attitude."

Lousma recalled the transition from weightlessness to four-G during reentry, as well as the unforgettable view: "Facing aft during entry, Al and I could watch our fireball. It was about four feet in diameter and about forty feet behind the CM. It was like flying in a cone of flame which extended from the CM to the fireball formed by ionized gases and particles from the ablative heat shield. The fireball would dance rapidly around its central location but would break up when the roll thrusters fired, after which it would quickly reform. There was a frequent, loud banging noise right next to our heads when the roll thrusters fired followed by frequent right and left rolling maneuvers to keep the CM on trajectory."

They soon began to feel atmospheric drag increasing, and eventually the smaller stabilizing parachutes opened, and then the three large main chutes opened to slow the Command Module down for a splash into the ocean. "Entry was very dynamic in terms of sound, sight, and physical sensations," Lousma said. "At 25,000 feet, there was a loud, clanging noise as the nose-cone ring was explosively jettisoned to expose the parachutes. It tumbled away, and we were jerked into our seats as the two, small, white drogue chutes were deployed on long lanyards above the CM to slow it down and stabilize it for main chute deployment. At 10,000 feet, the drogue chutes were cut loose. There was a rapid sinking feeling until the main parachutes unfurled into a partially open, 'reefed' configuration so as not to tear the panels in the parachutes. In a few seconds, the reefing cords were automatically severed to allow the main parachutes to open fully for the remainder of the descent into the Pacific Ocean."

Apollo Command Modules were designed to remain stable in the water in two different positions. The more preferable of the two was called "Stable 1" and involved the narrower nose end of the CM pointed toward the sky with the crew lying on their backs inside. The second stable floating mode, Stable 2, was the inverse of the first. In Stable 2, the Command Module settled upside down, with the heat shield on the wide end of the cone facing upward and leaving the upside-down crewmen literally hanging in their seat straps.

When the crew's Command Module landed in the water, it settled into Stable 2. Then a switch was thrown, inflating several small balloons near the apex of the spacecraft. As the bags inflated, they slowly tipped it back to an upright position from which it would eventually be lifted out of the water to the deck of the USS *New Orleans,* the recovery ship. The crew remained in the capsule while it was hoisted so that the flight surgeon could make measurements before they got out of the spacecraft.

"The frogmen were in the water immediately after splashdown," Lousma said. "One of them looked in my window to determine our status while we were still in the Stable 2 orientation and while we were pumping air into the three spherical air bladders on the nose of the CM to change its buoyancy so it could rotate nose up. 'Hanging from the ceiling' in one-G was uncomfortable after two months of weightlessness. The CM is not a good boat, either, especially upside down."

On the ship, Garriott, who had no interest in using the shower on Skylab, finally got his chance to enjoy the real thing. "I had my first real shower in two months and it sure felt good," he recalled. Although trained to take short Navy showers after three years of sea duty on destroyers, an exception was made for this one—long, warm, and pleasant.

He also found that when he turned off the wall light in his sleeping compartment, he realized that he could not walk to the bunk without falling over. His vestibular system was completely deconditioned, and only his eyes were of much use to determine what was up or down. "So, back 'on' with wall switch, go to bunk and turn on bunk light, then wall switch 'off' and back to the bunk," he said. It was several days before the otoliths could be trusted to provide a good sense of what was up and down in complete darkness.

After preliminary medical tests on the USS *New Orleans,* the ship steamed back to San Diego. Garriott recalled being greeted by a friendly face when the recovery ship finally made port. Throughout the mission the crew had complained about the tedious and, in their opinion, unnecessary constant calibrations they were required to make on the mass measuring devices. Garriott had made a note to complain vigorously to the principal investigator for these devices, fellow astronaut (and SMEAT crewmember) Dr. Bill Thornton, when he saw him.

"When we docked in San Diego, the first person I saw on the pier was Bill, carrying the biggest bottle of champagne I've ever seen and wearing a grin from ear to ear, a lengthy stretch of real estate," Garriott said. "My resolve evaporated in moments. Bill may not even know my original intent until he reads this."

"On the water, it was OK," Alan Bean said. "I felt heavy, but not especially weak or anything. And so they hoisted us out of the water, and they started taking us out. We had our G-suits inflated, which I thought was a waste of time until I stood up. And then they brought us out of the Command Module and helped us, which I didn't think we needed. I'll tell you now, I think we really needed it a lot!

"They set us down in chairs. And I can remember sitting in those chairs for a ceremony on TV, and I can remember thinking, 'I hope this gets over soon because I just don't feel good.' I didn't think I would faint, but I didn't feel right. So I wasn't into any ceremony; I was more interested in lying down. So we sat down with our legs apart. We were all sitting wide stance because of our lack of stability.

"We got through that. I felt like I faked it through because I didn't let anybody know how much I wanted to lie down. Then they had to walk us down to sickbay for tests. I can remember walking along with the doctors on either side and thinking, 'They don't need to be there.' But twice, maybe three times, during the walk, I suddenly pitched left to right, and they held me up, kept me from falling. And I can remember saying, 'Boy, this ship is sure rolling,' and they didn't say, 'No, it's you,' which they knew it was, but I didn't because it didn't make sense that I could suddenly pitch left or right. I never knew that was the problem initially. I don't ever remember having vestibular problems, ever again. It was later that I began to understand that the ship never rolled, it was me pitching off. So it wasn't that I was dizzy, it was like I suddenly lost my balance.

"And so we got down to the test facility, and the NASA doctors laid us on a table and started monitoring us, and boy, it sure felt good to lie down. After a while, they deflated my G-suit, and then they had me sit up for awhile and watched my blood pressure and pulse. I guess the blood pressure went down and the pulse went up, or whatever it does, they never said, because they didn't want to affect the data, I guess, but I could tell.

"Then they had me lie down again. I can remember going through this period and not really feeling good, wanting to lie down all the time. That's what I wanted to do. But they wanted to get me physically ready to ride the exercise bike again. So they sat me up again and looked at my vital signs. After a time they had me stand up. Well, my pulse and blood pressure didn't like standing up, so the doctors had me sit down.

"During this time, other doctors were performing the same evaluations on Owen and Jack. I could see both were further along in recovery than I was. That was motivation for me to do better, but there was nothing I knew to do. And we'd hold on, but I wanted to lie down. Finally they got me on the bike. I think I was the last one on the bike. But, I got on the bike and rode the bike. I'm sure I didn't do very well, but I didn't faint or anything; and I sure was glad to lie down again.

"We probably did the LBNP, which I probably had to punch off without fainting, because several times in orbit I had to punch off or nearly had to punch off [that is, relieve the negative pressure on the lower half of your torso, which tends to pool blood in your legs and may cause fainting]. For me, the toughest thing in flight was the LBNP. I dreaded that thing. Because I really had to concentrate almost like when you're pulling GS to keep conscious in aircraft acrobatic maneuvers. I've since found out that I'm a low-blood-pressure guy. It's just something that's good in a way to be low-blood-pressure, but it's bad in that way.

"I remember then for the next two or three days, not wanting to either sit up or stand up much, so every chance I got, in debriefing or anywhere else, I'd lie down. I'd get out of my chair and lie down on the floor and prop my head up and talk. It took me two or three days to finally feel normal. It probably took some time to get the LBNP and bicycle ergometer back to normal as well."

Lousma also recalled obligations dragging by after his return: "Upon return, we had a really long day. We had to get ready to come home and get picked up. We felt like going to bed when we got back, and the doctors wanted to keep us up and do all these medical experiments. I remember just really being up longer and feeling more tired than I imagined I would be, to get all the medical stuff done on the deck on the ship.

"The medics weren't always best friends with some of the guys, but I never felt that way about them. We cooperated with them no matter what it was, to do an experiment or to do some preflight test or postflight tests, whatever they wanted to do to get their job done."

As the crew's readjustment progressed, routine tasks occasionally took on new complications. Moving a suitcase on his first night back in his stateroom, Bean pinched a disc in his back and had to receive treatment for it. A couple of times, getting out of bed during the night to go to the bathroom, he fell to the floor while attempting a floating move similar to what he would have done in orbit. "I didn't get hurt or anything, but I thought, 'That's weird,'" he recalled.

Lousma said that it took between four days and a week for his vestibular system to fully readapt to life on Earth. "I don't remember having a big vestibular problem. I don't remember having vertigo or feeling dizzy. The vestibular response that took the longest was to walk in a straight line," he said. "Our muscles and our brains didn't work together on lateral motions, because we hadn't simulated any of this straight-ahead bicycling motion. We were strong, but we hadn't used those sensors that are used to do lateral. I remember getting back to the office in Houston in a big wide hall. I'd be going somewhere, and all of a sudden I find myself on the other side of the hall, and I didn't mean to be there. I wasn't falling over, but I meandered for three or four days, probably, something like that. Your whole sensory system recalibrates itself."

It took a similar amount of time for his body to return to something resembling the condition it was in before the mission. "The doctors said I was back in my preflight shape in six days," Lousma said. "That's overall. But when I got back, I felt lightheaded when we had to stand up. We had less blood volume, I think, and fewer red cells. For the first week or so when I went home, when there were things to be done, I didn't feel bad, I just felt lazy."

Other elements of the readjustment, though, took a little longer. "I measure myself on how fast I can run two miles, and I have that pretty well documented personally. I was running two miles between 12:30 and 13 [minutes]. I shot for less than thirteen minutes. I guess 12:25 was the fastest I ever ran, but I could usually come in around 12:45. I was under thirteen on a regular basis. If I wasn't, I was disappointed. It took three weeks to return to the same speed as I had left with. So it all depends on how you measure it."

Like Bean, Lousma had a moment or two when he forgot to take into account the effects of living in a one-G environment. "That first night on the ship, we were in sickbay, I guess. I was in a bed with rails on it," he said. Noticing that the door was ajar and letting in light, Lousma decided to get up and go close it. "I grabbed hold of those rails and was going to float over there, and I didn't go anywhere.

"One of the funny things that happened, after I was home for about five days or so, I was shaving one morning. I use shaving lotion, and got myself all shaved up." He picked up the shaving lotion with one hand and attempted to toss it to the other with the sort of quick push that would have done the job on Skylab. On Earth of course the bottle dropped immediately. "Pow, right in the sink. Smashed the whole bottle."

For Alan Bean the conclusion of Skylab II was not only the highlight of the mission, but also one of the proudest moments of his life. "It sounds strange, but for me, it was when we landed on the water. I felt like—and I still feel this way—that we had given the best we had for fifty-nine days," he said. "That meant a lot, and still does mean a lot. I felt like that mission was from my viewpoint the highlight of my career, as being the best astronaut that I could be. I felt like our crew was the best crew we could be because we had done the best we could. We got sick; we couldn't help that. We bundled along. And then we went normally, and then we went to overdrive to catch up, and then we passed. So we ended up coming with a great percent."

He said that he was very proud of a report published after the mission summarizing the crew's accomplishments, reflecting the fact that they had accomplished 150 percent of their assigned objectives. On 12 October 1973, the top headline of Johnson Space Center's "Roundup" newsletter read "SL-3 'Supercrew' Gets 150% of Mission Goals." It continued:

Although the Skylab-3 mission has been completed, scientists and principal investigators will be busy for years analyzing data from the experiments performed by astronauts Bean, Lousma and Garriott.

Kenneth Kleinknecht, Skylab Program Office manager, said at the post-flight press conference that the crew brought back to Earth more than 150 percent of their goal in scientific data.

"With the longer duration mission, the crew gets more proficient because of in-flight training and experience. . . ." Kleinknecht said.

Reg Machell, manager of the Orbital Assembly Project Office said that several new things which had never been observed before were recorded in this mission.

Among these new items are coronal holes, or voids in the sun's corona. Experimenters found that the velocity changes of the gasses and of the material moving across the sun were much higher than anticipated. Data was also gathered on major solar flares.

Over 10,000 frames were taken with the multispectral camera, 2,000 frames with the Earth terrain camera and 25,000 frames with the visual tracking system. The multispectral scanner, infrared spectrometer and micro wave sensors recorded over 90,000 feet of magnetic tape data. "The VTS *film turned out to be better in this mission than the previous mission from a standpoint of resolution and clarity of Earth sites. This Earth resources data is about three times the amount of data gathered on Skylab 2," Machell said.*

Also, the beginning and ending stages of tropical storm Christine were covered as were African drought areas, Mt. Etna—an active volcano and a severe storm in Oklahoma.

"I've always been proud of this," Bean said recently of the article. "That's why I have it in my briefcase, even though I haven't looked at it for a long time; I've had it there. We were called a 'supercrew.' We were. Nobody had done that. We did, compared to previous mission estimates, more than any crew had ever done in any program, and we started out behind. So we really were as great as we could be. I've felt good about that. That's the primary feeling I have about Skylab, is just 'Wow, we did what we wanted to do. We did the best we could do.' "

"You have to find a way to accomplish the goal. We were able to do that. We went fifty-six days, and three more. Even with all the thruster problems, we accomplished the goal."

Jack Lousma said: "Maybe the best way to characterize it for me was the final impression I had when we were rolling around in the Command Module on the water. I felt the most professionally satisfied I have ever felt, with the exception of the Columbia mission I commanded, about equal, I guess.

"That number one, we were alive, and number two, we did a good job. We'd not only done the best we could, but we got it all done and really did a good job. That was the most rewarding professional sense I ever had, was on both flights, and that professional satisfaction lasted a long time after the Skylab mission. If I had never flown another mission, I would have been a satisfied guy that I'd done a good job on my spaceflight and had been professionally rewarded."

Owen Garriott said: "I have asked myself, to whatever extent it is true, what are the reasons for our success on this mission? No doubt a commitment to doing the best one can was important and even Alan's 'positive mental attitude' was to some degree contagious. An adequate degree of competence is obviously essential.

"But the one overriding characteristic of our flight, even the whole Skylab program, is that of team spirit. We had it to a greater degree than experienced in any other group I've been involved with in my career. How else can the ten-day effort to 'Save Skylab' be explained, after all the problems that arose when Skylab was launched on May 14, 1973? The thousands of Skylab Team members had it too.

"I believe it was that unquenchable team spirit that was the most important single characteristic responsible for our success and that of the whole program. It should not be overlooked that this characteristic is definable and teachable in other situations for those who are willing to make the not insignificant commitment to maximum achievement."

10. Sprinting a Marathon

You lost a crewman? How could you lose a crewman inside a spacecraft?

Skylab III Mission Announcement

The third manned Skylab mission is scheduled for launch November 10
at 11:40 a.m. EST for a mission duration of 60 days or more,
William Schneider, Skylab Program Director, announced.
The mission will be planned as a 60-day open-ended mission
with consumables aboard to provide for as many as 85 days.
Mission extensions would be considered on the 56th, 63rd, 70th and 77th days of the
flight based on the medical well being of the crew,
consumables and work load. The extension of the mission to
85 days would substantially increase the scientific return.

NASA—*Press Release, 26 October 1973*

Skylab III was to break new ground in mission duration and accomplishments. And it would do it with an all-rookie crew. When they launched, the three crewmembers did not have a single day of spaceflight experience amongst them. But when they returned, each would have spent more continuous time in space than any other human being.

For the mission's commander, Jerry Carr, and its pilot, Bill Pogue, the Skylab assignment started off as practically a consolation prize. "I was tentatively scheduled to fly on Apollo 19," Carr recalled. "Our crew was to be Fred Haise, the commander; Bill Pogue, the Command Module pilot; and me, the Lunar Module pilot. We got started on that assignment and began our training program. Then if my memory serves me correctly, it was around 1970, early 1970 or so, when it was decided that Apollos 18, 19, and 20 would be canceled. So that was a bad day at Black Rock for the three of us. We had lost our opportunity to go to the moon.

"We moped around for quite a few weeks," he said. "Then Tom Stafford called me into his office and informed me that I was to be the commander of the third Skylab mission and asked, 'Do you think you can work with Bill Pogue and Ed Gibson?' And I said, 'Of course I can.' At that time they took us off our roles as the backup crew for Apollo 16 and put another crew in there, and we began focusing on the Skylab mission.

"I was delighted to get a seat, and I was absolutely floored that they would select me to be a commander because there hadn't been a rookie commander at NASA since, what was it? I guess it was probably Armstrong on Gemini 8. And so I was really flabbergasted to be selected and very happy to do it. What delighted me the most was that I was going to be working with Al Bean, Pete Conrad, and people like that again, which was really a wonderful thing."

Ed Gibson recalled: "When assigned to the mission, I knew I was in fast company. Bill Pogue initially appeared to be just an average mild-mannered mathematician, who he had been; but he was also once grounded for flying too low behind enemy lines, was an Air Force test pilot, and flew with the Thunderbirds. He is a sharp, aggressive guy. Jerry Carr had a good education in aeronautics and was a Marine aviator, which pretty much said it all. The all-rookie crew aspect didn't faze me. I was just happy to get a seat, and flying with guys I really respected. In retrospect, I lucked out. I got to do great science, be fully immersed in all aspects of astronaut activities, and fly high-performance aircraft. It just couldn't get any better than that!"

While the three rookie astronauts were excited to be getting their chance to fly, they had little idea that before Skylab III even launched, it already had two major strikes against it—the past and the future. The strike in the future was the next great thing looming over the horizon—the Space Shuttle program. Early development of the Shuttle was already underway by the time of Skylab, and the orbiter contract had been signed the month before the SL-1 launch with a critical design review scheduled for 1975. However, the program still faced opposition in Congress. A major part of the system was a one-shot pilot-controlled landing from orbit with no go-around capability. There were those who felt that landing would be too large a challenge, particularly if the pilot were suffering from space sickness. The Skylab III crew had been made aware of how important it was that they not give the orbiter's enemies ammunition against the program in that respect.

The past affected them in the form of the two Skylab missions that flew before them. Both Skylab I and II had been behind their timeline early in flight. In both cases there were obvious factors that contributed to these slow starts. The Skylab I crew had to deal with the high temperatures and power shortages on a crippled spacecraft. The Skylab II crew was slowed by motion sickness. Those obvious factors, though, obscured the fact that the major cause was simply that people had to get considerable on-the-job training to efficiently perform tasks in weightlessness—especially when large habitable volumes are involved. With repetition the second crew in particular became extremely efficient and was accomplishing more than their scheduled science work by the end of their mission. While the actual factors involved in the slow starts of the first two crews would become a major issue once the third crew was in orbit, the efficiency the second crew developed over the course of the mission had an impact on the third crew while Carr and his colleagues were still on the ground.

Upon learning of the 150 percent return of the Bean crew, scientists and mission planners saw an opportunity. Clearly, they had not sent enough work for the second crew to do—and they began making sure the third crew was going to have plenty of work to accomplish. "We got to Skylab III, which was going to be the last mission in the program," flight director Neil Hutchinson recalled in a NASA oral history interview. "The train was leaving the station, and all kinds of experiments and experimenters were running for a seat."

In addition NASA decided to use the third crew's flight to capitalize on another opportunity. In late December 1973 and early January 1974 the Comet Kohoutek would be passing through the inner solar system. Tasking the third crew with observing Kohoutek from Skylab would let the agency show off the potential of orbital astronomy by performing an unprecedented feat—no comet had been observed from space before. "Some other training we got at the last minute included that on Comet Kohoutek," Pogue recalled. "Early in the year it was discovered at the Hamburg Observatory in West Germany that this comet was headed toward the sun and was going to reach its perihelion about Christmas Day of 1973. There was a lot of talk about the period of the comet being about two thousand years, which led to speculation that it was actually the Christmas comet, the one cited in Biblical stories of the new star. At any rate, we did some studying and training for that experiment as well."

Press releases from the time illustrate the situation that confronted the third crew.

NASA-JSC ***Release No.: 73-107***

NASA today announced tentative plans to observe the Comet Kohoutek during the Skylab III mission which is planned for launch on or about November 9 from the Kennedy Space Center. The November date is the original planned launch date for Skylab III. The Comet Kohoutek was identified earlier this year and will be clearly visible from Earth. It is expected to be the brightest object in the night sky except for the Moon in late December and early January. Skylab's Apollo Telescope Mount instruments, designed to obtain data on the Sun, will observe Kohoutek during its nearest proximity to the Sun late in December.

Aircraft-borne Lasers to Profile Earth and Sea during Final Skylab Flight—

As Skylab's third crew collects data on the Earth's resources from 270 miles out in space, two aircraft from the Johnson Space Center (JSC) will skim near the surface using laser instruments to provide an exact profile of the land and water at more than a dozen sites. During the coming months, NASA aircraft will use laser profilometers over portions of the North Atlantic Ocean, the Gulf of Mexico, the Puerto Rican Trench, and the Great Salt Lake to support Skylab remote-sensing passes over the same areas.

Skylab Gypsy Moth Research Project—

One thousand gypsy moth eggs in two special vials will be launched aboard the third and final Skylab mission on November 10. The first moths in space are part of a research project sponsored by the U.S. Department of Agriculture's Agricultural Research (APHIS) in cooperation with NASA. Agriculture scientists are trying to find out if the state of weightlessness might be the key to altering the gypsy moth's life cycle. If weightlessness does prove to be the factor, the key point may be found in rearing insects by the missions and thus controlling a whole class of insect pests with similar life cycles.

Third Skylab Crew to Expand Knowledge of Earth's Resources—

Astronauts Gerald Carr, Edward Gibson and William Pogue will be well equipped to survey the Earth during a final Skylab mission that could last nearly three months. Their training included 40 hours of special lectures on Earth observations and they're taking along a detailed handbook for viewing Earth from space and the largest store of film and computer tapes ever supplied for a Skylab mission. Meanwhile, the 20,000 Earth photographs and 25 miles of computer tape obtained during the two previous Skylab flights will be undergoing extensive analysis by 137 Principal Investigators and their staffs in the United States and 18 foreign countries. Before the first EREP pass of the final Skylab mission can be undertaken, Science Pilot Ed Gibson, assisted by his fellow crewmembers, will attempt to repair the antenna drive system for the microwave radiometer-scatterometer-altimeter (S193). Gibson will work on the S193 instrument during the crew's first walk outside the space station, scheduled for the week following launch. Pilot Bill Pogue will join him outside.

Operation Skylab/Barium

Skylab's third and last crew of astronauts, now in orbit and embarked on a full program of scientific research, is scheduled to add another important data-collecting task to an already full agenda. In addition to continuing investigations of the sun, Earth resources, and medical effects of long duration space flight begun by preceding Skylab crews, the astronauts are going to participate in an experiment to trace geomagnetic field lines with barium ions. Beginning with the morning of November 27, Marine Lt. Col. Gerald P. Carr, civilian scientist Dr. Edward G. Gibson, and Air Force Lt. Col. William R. Pogue will join a widespread network of observation stations waiting for the launch of a National Aeronautics and Space Administration Black Brant IV rocket from the Poker Flat Range near Fairbanks, Alaska. The rocket payload is designed to create a high-explosive-driven jet of barium vapor and inject it into the Earth's magnetosphere. It is hoped that the barium vapor, ionized by solar ultraviolet radiation, will illuminate geomagnetic field lines and make them visible to sensitive optical equipment for many thousands of kilometers.

Special Camera to Photograph Comet Kohoutek from Skylab—

As the Comet Kohoutek streams through space at speeds exceeding 160,000 kilometers an hour (100,000 miles per hour), astronauts aboard the Skylab space station will use a special camera to photograph features not visible from Earth's surface. The camera, called a Far Ultraviolet Electronographic Camera and designated Experiment S201, was built by the Naval Research Laboratory (NRL) in Washington DC. Dr. George R. Carruthers prepared the instrument for use aboard the space station during a three-month crash program.

Release No.: 73-156

Skylab Science Demonstrations

A Skylab bonus of three unscheduled science demonstrations performed by the SL-II crew in their spare time has resulted in plans for expansion of this activity by the crew of SL-III. The demonstrations, to be filmed by TV, movie and still cameras, will require a degree of inventiveness from the crew and will provide a change of pace for them during the mission. The activities will also provide material for educational applications. In addition, NASA scientists believe that examination of the photographs and video data of these demonstrations will be of considerable assistance in designing even more valuable and complex science experiments onboard the Space Shuttle.

All of these new activities and others added to the mission at the last minute meant that crew training for Skylab III presented a real challenge. The crew had been working to prepare for one mission and suddenly found itself with little time left to prepare for a greatly expanded one. And its position as last in line made getting adequate and proper training even more difficult. "We didn't have a chance in the beginning to get much real simulator training because the two crews ahead of us were going to get it all, and all of us Skylab guys had to wait till the Apollo Program was over," Carr said. "So even Pete Conrad and his crew were only getting catch-as-catch-can training whenever the simulator was available. We were left playing with cardboard and other low-fidelity mockups to try to figure out what to do in flight."

Pogue agreed: "We were the last crew. The first crew dominated the simulators when they were training. Then obviously the second crew also had to spend a lot of time in the simulators. In addition, the backup crew required increased training when the potential arose of rescuing the second crew if their RCS [thruster] problem worsened. Thus, we were left doing only peripheral stuff. We'd go wherever they weren't getting trained. Whichever simulator or trainer they weren't using, we would use, if it wasn't down for maintenance. Then, of course, as soon as they launched, we finally got three months of relatively intense training."

Despite being low on the priority list for the simulators, the crew kept busy with training activities. "One of our main tasks was to help put together the training program for all Skylab crewmen, so we worked hot and heavy with people in the training department to help them brainstorm and get that sort of stuff out of the way," Carr said. "Since we didn't have any simulators to work with, and we couldn't do anything else, it was probably an excellent use of our time.

"We each ended up with individual jobs: Ed was the guy in charge of experiments, and particularly the solar physics experiments. He had recently written a textbook called *The Quiet Sun* and was the solar physics expert in the astronaut corps. Ed really focused heavily in that area. Bill managed a lot of the Skylab fluid systems and other experiments. My main focus was the Skylab navigational, guidance, and related systems. We structured our training so that all of us could operate anything, but if something went wrong, there was always one expert."

The crew put particular focus on preparing for the Earth observations tasks during their training. Jerry Carr explained: "We did not want to be in the position at a debriefing of having someone ask us about something we saw and being able to say nothing more than 'Yeah. We saw it. Sure was pretty.' We went to Ken Kleinknecht and said that we really wanted to be intelligent observers of the Earth when we weren't doing other things and asked if he could help us. They gave us forty hours of training time and promised to find at least twenty world experts on various Earth phenomena. Each of them was to come to the center to give us two-hour briefings on what's important, what they wanted to know, and how we were to look for it.

"That turned out to be probably the most exciting and rewarding of all of the experiments that we did (Ed would probably put the ATM first by a narrow margin) because it provided the opportunity to ad lib, and ad lib intelligently. The kinds of people we worked with included Lee Silver, who was an earthquake-fault expert from southern California; John Campbell, who was an expert on ice formation in the northern and southern latitudes; Bob Stevenson, who was an outstanding oceanographer from La Jolla; a desert formation expert; and several meteorologists. These people were programmed into our training, enthusiastically came to the center, and talked about what data we could acquire that would provide them with the best insights into their particular studies of the Earth.

"We thoroughly enjoyed those forty hours of training. They also gave us a lot of extra film, partially to make up for some of the film that got ruined by the high temperatures in the station early on. On balance, we were able to do pretty well with what we had."

While the first two missions left legacies that would create challenges for the third crew, there were some benefits as well. "We drew a lot of conclusions from what we saw on the first two missions," Carr said. "I think the most important one was that when the first crew came back after twenty-eight days, they were pretty wobbly, pretty weak. So the second crew and ours decided to bump up the exercise periods. Al Bean's crew doubled their exercise period from a half hour to an hour a day. Turns out that that didn't appear to be enough either, so we increased it again to an hour and a half."

"We were determined that we would stay longer and come back in better condition than the previous crews," Gibson said, "partially because we learned from their experience on how to best exercise to counter the effects of zero gravity."

Looking at the results of the second crew's mission, Carr saw the roots of a potential problem for his flight and took action to prevent it. "We watched the way experiments were being done, and some of our procedures were modified based on what the first two crews had learned," Carr said. "We noticed that the second crew was really hustling all the time. By the end of their mission their rate of activity was extremely high. We began telling some of the managers that we didn't think that rate of work was wise for a ninety or an eighty-four-day mission because we weren't sure that we were going to be able to sustain it. We thought that the workload should be slacked off some and there should be more rest. Everybody agreed to that, and the experiments were slowed and spread out quite a bit."

It was to be a short-lived respite, however. "Unfortunately, they then added a whole bunch of new experiments, and we allowed ourselves to get trapped into this new situation. All of these experiments that were added at the last minute came with a lot of problems that we didn't have the time to detect and take into consideration," Carr said. "So, when we got up there, we found that we were overcommitted just like the first crew and that we were going to have to sustain the high Skylab II work pace for eighty-four days instead of the fifty-nine that they experienced."

"The first crews really performed well and set pretty high standards for us to live up to," said Gibson. "But in critiquing their performance, we couldn't let them get swelled heads. Yes, the troops on Skylab I faced temperatures of 140 degrees and did a great job of making the space station useable. But after all, it was a dry heat!

"The second crew erected a larger sun shade that further lowered the temperatures down into the comfortable range except for one hot spot that formed when the station was in nearly continuous sunlight (technically, high beta angles)—at MY sleep compartment! At those times, I just floated my cot into the MDA and slept there."

Finally, launch day drew close. And then it was postponed. Skylab III's scheduled 10 November launch date had to be delayed when cracks were found in the fins at the base of their Saturn IB booster—something that could be blamed on the thruster problems of the second Skylab crew. The SL-4 booster had been transported to the launch pad during the summer to serve as a booster for the potential Skylab II rescue mission; it was thought that this additional period of resting on the fins had caused the cracks. After the fins were replaced, the final Skylab crew left the launch pad at one-and-a-half minutes after 10:00 a.m. on 16 November.

"I went to bed early that night knowing full well I wouldn't sleep worth a hoot," Jerry Carr said. "Several days earlier we had started trying to shift our circadian clocks to allow us to go to bed at something like six in the evening and then wake up at two or three in the morning. So at about four o'clock in the morning, Elmer Taylor, who was our flight crew systems coordinator, came into my room and said, 'The bird's waiting. It's time to go.' I had actually fallen asleep, finally, but then I awoke with a start and got up.

"The first thing scheduled was our physical. They took microbiological swabs from many parts of our bodies to find what kind of flora and fauna were living on us. They catalogued their findings as part of a long-term experiment to determine how much microbiological material we would leave on the spacecraft and what we would pick up, if anything, left by the crews ahead of us. It turned out that we did pick up some of the bugs left behind by the Skylab II guys.

"After our physicals, we went into the crew dining room and had breakfast with Deke [Slayton], Al [Shepard], Kenny Kleinknecht, and other managers. It's interesting that our meals at the crew quarters were always steaks, eggs, and all those good things that are just wonderful for cholesterol. In the subsequent years, my wife and I have totally modified our diet so that now we don't touch any of these foods mainly because of their high cholesterol and fat content. It's amazing that dieticians in those days thought that lots of steak and eggs was the best thing in the world for us.

"After the meal, we began suiting up. I put a watch on my ankle, although I was not supposed to be taking anything extra up. But I had this Movado, which was a self-winding watch with one of those little counterweights in it. I was very curious to find out if this self-winding watch would still work in a weightless environment or whether the weightlessness would inhibit the motion of that little counterweight and keep it from winding the watch. Our official watches, Omegas that we wore on our wrists over the pressure suit, were regular hand-wound, plain old mechanical watches. So I put the Movado on my ankle and finished suiting up.

"Launch went off perfectly. It was a beautiful, clear day. I remember when the escape tower was finally kicked off, and it took the shroud with it. The light that came in the cabin was just blinding for a minute. It was incredible. I tell a lot of people that riding on a booster like that is kind of like riding on a train with square wheels. You've got lots of noise, lots of vibration. Then sure enough, when you hit that first booster shutdown, staging, and then the next booster kicking off, it's just exactly what everybody has called it: a train wreck. I thought that was very apt.

"We got into orbit without any problems. Everything worked just fine. Eight minutes and twenty-eight seconds later we were on orbit and things were beginning to quiet down. Looking out the window for the first time, I was totally disoriented. I didn't recognize a thing. Suddenly, somewhere in the first thirty minutes or so, I saw Italy, and I said to myself, 'Italy really is shaped like a boot.' I've never forgotten that particular experience."

Ed Gibson described the experience: "Liftoff is an exciting time, and any crewperson who is not excited doesn't really understand what's about to happen.

"On that crisp cool morning of November 16, we rode in the standard NASA van out to the launch gantry, a thirty-seven-story building, and our Saturn IB booster resting on a structure that brought the Command Module hatch to the same level as if it were on top of a Saturn V booster, a structure that resembled the world's largest milking stool. As we rode, the big blue eyes of Al Shepard bored into each of us looking for any sign of weakness, any indication that one of these rookies was not ready to go. I looked back with a defiant smile, 'Not you, Big Al, or anyone else is getting my seat!'

"Then we took an elevator to the top floor of the gantry, walked along a narrow but exposed hallway, and waited to get strapped into the Command Module. Since I sat in the center seat under the hatch, I was the last one in, which gave me a chance to just stand outside and gaze at the vehicle. For most of the preflight time we were busy and didn't have time to reflect. But then I had about twenty minutes where I could just stand back and drink it all in.

"It was dark, but the booster was brightly illuminated by search lights on the ground. Because it had just been fueled, it was creaking, popping, and groaning from the weight and frigid temperatures of the liquid hydrogen and liquid oxygen, which caused continuous shrinking and readjustments of the metal. All of the electrical systems were up, gases were venting, and lights were blinking unlike what we had ever seen before. No longer just passive metal, the vehicle had taken on a life of its own—it was alive!

"I found it difficult to get the wide grin off my face as I was strapped in. It was an exhilarating few moments of anticipation that to this day I highly value and feel fortunate that I had, an experience similarly noted by the previous two science pilots on their missions.

"As we waited for launch, we learned who was really in charge, who would have the last word. A few days before launch they discovered cracks in the fins on our booster. Because we were eager to go and not happy with the five-day delay required to replace the cracked fins, we started to refer to the booster as old Humpty Dumpty. Well, somehow that got out in the press and of course didn't sit too well with those good troops who were working around the clock to get the booster ready in time. But, much to their credit they said nothing . . . at least not until twenty minutes before launch when we got a message, 'Good luck, and God speed, from all the king's horses and all the king's men.'

"Finally we heard launch control start counting backwards from ten, then a tremendous sucking sound as propellants got ripped into combustion chambers, a noise Bill later said 'sounded like they had just simultaneously flushed every toilet in the Astrodome.' Far below and lasting less than a second, we felt eight engines ignite in a ripple fire, and we crept off the pad. The front of my mind was focused on gauges and abort procedures, even as a little whisper spurted up from the back of my mind, 'The basement just exploded!'

"The ride on the first stage was noisy and rough, like a Hummer doing eighty miles per hour over moguls. At about one minute into the flight, we went through the speed of sound and also reached the maximum of the aerodynamic forces and turbulence that built up as we rammed through the wall of air resistance ahead of us. The vibration became severe; I felt like a fly glued to a paint shaker. Then it smoothed out a little until staging at two minutes, which jolted us like a head-on crash quickly followed by a sharp impact from the rear."

Bill Pogue recalled the incredible noise and vibration of the launch: "The noise caused by airflow over the booster had been building all during the first minute of launch. It was so loud that it was difficult to hear the intercom between our suits. Once we were supersonic, all the outside noise ceased because the air noise couldn't penetrate the shock wave attached to the Command Module. Then we could hear the creaking and groaning of the structure as it responded to abrupt swiveling and gimbaling of the engines. We also heard liquid propellants rushing through feed lines.

"Because of the intense vibration, I had difficulty reading my hardcopy checklist, which I was supposed to use to compare the predicted performance against our actual performance as indicated on the computer display."

Ed Gibson said, "The second stage reminded me of a long, smooth elevator ride that accelerated ever faster as the mass of the propellants burned away. Eventually we weighed three times our normal weight, which was not bad because our hearts were at the same elevation as our heads so graying out was not even a possibility. But it was hard to lift a hand, and I noticed my cheeks and ears sliding towards the back of my head.

"Then, at a little over eight minutes, the engines cut off—sharply! Immediately, everything floated. Our spacecraft, which they tried so hard to keep clean at the Cape, filled up with small dirt and debris that floated up from its hiding places on the floor. In short order the air conditioning system cleaned it all up.

"Outside I saw the curved horizon and the coast of Florida receding. This was the best simulation yet! I looked back in to study the gauges and threw a few switches as we reconfigured the spacecraft for rendezvous with Skylab.

When I glanced out again, Italy going by and I understood what it's like to travel at five miles a second. After a presentation when I got back, a highway patrolman stepped forward and presented me with a ticket. Said he'd clocked me at 17,682 mph . . . in a 40.

"After several orbital maneuvers, a distant speck expanded into Skylab. It was missing one wing and a micrometeoroid-thermal shield, and it was covered by two jerry-rigged sunshades. I felt a warm glow—we had arrived at our new home. This was going to be great!"

After docking, the crew was to spend the night in their Command Module before moving into their new home. The delayed entry was prompted by the problems the second crew had encountered with space sickness upon their arrival. Mission planners decided that in order to try to avoid the adaptation problems the second crew had encountered the third crew should spend a night in their Apollo capsule, giving them time to adjust to weightlessness before moving into the open volume of Skylab and getting to work. So after arriving at Skylab, instead of going inside the crew worked late stowing equipment in their Command Module.

"About that time," Jerry Carr said, "Bill was saying, 'I'm not really feeling too terribly well,' So we talked about it, and I said, 'Well, best thing to do, probably, is to eat. You'll feel better.' So we went ahead and ate our dinner. One of Bill's items was stewed tomatoes. He ate them down, waited for a while, then said, 'It's coming back up.' So he got out his bag and barfed.

"The day before we left JSC, the doctors said, 'Now, we're real concerned about this space-sickness thing. We want you to take medications.' In the medical sensitivity tests they'd done on us, they found which of the antinausea medications were best for each of us and which had the least side effects. The doctors said, 'Jerry, we want you to take something. In fact, we want all three of you to take something.' I said, 'Wait a minute. I'm driving this multimillion-dollar vehicle, and I'm not even allowed to drive an automobile or fly an airplane when I take Scop-Dex [one of the medications]. Why do you want me to do it now?' They said, 'We don't want you to get sick.' I said, 'I'll take the sickness rather than the disorientation,' and decided not to take the medication.

"Well, Bill wanted to be a good patient and said, 'Okay. I'm not driving, and I'll be able to manage fine, so I'll take the Scop-Dex.' What surprised us was that Bill was the one who got sick. Whenever Bill and I went up in a

T-38 to do acrobatics, I was usually the one that turned green, not Bill, and he had taken the medication!"

Gibson agreed: "We called Bill 'Old Iron Ears.' You could never make him sick on the ground. Put him in a rotating chair, and he'd never get sick. He used to fly for the Thunderbirds, so you figure that if there's anybody going to get sick, it'd be me, the real novice, or maybe even Jerry, but not Bill, which showed us that we didn't really understand the problem."

Bill Pogue said, as his experience proved, "There's not a direct correlation between who suffers from motion sickness on the ground and who has problems in space. I've observed that people who are susceptible to motion sickness, particularly susceptible, on Earth tend to not be in space, and vice versa. Clearly someone like me who went through the full limit of head motions at the highest RPM in the rotating chair at the Pensacola naval facility and could have continued indefinitely is, by definition, highly resistant to motion sickness on the ground. They never could make me sick. But who got sick first on Skylab? I did. It's sort of an inverse relationship."

Faced with Pogue's sickness, the crew discussed what to do about it. One of the biggest things on their mind was the burden they carried for the future. Right then, down on Earth, work was beginning on the Space Shuttle. Right then, also down on Earth, there were those in Congress who were opposed to the program. The success of the Space Shuttle depended on astronauts being able to make that one-shot glider landing. Sick astronauts, the Shuttle's opponents would argue, would not be able to make that landing. The future, it seemed, was resting on the third crew proving that there wouldn't be a problem.

And so with the future of spaceflight in mind, they decided what to do about Pogue's sickness: "With all the pressure they were putting on us not to get sick, Ed and I said, 'Well, look. Maybe we just won't say anything,'" Jerry Carr recalled, "In fact we thought it might even be best to toss the vomit down the Trash Airlock and not to report it. That way we wouldn't get people all fuzzed down on the ground, and we could get the mission off to a smooth start. We knew we had a lot to do. So we said, 'Okay. That's what we'll do. We hope Bill will feel better tomorrow, and we can press on.'

"Well, unfortunately Bill, being the sick one, was also the guy in charge of the communication system, and he had left the switch on to the equipment that was recording all the intercom conversation. So while we slept that night, people on the ground played it back and heard all of our previous conversations. The next morning, Al Shepard came up on the Capcom loop and said something like, 'You guys have made a mistake here, and I hope you haven't destroyed the vomitus bag.' I said, 'No, we haven't done anything like that, and I agree with you. It was a dumb decision. We'll put it in our medical report, weigh it, do all the necessary things, and go from here.

"So they discovered that we were trying to conceal information, which we felt pretty bad about. But that was our motive: we didn't want to fuzz things up anymore on the ground. It was dumb. Yet we did it, we wish we hadn't, but we did."

That mistake behind them, it was time for the third Skylab mission to truly begin. After a night's sleep, the crew awoke and prepared to move into the space station. "The next morning, Bill wasn't feeling great but he was feeling better," Carr said. "Ed and I were both okay. I had a feeling in my stomach that was kind of like a big knot, but I wasn't sick. Ed just didn't have any problems at all. We always thought that was kind of a marvel. Ed, the one who had the least flying time, was the nonsick one."

The crew opened hatch and entered Skylab. Upon moving in, though, the crew found that they were not alone. Three figures, wearing the unmistakable brown Skylab flight suits were waiting for them in the workshop. Before their departure, the Skylab II crew had stuffed the suits and posed them at various work locations on the lower deck. "When we arrived, we found three dummies that had been packed and put there by three previous dummies," Carr said. "It was quite a surprise to roll down through the tunnel and come across three other people in the spacecraft that we weren't expecting."

"Because we were really rushed at the beginning," Ed Gibson said, "we left the dummies where they were for quite a while. Every time I was down there, I felt them staring at me, inspecting everything I did, but not lifting a hand to help—eerie.

"During those initial days, there was a real adaptation to zero gravity that had to take place. When launched, we were literally thrust into a whole new environment. When I looked in the mirror, a pumpkin looked back, a round red head with bright red eyeballs. No longer countered by gravity, my heart and arteries continued to ram blood up towards my head. It felt like I was lying down back here on Earth with my feet a little over my head. But after a few days, I lost about three pounds of water as did Jerry and Bill. Jerry and I then felt pretty good, but Bill continued to suffer.

"After working hard to become efficient, it all started to seem so easy, so effortless—from a physical standpoint. That's because one of the real problems with the stresses of spaceflight is that there were none. With no gravity to work against, our muscles weakened if we didn't exercise enough, and our bones slowly lost calcium and also weakened, just like bedridden patients down here.

"But we had learned what exercises to do from the previous crews, and we lengthened our workout durations 50 percent above those of Skylab II. We wanted to not only walk out of the Command Module at the end of our flight under our own power, we wanted to be in better condition than the previous crew, even though we would be in zero gravity over 40 percent longer. We dedicated ourselves to that goal and continued to aggressively pursue it through strenuous workouts throughout the full duration of the mission. And we succeeded."

Between the missions on Skylab, ground control had dumped the pressure in the station down to a quarter of a psi. They had then repressurized it to provide pure, clean atmosphere. "I recollect that when I first entered Skylab," said Pogue, "my first impression was, 'Boy, it's cold in here.' But it felt really good, especially after having the nausea event the day before. Of course I also knew it was going to be big, but after entering, I felt, 'This really IS big!' Our immediate problem on entry into Skylab was trying to find all the right books and other things that we had to use. We worked till about 10:30 p.m. Houston time that first day just trying to get caught up."

The enormity of Skylab created a situation never encountered before on a space mission. "Skylab was so large that they actually lost me one morning," said Gibson. "Skylab had many different compartments, and I was in the Orbital Workshop trying to find some of the old procedures that the previous crew had left. I was buried deep down behind the freezers where they had stowed most of the previous mission data. When Jerry and Bill started looking for me, they just glanced in the workshop and didn't see a soul. Then they looked outside and said, 'Hey, the Command Module's still here. The hatch is not open. Guess he hasn't left. So then, where is he?' When I finally floated into view they said, 'Where the heck have you been?' So, it was possible to get lost in Skylab."

Also the use of the same spacecraft by different crews created problems. Items got misplaced or totally lost, making it harder for each successive crew to operate.

"Skylab gave our nation its first experience with long-duration spaceflights in large spacecraft," said Bill Pogue. "It had an internal volume of 12,500 cubic feet, the volume of a three-bedroom house. The huge forward compartment was twenty-one feet in diameter and over twenty-five feet high. This spacious volume, numerous stowage lockers, and our longer missions led to some problems we had not encountered before. Some were amusing, but others were downright aggravating.

"Floating through the forward hatch of the forward compartment I saw Ed floating a few feet off the grid floor twenty-five feet below and obviously out of reach of any handholds or other structure. I lunged toward him, gave him a shove, and, like two billiard balls, we went flying off in different directions toward the walls where we could grab something. We were both laughing as we went back to work.

"In other instances, the multiplicity of lockers and stowage locations led to frustrating problems and delays. One evening my flight activity message for the next day directed me to recharge the fluid level in a water loop used to cool an electronics package. The job looked simple: get a couple of tools, a flashlight to observe the accumulator, and a long hose that stretched from our water tanks to the work site, and then follow the procedure and restow everything. A piece of cake? Well, not quite.

"The hose wasn't where it was supposed to be. No problem! I'll just call ground and get some help, but it would be another twenty minutes before I could call Houston (no relay satellites back in '73). I started looking in lockers adjacent to the one designated in the procedure and anywhere else that seemed like a logical place to stash it, but it was all to no avail.

"At the next AOS, I explained the problem and asked if they could get in touch with Jack Lousma to see if he could remember where he put it after the last use. Jack is a highly disciplined individual, and I was confident he could tell me right where to find it. Jack was busy mowing the lawn at his home in Friendswood a few miles from JSC when he got a call from Mission Control. He wiped the some of his sweat off and said he did remember using it, but if it wasn't in the designated stowage location, he didn't have the foggiest notion of where it might be.

"When I learned that Jack couldn't help, I really felt defeated. However, Capcom had an alternative approach and told me where I could get two shorter hoses to connect together that would span the distance. I did, it worked, and I was able to finish the servicing task, exceeding allotted time by only a factor of five. Incidentally, I never found the hose. We had a stowage book, which was generously cross-referenced, but the book only told us where an item was supposed to be.

"We had other cases of mysterious disappearances. Once a set of calibration weights just vanished, and I spent four hours on my day off looking for them. They never did turn up. A systems checklist apparently floated away and was missing for weeks until Jerry flushed it out from its hiding place with thruster blasts from the maneuvering unit [the Manned Maneuvering Unit prototype tested inside Skylab].

"We lost other items, some of which eventually did turn up. One day when I whirled around to get a camera to take a picture of Hawaii, my eyeglasses flew off. I heard them bouncing around through the experiment compartment as I was taking the picture, but when I went to get them, they were gone. Three days later, Ed found them floating near the ceiling in his sleep compartment.

"Frequently our tableware, usually a knife, would get knocked off the magnetized surface on our food trays and get caught in the airflow, which gently wafted it to the intake screen of the air duct system. It would hover there on the surface until retrieved. The screen became our lost and found department and the first place we looked whenever something was missing.

"Fortunately, today there is technology that can solve the problem. Tags placed on stowed items respond to an interrogation device and reveal their location using RF energy from the locator. It's just what we need. Let's hope it's implemented on [the International Space Station], which ultimately will have more volume and surface area than Skylab. Otherwise, it's back to the Skylab mode of operation: If it isn't there, then happy hunting."

The large volume also provided some interesting opportunities. Ed Gibson said, "One night I could not resist the temptation of Skylab's large open volume, and I tried sleeping out there floating completely free. It was the ultimate in relaxation—no pressures on your body whatsoever. Once I relaxed, my knees would bend slightly and my arms would float out straight, just like the position I had assumed floating in water many times on Earth. After a few minutes, I would drift off into a nice . . . relaxing . . . quiet . . . WHACK!

"I had drifted into a wall that jarred me awake. During all subsequent tries, I remained poised just wondering when, where, and what I would hit again. It just didn't work. Once I even ended up on the air intake screen in the OWS, our lost and found department that usually rounded up considerably smaller objects. Eventually, I discovered that all I had to do was to slip an arm or leg under a bungee cord, and I could drift right off to sleep.

"Sleeping turned out not to be difficult at all without gravity in our sacks, especially early in the flight when we all were exhausted. If we did have trouble turning off because we worked right up to the time we floated into our sacks, reading was usually a good sleep aid. This situation was about the only time we did pull out a book. Time in space was too valuable to use for things we could do on the ground, a sentiment previously stated by Owen on Skylab II.

"The fifteen sunrises and sunsets a day that we experienced could present a problem when trying to turn off and go to sleep. If you made the mistake of sneaking out of the sack to look out the window, you might see China at high noon, and you then had the difficult task of convincing your mind and body that it really was time to sleep. Also early in the mission if you were clumsy in your sneaking, the guys watching Skylab's rate gyros on the ground could tell you were up and might just call up and ask you to do 'just one more thing.' Later in the mission those 'one more things' got ruled out."

With Pogue still suffering somewhat from space sickness, the crew tried to compensate for the reduced manpower available. "Bill and I decided to change jobs because my job was a little more sedentary than his," Carr said. "So we swapped checklists and went on. Bill was able to stay quiet and get my work done while I did his. It worked out well. For the next couple of days, when Bill got to feeling a little funny, we would swap jobs. But for the most part, Bill was able to pick up and carry his load without any trouble at all."

Despite their best effort, however, the crew began to run into what would become the second major problem of their stay on Skylab. "The schedule caught up with us," Jerry Carr said, "We found that we had allowed ourselves to be scheduled on a daily schedule that was extremely dense. If you missed something, if you made a mistake and had to go back and do it again, or if you were slow in doing something, you'd end up racing the clock and making more mistakes, screwing up more on an experiment and in general just digging a deeper hole for yourself.

"The schedule was very tight, and we were hustling each and every minute just trying to meet it. That went on for many, many days. It was hard on morale. We were rushed and not able to get things done and experiments completed. We knew, we were just sure, that the experimenters on the ground were grinding their teeth when we had to report, 'Well, I didn't get your experiment done because, in my rush, I put the wrong filter in, or I made another error.' We found that it was almost to the point where you had to schedule time to go the bathroom.

"Then we discovered that we had been scheduled at nearly the same rate that the second crew had achieved at the end of their flight! That explained why we were having so much trouble keeping up. But by the time that was finally recognized, we had achieved a skill level that was adequate to get the work done.

"After the first few days, we realized that eating three meals together was not an efficient use of time. However, we did have dinner together so that we would make sure we were functioning as a cohesive crew, and we each also needed that bit of social contact. It turned out to be a great decision. But after dinner, we'd go right back to the experiments and work till probably nine o'clock at night when it would be time to wind down and go to bed. So at ten o'clock, when we were supposed to be in bed, none of us were ready to go to sleep because we still had things to pick up and put away and other things to do. Our minds were still moving too fast to rest. So, we just weren't getting the right kind of rest and the right kind of leisure time that would allow us to do things right.

"Finally we began to get a little bit testy. In order to make up time on some of the experiments, to account for some of our fluffs, they had to redouble efforts to tighten the schedule even more. They were juggling our exercise around, and we ended up in several cases having to exercise right after a meal. That's no time to be exercising, particularly up there where you couldn't belch because with your food floating around inside you, you were liable to get it back with your belch.

"So we started grousing at them about that, they were working hard trying to keep us up with the schedule, we were giving them a hard time, and they were giving us a hard time. Finally we reached a point in the mission when we just had to take a day off. We had set up a ten-day week with the tenth day as a day off when we could do what we wanted. That was also to be the day when we could take a shower in our makeshift shower. But we gave back our first two or three days off. We said, 'Go ahead and schedule us, and we'll do some makeup work.' Well, we got to the point where our morale was low, we were feeling lousy, and we were really getting drained. So we said 'Let's take our day off and get a good day's rest. It'll get us back in good shape again, and we can begin to maintain the pace.'

"So we took our day off and did what we wanted to do. We each took a shower. Bill and I did some reading, looking out the window, Earth observations, photography, and other things. Ed worked his own schedule at the ATM panel, did some relatively simple experiments, and made some ad lib observations. We had a good day."

"Though we didn't understand it at the time," Ed Gibson said, "we and Mission Control were about to learn some valuable lessons for the future—lessons that had to be learned sometime, and each of us, playing our respective roles, were the unsuspecting students.

"We found it disheartening to be in a situation where you could never catch up; it's only a question of how far you are behind. We just pushed the buttons as fast as we could and moved on to the next. We were not used to working in that mode, and we didn't plan on it being that way. An image of my high-school track coach flashed into my mind. With a wide grin he gave me a tip, 'If you want to win the quarter-mile race, sprint the first hundred yards then just gradually increase your pace.' 'Thanks for that bit of wisdom, Coach.' And that's exactly what we're trying to do here.

"Early in the mission we used our time at night and other open times to work to catch up. Later on I used these times to perform ad hoc experiments, such as the study of fluids in zero gravity, or when several open hours appeared, I'd go to my favorite spot, the ATM control panel, where there was no end to challenges and opportunities to learn and contribute. I remember these open times the most, times when I had a chance to use some creativity; the rush to continually catch up is remembered as just a blur.

"Of course our rushed pace caused mistakes, and I still chuckle about one of them: the televising of an experiment or other event. The switch to turn on the video tape recorder was not controlled by the camera but located in the

MDA, which was usually far from the subject that we were televising. More than once and always in a rush, I got the subject all nicely prepped, the mike and camera turned on, and started to record, or so I thought. Eventually when I'd realize that the video recorder wasn't on, I'd drop every thing and streak into the MDA muttering some rather creative profanities as I went. All too late I also realized the voice recording was on. 'Oops, sorry ground.'

"The situation was compounded a bit because people had not yet fully come to grips with the fact that Skylab was a different animal than all the relatively short missions to date. As in ascent, reentry, EVA, or hazardous aircraft operations, which preceded spaceflight, it is absolutely essential that nominal and malfunction procedures be spelled out in detail, simulated with fidelity, then followed precisely. It's a mindset that keeps people alive. However, once the hazardous operations give way to a normal day-to-day type of operations, like we usually experience here on Earth, it's time to back off the rigid specification of every action, set goals and objectives, and let the people on the spot use their intelligence to perform to the best of their abilities.

"Because of everybody's heritage and life-long conditioning, it was a tough mindset to break. As we began to get behind early in our flight, Mission Control, God love 'em, tried to help us as best they could in the only way they knew how: plan to the hilt and specify the procedures in detail. One morning we got a teleprinter message enumerating that day's activities that stretched from the Command Module down to the Trash Airlock, a distance of sixty-five feet!

"We wanted to be given some latitude in how we applied the brush strokes to the canvas; Mission Control, in their sincere efforts to help us, wanted us to continue painting by the numbers and in areas of ever decreasing size.

"I believe another contributing factor was that we lacked adequate integrated training with the Mission Control team. This team and ourselves never really understood what the other was thinking and planning before launch. Usually integrated training is done as much to train Mission Control as the crew, but they'd been through it all with the first two missions and weren't eager to revisit that 'demanding boredom' more than absolutely necessary. So when they came to us with a set of procedures, we simply said, 'You've been through it all before, and we haven't. So, we'll just do it.' But that didn't allow us to develop much interaction, communication, and real rapport with the Mission Control team before we reached Skylab. This lack of flight experience and the time crunch led us to just accept almost all suggestions presented to us without question or resistance even when it really would have been appropriate.

"Lastly, the situation was further compounded by lack of open communication after liftoff. You couldn't just call down and say, 'Hey, guys, let's talk this out,' because everything had to be open for the whole world to hear including the sensationalism-seeking press. So we thought, 'Okay, we'll just work through it.' But that stoic approach didn't work."

As with the previous two crews, one form of open conversation was relished by the crew of Skylab III. "Every third day we had a link to the real world when we each got to talk with our families for ten to fifteen minutes," Gibson said. "We really looked forward to those talks. Once I was describing the awesome beauty of fires that I could see all along the African coastline, a result of the farmers' policy of slash burning. I pictured my family hanging with breathless anticipation on every word. Then I heard Julie, my youngest daughter, say, 'Mommy, can I go out and play?'"

"We each really looked forward to talking with our families," said Pogue. "However, the news wasn't what we wanted to hear. Before our mission, I went by the office that handled government employees' life insurance at JSC and asked if I could pay three months ahead to cover the time I would be on Skylab. I was told that a prepayment wasn't possible but that the policy would be held effective until I returned. The bureaucracy didn't coordinate too well within itself (or maybe it knew something that we didn't) because my wife told me that we had just gotten a letter informing us that my policy had lapsed and was about to be canceled."

"One day in the midst of all our efforts to get back on schedule," said Gibson, "we were each working hard and lost in our individual worlds when we heard 'BANG! . . . BANG! BANG!' The attitude control system thrusters, for the first time on our flight, had fired to help the Control Moment Gyros counteract the gradient of gravity trying to torque Skylab off target. As we worked inside the huge OWS tank, it sounded like someone was outside working over the tank with an equally huge sledgehammer. Now I know what it would be like to live inside a drum."

While the crew bore the burden of getting back on schedule, they should not have borne the blame for being behind. According to lead flight director

Neil Hutchinson, "If you've read anything on the third manned flight, you know 'we,' the ground, and I who was right in the middle of it, were on the wrong side of the work scheduling issue. It was clearly a mistake on my and the control center's part. We expected those three guys to pick right up where the Skylab II guys left off. We did not give them one ounce of zero-G time to get used to it; that is, to do the task a few times and then schedule it tighter. When they got up there, Bill wasn't feeling very good, which is another thing we've now come to accept as well. Yes, it happens, so what. But it was still kind of spooky back then when these guys were getting sick, which was not the fighter pilot image. Oh, what were we going to do?

"Once the guys got up there, I went through the activation, they did a terrific job. Then on the third day we sent a flight plan up that was like the day after the last flight plan of Skylab II, which we didn't get done when the last flight crew returned. Of course we had practiced some with them on the ground before they launched. We had simulated between the umanned missions with each upcoming crew while we were unmanned for a while. Still it was a serious mistake on the part of the control center because we just expected Bill, Ed, and Jerry to just jump right on the bandwagon and take off.

"On their side of the equation, there was not enough communication early on to let us know that we were getting them in trouble. They were pretty quiet about it. Again it was the fighter-pilot mentality. 'I'll be damned if I'm going to cry "uncle." I'm going to just keep trying to get this done. If they keep sending me a flight plan I can't get done, I'm just going to try again.'

"Of course as we continued to press them, more mistakes begin to be made, more than we had seen with the other crews. And then you began to wonder, 'Hmmm, what's going on here?' I think it might have been even a year or two later that I sat back, looked at that whole thing, and said, 'You know, we really did something stupid. They didn't cry "uncle" soon enough even though they had an absolutely valid reason for doing so. The control center had fouled up, and we just kept fouling up until we got them all fouled up too.'

"In the end of course they turned out to be every bit as good as the other guys. They really turned out the stuff. You wouldn't have believed that they were up there for nearly three months.

"It's funny, one of those guys, Ed Gibson, has since become a very good friend of mine, and he and I have chatted about this off and on. He knows a lot about what went on there. It was clearly a case of the control center not recognizing that people needed some zero-G adjustment time before they could really be productive. There was just no point in pushing them early on, because they weren't going to get the job done. We don't do that these days on the Shuttle. We let them get really organized first."

Public relations were also impacted by air-ground communications: "In an effort to increase our efficiency," said Gibson, "we occasionally would have only one of us listening to the voice traffic from the ground and responding to it while the other two of us turned off our radios and worked without interruption. We each signed up for an orbit as the radio-response guy. Well one day we made a mistake and for a whole orbit we all had our radios off!"

"When we came up to AOS over one of the sites," said Carr, "the ground called us, and we didn't answer them for a whole orbit. Regrettably that caused a lot of concern down on the ground. And of course the press just thought that was wonderful. They said, 'Look at that. These testy, crabby old astronauts up there won't even answer the radio now. They've turned it off and won't listen to the ground anymore.' We've had to live under that stigma they falsely created ever since."

"Problems that surfaced early in our mission were created by competent, well-intentioned people," said Gibson. "The exceptions were the dramatic stories fabricated by the media and later repeated and exaggerated in a book on Skylab and a Harvard Business School study. There was no 'strike in space' by any stretch of the imagination. What could we threaten to do, go live on the moon? If any of these writers had gotten their information from just one of us, the crew or other people directly involved, responsible reporting and validity would have prevailed over expediency and sensationalism."

While finally taking a day off gave the crew a much-needed break and helped relieve some of the stress they were under, it didn't really change the situation. "Right after our real day off," Jerry Carr said, "we got right back onto the treadmill, and things weren't getting any better. Finally after several weeks into the mission, it all came to a head. After dinner we always had a medical conference with the flight surgeon where we would tell him how we were doing physically, and we give him the readings for the food that we'd eaten and the water we'd drunk and all other data that they needed for their metabolic analysis. I said, 'You know, I think we need to have a séance here.' I told him about our situation, that we weren't too terribly happy and that we were quite sure the ground wasn't happy either. 'It's time for us to have a discussion, a frank discussion. We can do it on this channel if they want.'

"That request went down to the doctors, they passed the word, and, when the press got a hold of it, they raised Cain. So Mission Control came back and said, 'We're going to have to do it on the open circuit.' I said, 'That's fine.'

"So one evening we started talking with ground as we came up over Goldstone [California]. We had the whole U.S. pass, essentially, for me to tell them all the things that were bothering us. 'We need more time to rest. We need a schedule that's not quite so packed. We don't want to exercise after a meal. We need to get the pace of things under control.' Then we said, 'Okay, now, next pass over the U.S., you guys please tell us what your problems are.'

"So during the next U.S. pass, they bent our ear with all of the things that we were doing, including our rigidity that made it difficult for them to have the flexibility to schedule us how they needed to. We came back with, 'Let's think about it overnight and try to come up with a solution by the morning.'

"The next morning they sent a teletype message in which they recommended quite a few things. The most important one was to take all of the menial, routine housekeeping chores out of the schedule and put them on what we called a shopping list. They were things that needed to be done that day but not at any particular time. Of course, they still had to hard schedule those activities that were required at a specific time or location in orbit. By opening up the schedule that way, they really took the pressure off. We were no longer racing the clock to get things done. It solved the problem.

"They also said, 'We're not going to hassle you anymore during meals or give you any major assignments after dinner. After dinner is relaxation time for you. Do a few things like some student experiments, but we're not going to have any major experiments after dinner.'

"We said 'That sounds great. Let's go with it!' And it worked beautifully. It's a testimony to the human condition. Henry Ford probably learned it on his assembly line. The line can only go so fast before you start making mistakes.

"We also felt that the extra time was needed to do some creative thinking. As a result of having all that extra time, we were all able to gin up some experiments that we had wanted to do and put on TV. Some of the results are being used today in schools such as short physics experiments and experiments with water in zero gravity. The loosening of the schedule really solved the problem. We got the more important experiments done immediately or at a required time, and everything else got done when we could. That flexibility gave us some control, put us in positive frame of mind, and increased our productivity. Everybody won!"

"After the crew came back and we had gotten through the debriefing process," Neil Hutchinson recalled, "it was pretty obvious that we had had some real scheduling and performance problems at the beginning of the flight. There have been a couple of books written that stated that there was a strike in space even though that was clearly not the case. There is even a Harvard Business School case about it. If you get an MBA at Stanford or somewhere, you're likely to get the Harvard Business School case about Skylab III. They talk about people's expectations and miscommunication as part of a management process. I don't know if it's a good example or not.

"I just look on it as a time when we just weren't thinking straight. We should have seen it even though it was very insidious because the mistakes were little at first. Just every once in a while you kind of caught in somebody's tone of voice that he was irritated. It was not a good scene, but yes, good lessons were learned."

Ed Gibson noted that, long before Skylab III, he had experienced slow starts: "As a little kid, I was slow, a lethargic dreamer. One of my earliest memories is that of lying on the living room floor, drawing pictures of the solar system, and dreaming. I sensed a fascinating and never-ending world in the night sky and inherently knew that, somehow, I had to become part of it.

"However, at an early age, I had contracted osteomyelitis, a bone infection in my leg; and amputation, the standard treatment, was contemplated. That would have really slowed me up. But first, my doctor thought a newly developed drug called 'penicillin' was worth a try. It worked.

"Then I encountered another roadblock: me. Dreamers make poor students, and the kindest thing I can say about my early academic career is that I was president of my first-grade class—two years in a row! Fortunately failure was not in my dad's makeup, and he was determined it wouldn't be in mine either. My performance in high school rocketed up to mediocre. At the University of Rochester, the only school that would accept me, I decided it was then or never, and I got to work.

"The world of high-performance aircraft and rockets, steps towards the stars, fascinated me, but because I once had osteomyelitis, I could never pass a military flight physical. I had to accept my destiny as a 'ground pounder' and developed the skills to design what I couldn't fly. It was a slower paced life than I wanted. That's when my wife read me the article in the *L.A. Times* that ultimately led to my presence on Skylab. Julie has always been my most ardent supporter and constructive critic. Anything I've been privileged to do would never have been possible without her support at every step along the way.

"I guess slow starts are in my blood."

Others on the ground reflected on the situation. Skylab II commander Alan Bean said that the failure of NASA to shift gears after his mission was a major factor: "I think Mission Control should have gone back to how they started with us. I believe that they started them out near where we ended, rather than maybe 10 to 15 percent less. Kraft called Pete and me over to talk with him and his managers. I told them, 'Mission Control plans to lighten up on these guys, but they don't ever do it. They have to lighten up and let these guys catch their breath.' Then finally Jerry Carr said 'We're not going to do this anymore, because we can't.' And he was right. They couldn't. We couldn't do it on Day 1—or 2 or 3 or 4 or 5—either."

Bob Crippen, crewmember of SMEAT and Capcom for all three Skylab missions said: "I can see how the situation developed over the course of the three missions. On Conrad's crew, most of that time was spent repairing Skylab so that it would function, and we didn't really work that hard on the experiments. Then Bean and his crew went up, started off at a slow rate, and then kind of built up speed and got more efficient, and we accelerated after them. At the end of that mission, we on the ground were used to operating at about that pace. And then here comes the new crew, Jerry Carr and his guys, and we started scheduling things at about the same rate that the last guys had ended up with.

"Part of what my job as Capcom was to try to sense what was going on, and truthfully, they were having some problems here and there, and we tried to scale back a little bit while we were doing scheduling at night. It was not until Jerry finally requested the conference to work things out that

Mission Control really understood what was happening. It took that to hit us on the head.

"But that's also the job of the crew because when you're sitting on the ground and trying to communicate only over the radio, it is hard to put yourself in their position in orbit. That's one of the responsibilities of the commander—to come back and say, 'Hey, this doesn't work and that doesn't work.' They have to let us know what's really going on.

"The ground controllers, my flight director [Don Puddy], and I were upset because we had not seen the problem coming on as big as it did and had not appreciated the extent that it was actually affecting the crew. They just kept trying to make things work without telling us about their difficulties.

"Even though we all initially got off on the wrong foot, Jerry, Bill, and Ed did super once we got things back on track. And no, there was no rebellion. I think the rest of the flight directors and the Capcoms would certainly say the same thing."

With the scheduling problems worked out, the contrast was sharp. No longer held back by these difficulties, the crew's performance accelerated rapidly. The slow start was behind them. "As it turned out, when the mission was over, we had completed every one of the experiments that we needed to do, plus a lot of extra ones that we dreamed up," Carr said. And although it was not obvious to everyone at the time, valuable progress had been made in moving America's space operations experience forward.

"As our mission progressed, Mission Control and we learned together how best to achieve the highest performance," said Gibson. "They were hard-won lessons, and because of past history and philosophy of operations, they were inevitable lessons that had to be learned either right there on our mission or ultimately on early space station missions."

Throughout the whole mission, ATM (solar observations) was an area that received considerable attention. "When we studied the sun," Ed Gibson said, "we used the ATM panel to monitor and control seven different instruments that 'looked' at the sun in visible light rays all the way down to X-rays. Even though I helped design the panel, it was a still a highly demanding and sometimes humbling task. Choices had to be continually made in space, time, and wavelength, sometimes within seconds, for experiment observations and then translated into panel switch actuations. The Joint Observing

Programs helped quite a bit, but the real value of having a human at the controls was when targets of opportunity arose and we'd have to put the sheet music away and play more by ear.

"I had a background in plasma physics from Caltech (the sun is one big ball of hot plasma), and I also had studied solar physics ever since I knew I had a chance to fly Skylab. I used the writing of a textbook as a way to focus my efforts and gain more credibility to help put my body into one of the three front-row seats on launch day. I still found the ATM a major challenge and empathized with other crewmen whose expertise lay elsewhere. However, after being an operator of Skylab and the ATM for eighty-four days, I feel strongly that mental challenges of this magnitude are essential to maintain sanity on future long-duration missions.

"The most demanding task was trying to capture the birth of a flare, which lasts only a minute or two. Understanding a flare's triggering mechanism is essential if we are ever going to be able to predict when and where flares will occur. The difficulty came in because almost all instruments used film to record their data, a limited supply of film that could be rapidly consumed during the high data acquisition rates required by flare observations. We had only so much film onboard and [a limited number of} EVAS to replace it. Thus, we were in a Catch-22. How do you know when to go into flare observations until a flare is well underway; that is, past its birth and well into its teenage years? It took a while to get the hang of it, and the extreme ultraviolet light monitor was indispensable.

"It was in the XUV monitor that one could see an active region start to simmer. It was almost like watching a pot of water getting ready to boil. When were the releases of points of XUV radiation (like the formation of little bubbles on the bottom of a pan) rapid and intense enough to predict the eruption of a flare (like large bubbles exploding upward to bring chaos to the water)?

"Late in the mission I intently stood guard over the ATM panel during my scheduled times of ATM operation or any of my free time. After many hours of concentration and a few cases of infant mortality, I did catch a flare very early in its life (maybe even still just a toddler). It was much earlier than we'd been able to get data up to that time!

"I'm confident that given high resolution displays of the high energy emissions from the sun (XUV and X-rays) and the time to really study them, the true birth of a flare could be observed. Of course, these days the problem can be brute forced by continuous acquisition of electronic data on active regions at ultra high rates."

A NASA press release at the time explained, "A solar flare recorded on January 21, 1974 by the Skylab SL-III mission has created considerable excitement within the worldwide solar physics community. The flare was not large by comparison with those recorded on previous Skylab flights. Ground observers classified it as a medium sized flare. The excitement stems from the news that for the first time in the history of the Skylab missions, a solar flare has been recorded from its beginning through its expiration."

"Also on our mission," Ed Gibson said, "the liftoff of a huge prominence on the limb was observed by the coronagraph instrument. The resulting data yielded one of the classic pictures resulting from all of the ATM missions. The solar observatory in Hawaii saw the prominence start to liftoff and notified ATM scientists in Houston. It was night for us so all three of us were fast asleep. Fortunately, the coronagraph was one instrument that could be operated remotely by the ground.

"Like on previous missions, we also observed the sun hurl out massive amounts of material called coronal mass ejections or CMEs. The light from the corona is usually very faint. In contrast, CMEs are seen as tight, ragged-edged knots of very intense light that explode outward at tremendous speeds. If conditions are right, some of the CME material can impact our upper atmosphere, our magnetosphere, and play havoc with our communications and electrical power grids on Earth. These events are commonly called solar storms.

"Although lunar geologists and space doctors would give me an argument, I maintain that the ATM was the best application of a human's scientific knowledge and judgment in space ever accomplished.

"On the previous mission, ATM operations also set a precedent when principal investigators were allowed to talk directly to the crew. The first time out of the box it was a pressure packed event but Bob MacQueen, an ATM experimenter, did an excellent job. On our mission we also had a few discussions with those who were ultimately responsible for the scientific return from several experiments, but we would have liked more.

"Several years after our flight, I talked to a cosmonaut who had flown much longer than we did on Skylab III. They had a somewhat looser operation. After gaining some familiarity with an experiment before flight, they

39. Ed Gibson at the ATM console.

would have a private one-on-one discussion with the investigator the night before it was to be performed. No end-to-end rigorous detailed procedures and timelines were usually created or desired. I believe that a middle ground, the Goldilocks solution, will achieve the highest scientific return."

With the crew performing at high efficiency and rapidly catching up with and surpassing the tasks that had been planned preflight, they found time to laugh at themselves. "About half way through the mission," Gibson recalled, "we all noticed water collecting on one on the panels in the Lower Equipment Bay of the Command Module. We thought it would be a bad situation if that water ever seeped into a compartment full of the electronics.

"Jerry took the initiative to get some cloth towels and soak up the water. He did a very neat job. Not wishing to waste the towels, he hung them out to dry in the OWS. We all slapped our foreheads when the water evaporated from the towels and went straight back to the coldest spot in the station—the Lower Equipment Bay—from where it had just come! We'd just found another way to keep a Skylab crewman busy."

"Ed and Bill dreamed up more experiments than you could shake a stick at," Carr said. "I think one of the funniest pieces of footage I've ever seen is from one of Bill's experiments. He wanted to demonstrate that, although air is a fluid, a medium just like water, it's a lot harder to kick, paddle, or swim to get somewhere. He made some big cardboard fins for his hands and feet and put on a crash helmet with big bubble eyes, which made him look like a huge bug. When he drifted out to the center of the workshop and started flapping his paddles, he actually started moving, although very slowly, demonstrating that air is a medium just like water in zero gravity, and you can move around in it—if you the have the right kind of tools and the patience. While Bill was really putting out the energy flapping around, Ed and I got a good laugh."

The crew was also hitting their stride in terms of the experiments they were performing. They all became much more proficient at Earth observations as the mission continued not only because of the time on orbit but also because of the extra training and emphasis they had put on it.

Ed Gibson recalled, "When we first got up there, we would say, 'I guess we're over Africa because it looks like the coastal outline of Africa. But after a while we could just look at a little patch of our planet and say, 'There's a red windswept desert. We must be over north Africa.' or, 'There's an ocean current, and we can tell by its color and the way it's meandering that it's the Falkland current right off the east coast of South America,' or, 'There's a little round patch of clear ocean surrounded by a ring of clouds. That's where cold water is welling up and quenching cloud formation. Fishing must be good down there.'

"Whenever I had the chance, I would study and photograph my hometown of Buffalo. In December and January, it displayed its standard winter color: white. No matter which way the wind blew in from Canada, it picked up moisture over Lake Erie or Lake Ontario, then, when it hit the cold land, it all dumped out as snow onto Buffalo. However, the sight of it still warmed my heart. The people are great and ultimately responsible for what I'd done in life including flying over them 270 miles up (GO BILLS!)."

"The diametrically opposite corner of the United States, southern California, received considerable attention from all three of us primarily because of the interest of Lee Silver at Caltech in the multitude of crisscrossing fault zones there," said Pogue. "From our data he discovered a fault that propagated through the airfield at Palmdale, which was the construction site of the Space Shuttles. Fortunately, it has remained inactive. The San Andreas fault and others of less prominence could clearly be seen by eye."

"On the other side of the Pacific plate," said Gibson, "the Alpine fault looked like someone had scribed a long, deep, straight line from head to toe on South Island, New Zealand. It was boldly visible, especially at low sun angles."

"I found weather observations equally interesting," said Gibson. "One night I watched an extensive series of thunderstorms over the Andes. It was clear that the flashes came in groups as if one flash set off others perhaps through electromagnetic triggers transmitted through the ionosphere. A charge would build up, then ripple fire across the total system, not just at one location. At the time it all seemed so obvious and natural. It was enjoyable to see and acquire an instinctive feel for some of the natural forces on the Earth. However, since then the results have been difficult to reproduce with instrument observations."

With the longer duration of the third Skylab mission came more opportunities for its crew to venture outside. When the Skylab design was in the requirements phase, the only way an EVA capability could be justified was by the film installation and retrieval to service the ATM experiments. It became a classic case—in space—of "If you build it, they will come." The first of the mission's four EVAS took place on Mission Day 7, better known on Earth as Thanksgiving Day. On this EVA, as well as all the remaining ones, considerably less than 50 percent of the crew's time was spent on the installation and retrieval of film. The rest of the time was spent on repairs and deployment of other experiments.

Pogue was feeling much better after his initial bout of space sickness by the time he performed the first EVA with Gibson. They installed ATM film, repaired the microwave antenna, placed an experiment on the ATM truss, and took pictures during the six-and-a-half hour EVA.

EVAS were savored by each one of the Skylab crewmen.

"EVAS were good hard work that always left a feeling of accomplishment," Ed Gibson said, "as well as some stimulating and lasting visual images. Our training at the neutral-buoyancy tank at Marshall was excellent. Working in the water was always a bit more difficult because of the water resistance and the fact that you could never get weighted out perfectly, which left forces and torques on your body that didn't exist in space. If you could do it in the tank, you could do it in space.

"Over the years I spent a lot of time at Marshall, not only in training but also in the development of procedures and hardware. In fact the first time

40. The third crew performed a total of four spacewalks during their eighty-four-day stay on Skylab.

Bill and I went out the airlock hatch, part of me expected to see the eyes of safety divers magnified behind their masks ready to assist and big bubbles streaming past my helmet then breaking up, flattening into mushrooms, and turning to white spray at the surface. Instead, it was all clear, no divers, no bubbles. Nothing was outside the hatch but our space station and the Earth 270 miles below.

"Three times I went out that hatch into the 'truly great outdoors.' When I was out there, it was a silent world, except for the whispers of my own breath. Sometimes I felt totally alone, like the world below didn't even know I was there. But then I thought of the many people on consoles in Mission Control who monitored everything on the station, including my every breath, word, and heartbeat, and I realized that I was being fully supported in the most extensive way possible."

Jerry Carr described the spacewalk: "On the first EVA, Bill and Ed went out and did a lot of repair work. We had a microwave antenna on the side of the spacecraft that faced the Earth that needed some diagnostics and repair. Unfortunately there were no handrails or foot restraints on that side; so when we trained for it in the neutral-buoyancy tank, we had to figure out how we were going to get it done.

"Basically, we found a place on a truss where we could fasten foot restraints. Bill got into these restraints and held on to Ed's feet while he reached up and made the fix to the microwave antenna. It was ad hoc, very difficult, but it worked."

Ed Gibson said, "Removing the cover from the microwave antenna electronics box turned out to be exceptionally difficult. On one side of the box, four screws had to be removed. On the ground it was easy. But in flight because the real box had a metal lip that closely overhung the screws, it was anything but easy. The small screwdriver that fit the small screws had to be inserted into the slots from the side of the screw heads rather from than the top, which was extremely difficult in our large bulky EVA gloves.

"Bill and I both gave it a shot. I remember thinking on my last try, 'Our success here is limited only by something physical. We're just not going back in until this little hummer is fixed!' After the better part of an hour, we got the top removed and the work done. It felt good to achieve something difficult, even though most of my fingernails had turned purple from the intense and prolonged squeezing of the screwdriver.

"To get to the antenna electronics box, many layers of aluminized Mylar insulation had to be cut away with a scissors. Most of these highly reflective pieces floated free and were blown away from Skylab by the gases venting from our suits. It happened at sunset so that the red light of the setting sun reflected off these tumbling reflectors in the distance. We commented on the cloud of red flashing lights that appeared to be following us. One of the tabloids picked up on what we saw and of course did not give the real explanation. Clearly, we were not chased by flashing red UFOs guided by extraterrestrial intelligence."

The new holiday EVA tradition continued on Christmas Day. It was the first time a NASA astronaut had been in space on that day since the Apollo 8 crew was on their way back from the moon five years earlier. (Just a week earlier, another first had been marked—the 18 December launch of Soyuz 13 meant that for the first time, U.S. astronauts and Soviet cosmonauts were in space at the same time.) Among the tasks Carr and Pogue performed during the second EVA was to take pictures of Comet Kohoutek.

"EVAs were spectacular," Carr said. "The second EVA was on Christmas Day, and Bill and I were out for seven hours. I was amazed when I got back in because I expected that I'd have to go to the bathroom something fierce, but I didn't. Apparently, I had gotten rid of a lot of fluids in the form of sweat through my pores. When I got back in, I was really sweaty, but I really didn't have to urinate. I was just amazed that, after seven hours, I wasn't pretty interested in streaking to the urinal."

"On Christmas of 1973," Bill Pogue said, "Jerry and I were suited up for an EVA to do routine servicing of the solar observatory (removing and replacing film magazines). I was to also set up a camera to take imagery of Comet Kohoutek, and Jerry was to repair one of the solar telescopes using procedures developed by ground personnel. Jerry and I got into the Skylab airlock surrounded by two seventy-five-pound film magazines, a special camera for taking pictures of the comet, and tools for Jerry's repair job.

"We closed the airlock hatches that sealed us off from the rest of Skylab and dumped the pressure from the airlock. When the pressure dropped our suits began to inflate and stiffen just as we expected. At this point I was curious to see how our inflated suits and all the hardware were going to fit in the confined space of the airlock. It turned out to be no problem.

"When the airlock pressure dropped to a vacuum, we opened the hatch and began the first task. Jerry went hand over hand out to the end of the solar observatory while I got the replacement film magazines ready. I operated an extendable boom to transfer the first film canister to Jerry; he removed it and loaded the exposed canister to the boom; I retracted the boom while Jerry loaded the fresh canister to replace the one he had just removed, and when he gave me the OK, I sent the second canister out, we repeated the procedure and were finished in record time.

"I went into the airlock, grabbed the comet camera, and left the airlock as Jerry was returning for his tools. Everything was going like clockwork. I mounted the comet camera on a round Skylab strut and positioned it so that one of the solar arrays just blocked the sun. I couldn't see the comet but ground had sent me a diagram by teleprinter. The instructions were clear, and it was a fairly easy job. I turned on the camera, and I was finished.

"Because this was my last EVA, I decided to make the most of it. I crawled all over the accessible parts of Skylab. It reminded me of when I was a kid doing a mud crawl in a four-foot-deep stock tank used for watering cows and horses. The animals didn't appreciate it, but very few people had swimming pools at that time, and the stock tank was one way to get cool during hot Oklahoma summers.

"My adventurous foray over Skylab ended with me at the sun end of the solar observatory. I was positioned where Jerry had been earlier and the view was breathtaking. When I leaned my head way back I could see the Earth below with no intervening structure in my line of sight. As others had described to me, I had the feeling I was doing a slow swan dive through space. My euphoria was suddenly dashed by comments from Ed who was holding down the fort inside. After I listened to Ed describe the problem that now occupied Mission Control, he asked where I was. I said, 'The sun end of the ATM.' I quickly deduced that I had stayed too long at that location and got moving.

"On Skylab we had three large gyroscopes to maintain attitude control. It was obviously important to keep the telescopes pointed toward the sun during solar observations, and the gyros did a great job. Unfortunately one gyro had failed while Ed and I were out on our first EVA (Thanksgiving of 1973). Theoretically the remaining two gyros were supposed to be adequate, but in fact we frequently had to perform a special procedure to keep them working properly.

"I was really embarrassed. I had unintentionally caused the current problem, and it didn't take a rocket scientist to figure it out. Our suits were fed oxygen from inside Skylab, and there was no recycling of the air. It automatically fed in near the back of my head, flowed down across my face, and then escaped out the front of the suit near my waist. The outward airflow had acted like a small thruster, like letting the air out of a balloon. Although the force from the escaping air was small, my position at the sun end of the ATM magnified the thrusting effect because I was about thirty feet from the centerline of Skylab. In other words this lever arm was giving the force of the escaping air a lot of leverage. The airflow from my suit was rotating a one-hundred-ton space station!

"Jerry called asking for help, and I was more than happy to accommodate him. He couldn't reach a critical location, so I got into his work position and held his legs under my arms to extend his reach. It took a while but Jerry finally finished, we tidied up everything, I retrieved the comet camera, and we ended the EVA. When we got back inside, Capcom informed us that we had set a new EVA record of just over seven hours. What a blast!"

"That evening after the EVA," said Carr, "we did a TV presentation for the people on the ground. The three of us observed what it was like to be up there, what we saw on the ground, and how we felt about it. We had built a Christmas tree out of a bunch of food can liners from our kitchen and fashioned them into what looked like a little aluminum cedar tree. Then we had taken several kinds of decals, orange and red and green decals, and stuck them on the tree for decoration. Lastly in honor of Comet Kohoutek, we made a silver foil star with a tail and put it on the top. That was our Christmas tree."

Four days later Carr and Gibson each made their second EVA, taking more comet pictures. They also obtained a sample of the Airlock Module's micrometeoroid cover, which was to be studied to learn more about the effects of space exposure.

"In terms of brilliance, Comet Kohoutek was a disappointment," Carr said. "We and everybody on the ground thought that it was going to be a beautiful, brilliant comet. It turned out to be beautiful all right, but it was so faint that we really had to work to find it. Once we did find it, we observed a gorgeous thing: small, faint, but gorgeous! Although we took as many pictures as we could, I don't think our film was sensitive enough to really record good data. I believe the only decent picture taken was with the coronagraph on the ATM. The people on the ground got better pictures of the comet than we did."

Though the pictures were disappointing, their observations weren't a complete waste. "In order to best describe what we saw, we made drawings," Carr said. "Ed was the point man on that. With Bill and I looking over his shoulder, he made the drawings and then did a TV report showing the drawings and describing the colors that we saw. There was a little beak on the front end of the comet, which he described as well as [giving] its significance. These pictures are now in the Smithsonian Air and Space Museum in Washington DC. The comet's low brightness was a disappointment, but it was exciting to look for and find it. We also set up experiments outside to try to capture it after we went back inside. We then brought them in on a subsequent EVA."

The crew's fourth and final EVA, on 3 February 1973, was once more performed by Carr and Gibson. It was the last to take place on Skylab and its purpose reflected that finality. They were to gather everything outside that was going to go back to Earth, including the last of the ATM film and also try out a much simpler equipment transfer device than the extendable boom used on all other Skylab EVAS.

"On our last EVA," Ed Gibson recalled, "Jerry and I tried out the clothesline that had been proposed as another way to send the large ATM film packs

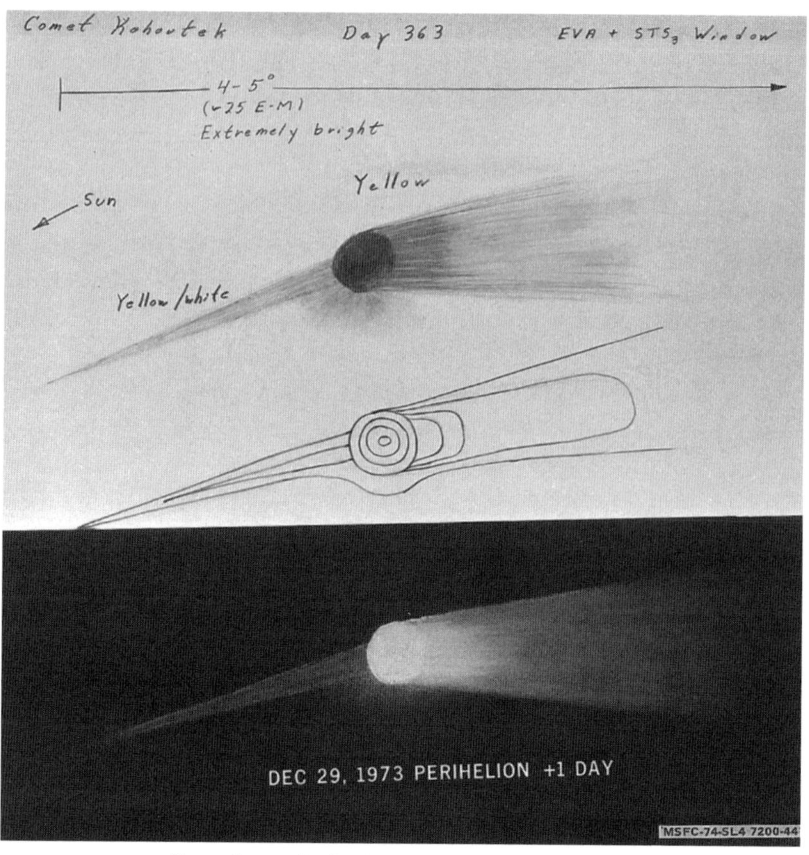

41. To supplement the disappointing ATM imagery of Kohoutek, the crew made sketches from their own observations.

back and forth between the airlock and the sun end of the ATM. An image sprang to mind of the clothesline [being] outside the station holding the wet towels that we had employed unsuccessfully to clean up the water in the [Lower Equipment Bay] of the Command Module, an image of 'wash day on Skylab' that we quickly censored and got back to work."

"The clothesline could not have been more simple: a closed loop of rope sliding over two polished cylinders attached to hooks on each end of the line and two hooks about two feet apart clipped to the rope. Aside from some oscillations of the objects being transported, which we easily controlled, it functioned well.

"However, after the clothesline had served its purpose, it was left up and led to some congestion in the workstation at the airlock, which caused me a problem. In subsequent work there, the rope got entangled with the umbilical connection to my suit and actually disconnected it. I could not feel it happening because the suit, once inflated, is a good insulator from all subtle contacts with the outside world. My secondary oxygen pack had picked up the task of keeping my suit inflated, but the fluid line leading to my liquid cooled garment was hosing out a water-glycol solution into the vacuum where it immediately turned to yellow ice. Once alerted by the ice fountain in front of me, I immediately remade the connection and everything returned to normal, except for the heart rate of the controller in Houston monitoring my suit."

Skylab III served as a bridge between two eras of EVAs. Not only did the crew perform the last EVA of the Apollo era and the last using a Gemini hatch, they also paved the way for future EVAs of the Shuttle era. Though it was not used outside, the third crew, like the second, tested an early version of what would become the Manned Maneuvering Unit (MMU). In 1984 the backpack would allow astronauts to perform EVAs floating untethered in space, essentially its users became self-contained human satellites orbiting the Earth.

"The MMU was a lot of fun," Bill Pogue recalled. "We flew it both shirt-sleeved and suited. Towards the end of our flight, we were running low on nitrogen gas that was used as the propellant, which meant that this had to be our last run. It got a little tense and exciting on Jerry's last suited run. I was observing and taking pictures. We didn't have any kind of remote radios or other types of communication with each other. But I could see the gauge on Jerry's oxygen bottle on his right leg, and it was running low. I kept pointing to it, and he kept gesturing that he wanted to finish.

"He kept going, finally got real close to finishing, but was getting red in the face. I slammed him down, pulled the release on his helmet, and popped it off. He was really sucking air but determined that he was going to finish, which I think he probably did. I was sweating bullets too because it looked like he was in CO_2 saturation. Actually, it was no big problem because at any time I could pop his helmet. I was just mother-henning him to death while he was sweating and puffing."

Medical experiments and observations took an increasing priority for the crew that would set a new world space endurance record.

"The bicycle ergometer was a great exercise tool as well as a good experiment," Ed Gibson said. "Especially early in the mission, it was a relief to have

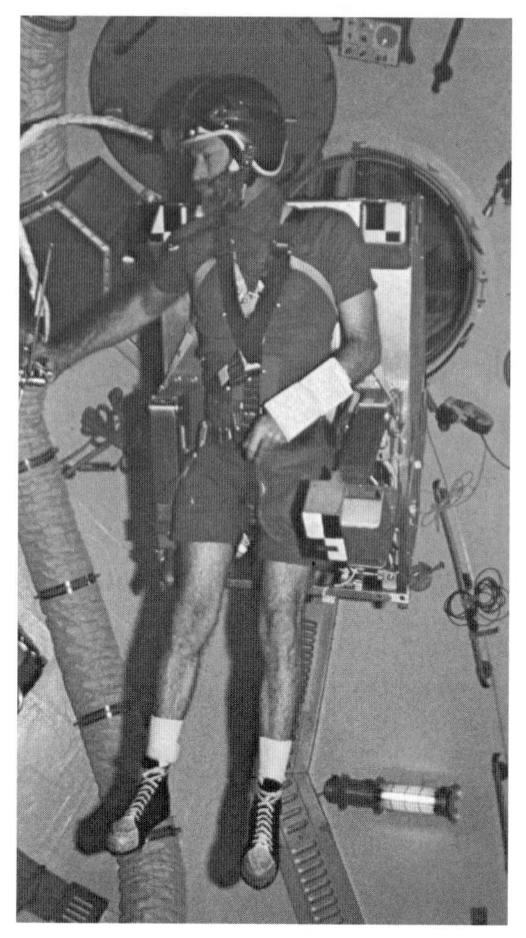

42. Carr pilots the maneuvering unit inside Skylab.

the blood pulled down into our legs to support our exercise, which relieved some of the fullness in our heads caused by the zero gravity and resulting upward fluid shift.

"Once I got pretty cranked up and developed a good sweat, a considerable amount of water clung to my back in a sheet and oscillated like Jello as I peddled. If a towel was not available, the shake-like-a-dog procedure usually worked. In zero gravity we couldn't use the seat on the bike, the straps to hold us in place caused too much chafing, and my arms got tired of holding me stable at high workloads for forty-five minutes. Instead I used my head. I taped a folded towel on the ceiling and put the top of my head against it to stabilize my body while I peddled. It worked!

"We also had something onboard that the previous crews did not, a device called 'Thornton's Revenge,' named in honor of Bill Thornton, the astronaut-physician who had a knack of doing highly beneficial things in clever and simple ways. Previous crews reported that they could have used some form of exercise that maintained the strength in their leg muscles that they used for walking and running upon return to Earth.

"Bill again came to the rescue with a poor man's treadmill. It consisted of a thin sheet of Teflon about a foot wide and three feet long and bungee cords that went over our shoulders to hold us down against the Teflon with a force equal to approximately our own weight. With only stocking feet against the Teflon, we could simulate walking or running by forcing our feet to slide over the Teflon one after the other, or we could just bounce up and down.

"Use of this exercise equipment was one of the few times I ever worried about what and when I ate before exercise. Eating some fire-hot chili before exercise is bad enough on Earth, but in zero gravity it's doubly bad, a real killer. That's because without gravity it bounces against the top of your stomach as well. Mixing chili and a treadmill aside, it was enjoyable exercise and definitely helped maintain leg strength. Thanks, Bill!

"Because of the extra requirements placed on the food system by the mineral balance experiment, this system was as much of a medical experiment as it was a crew habitability system. Despite having to do double duty, we found the food to be great. Many people picture tough astronauts in space surviving on food from squeeze tubes. That's the wrong image. Try the image of filet mignon, lobster Newburg, and strawberry sundaes."

"Our crew also broke new ground in the annals of spaceflight with the first full set of condiments in space," said Bill Pogue. "Rita Rapp developed them for the second crew after Pete had really railed about the 'yucky' bland taste of their food. Imagine, they had no condiments! The second crew took up only regular salt and pepper. But we had deluxe treatment: liquid salt, liquid pepper, hot sauce, horseradish, and garlic! Life couldn't get any better than that."

Between luxuriously seasoned meals, work continued. Ed Gibson recalled: "We also performed an experiment to nail down previous crews' observations. Light flashes had been observed by dark-adapted crewpersons when outside the van Allen radiation belts [lunar flights] and in Earth orbit when going through the South Atlantic Anomaly (SAA) where the inner radiation belt dips down lower than at all other locations around the globe. Even though a rough correspondence between the occurrence of the light flashes and presence in the SAA was observed on the two previous Skylab missions, no exact correlation had been made. Bill Pogue was selected and enthusiastically performed this arduous experiment.

"His task was to float in his sleep compartment wearing a blindfold and speak into a tape recorder every time he observed a flash. When the frequency of flashes was plotted against Skylab's position in orbit, a well-defined bell shaped curve resulted that was centered exactly over the SAA. Jerry and I praised Bill for his Herculean effort in the name of science.

"After a few weeks into the mission, something happened that made me think Skylab had a heart. I was looking out the wardroom window watching the spider web of lights blanketing the U.S. slide underneath while I held onto a handhold with the fingertips of one hand. Then I felt it—the station had a pulse, a heartbeat. I felt a beat just as real as a pulse in anyone's wrist!

"Of course I understood the absurdity of my observation, but it took me a few seconds to realize that I was really feeling the surges of blood through my own arteries and the accompanying deflection of my arm and fingers. Normally, on Earth these forces are unnoticeable because they are swamped by gravity forces. We really did live in a world of fingertip forces.

"By the time the third mission rolled around, Goddard Space Flight Center had gotten pretty accurate at pointing lasers at Skylab. Using lasers of *only a few watts,* they provided a point light source of various colors that we could track by eye from right over GSFC to almost one thousand miles out to sea. We thought it amazing at the time, and we still do."

Especially on the last of Skylab's three missions, cleanliness became a bigger challenge than ever. "As on Earth," Bill Pogue said, "a lot of trash accumulated during the day including food packaging, tissues, wet wipes, dirty towels, and washcloths. Most of this trash was immediately shoved through a push-through slot into a waste container. However, bits of skin, fingernails, hair, food crumbs, odd pieces of paper, and the like tended to drift around and eventually were sucked up against the air filter screens—our lost and found department. We used vacuum cleaners to clean off these screens, which took care of most of the problem.

"The worst mess was in the area where we ate. Small drops of liquid from our drinks and crumbs from our food would float around until they stuck on the wall or in the open grid ceiling above our food table. This grid and the area above it became quite dirty after three missions. Although we could see into this ceiling area, we couldn't get our hands in to wipe it clean, so it became progressively worse throughout the missions. Near the end of the flight, it began to look like the bottom of a birdcage. I just stopped looking at it.

"Every two weeks we wiped down the walls and surfaces of the toilet with a biocide [disinfectant] to prevent a buildup of microorganisms such as germs or mold. Periodic cleaning of this type will be required for the International Space Station to prevent a gradual buildup of biologically active contamination. It will be a time-consuming procedure but essential to preserve a healthy environment for the crew."

Like the other crews, Skylab III crew used the shower onboard. "Although we found that a washcloth, soap, and water followed up with a towel were perfectly good for maintaining satisfactory hygiene in zero gravity, we also tried out the shower that Bill Schneider, our Skylab program director, had worked so hard to get onboard," Gibson recalled. "He and others deserved that we each give it a fair try and evaluation.

"Granted it took a lot of time to set up and tear down, but I found it both interesting and refreshing. Because of its limited hot-water supply, it was like taking a shower with a Windex bottle. A smidgen of hot water was used to get wet and soaped up; the remaining smidgen was used to try to rinse off. The little hand vacuum, which was supposed to be used to remove the liquid, was awkward and difficult to use to reach all body parts. So I tried shaking like a dog, which sprayed most of the liquid to the inside surface of the shower enclosure, and then using the vacuum to clean it all up.

"I concluded that the whole procedure had to be made simpler and faster, analogous to passing through a car wash in two or three minutes if we are to have a shower on future stations. Nonetheless, we were appreciative that it was onboard and we had a chance to use it."

Several challenges to the crew grew progressively more severe as the last of the three missions progressed because of the gradual decline in the station's condition. Maintenance and repairs had been a part of the crew's duties even before the first crew ever docked, and there were always concerns about the potential effects of further failures.

"Below the hydrogen tank in the third stage of our Saturn V, our pressurized habitable volume," Bill Pogue said, "was the liquid oxygen tank or LOX tank, which was about the volume of a one-car garage [2,500 cubic feet] and served as the Skylab trash dump site or dumpster. Without it life onboard Skylab would have been altogether different, just as life in our homes on Earth would be different if we had to keep our trash inside, had no garage, and our trash pickup stopped. There was the constant threat that we would lose access to our dumpster, and our habitable volume would gradually fill up with our trash, which included biodegradable garbage and waste [food residue and urine bags].

"Our access to our dumpster was through an airlock, the Trash Airlock. We compacted our garbage as much as possible, placed it in a special bag, put it into the TAL, closed the lid, opened the TAL to the vacuum of the LOX tank, shoved the bag out and into the tank, and then repressurized the TAL to the pressure of our habitable volume for the next use.

"The lid on the TAL began to cause difficulties on the second mission. The hatch became more and more difficult to latch in the closed position. On our mission, the problem became more severe, and we were desperate to keep the TAL working.

"We finally worked out a system whereby Jerry would load the trash bag in the bin of the Trash Airlock, and I would float above holding onto the ceiling. As he pulled the lever to lock the hatch closed, I would push myself down sharply and stomp on the hatch lid while Jerry closed the locking lever.

"Voila!

"Was it a barnyard procedure? You bet, but it worked!"

Throughout their eighty-four days without gravity, the crew observed and thought about their reactions to this new mode of living. "Do we sense—or even need—up or down without gravity?" Ed Gibson recalled. "Early in our mission, our new world of zero gravity became familiar, then just plain comfortable. From many hours in a water tank, viewing films of previous crews, and actual zero gravity experienced for short times in aircraft, I came to picture a large switch on my forehead with two positions: one-G and zero-G. It got automatically thrown at booster engine cutoff from the first to the second position.

"Of course there was a lot to learn about the techniques of working without gravity, but zero gravity seemed familiar even on the first day. The harder we worked, the more efficient and confident we became. We soon realized at the gut level that space and its zero gravity is not foreign, not hostile. Rather it became just as friendly as gravity on Earth once we adapted. I do not know why I adapted so quickly and relatively painlessly. I was just lucky. I have always been able to visualize and think in three dimensions. Thus, as soon as we entered Skylab, I felt that my life had taken on another dimension, literally. No longer was there an up or down, except for visual references on panels or faces, but all dimensions became equal. Every motion was across, regardless of its direction inside or outside of Skylab.

"Yet my physiological responses did not forget gravity entirely. Some engineers came up with an 'experiment' for us to try out at our leisure. It was the 'Dynamical Acceleration Reference Trajectory Studies (DARTS).' When we tried out these Velcro-tipped darts, we were in for a surprise. Without concentration, a thrown dart would fly twenty to forty degrees up relative to the thrower, and far off the intended target. We lob things down here when we throw them to counteract gravity. Up there lobbing is not useful. Only by 'pushing' the dart out and away from my body was I able to achieve some accuracy. I tried to imagine what it would be like if I grew up in zero gravity, then came down to Earth and tried to gently throw something. In this case, gravity would be viewed as the exception, not the rule, and a real inconvenience.

"A related series of observations were made by the Skylab II guys who experimented with fish. Normally they swam with their bellies towards a surface or their backs toward a light. But when they were excited, they swam in what aviators call outside loops. However, when their offspring were born, they considered three dimensions natural and didn't favor one over another; that is, no up or down was recognized or needed. I felt a bit like them. But will it really be so easy to shed millions of years of human evolution by stepping down one generation that has never experienced gravity?

"Pete's and Al's guys made the most of the third dimension when it became available to them especially when they set up their own version of the Indy 500, streaking around the dome lockers. But by the time we got up to Skylab, the Control Moment Gyros were showing signs of real wear. Eventually one died and a second one was pulling back the covers on its deathbed.

Thus, running around the dome lockers was verboten because of the stress it put on the CMGs.

"We had to find other ways to enjoy zero gravity. I found that if I lay on my back on the grid floor of the OWS and used my wrists only to put some rotation and just a little translation into my body, I could go into a tuck position and spin exceptionally long times before I clanged into a wall. After reviewing the video, I asked Jannet, my oldest daughter, who was a diver at the time, if she could match one of my feats—it turned out to be a ten and one-half gainer in tuck position followed by a two and one-half forward in pike position.

"This tumbling exercise and the many others carried out by all Skylab crewmen illustrate the insensitivity to gross stimulation of the body's vestibular apparatus (semicircular canals) that developed in zero gravity. In my tumbles, I would develop severe nystagmus, or twitching of the eyes, as my eyes tried to catch up with the fluid racing through my semicircular canals, but none of it ever coupled into the gut to create nausea. I just passively spun and then watched the world flicker by for ten to twenty seconds."

The sensation of height proved to be inconsistent and elusive to Gibson: "There were a few exceptions in my ability to think of everything as just 'across.' One day after looking out a window in the MDA for almost fifteen minutes to watch the new and interesting features that never stopped coming over the horizon, I glanced back inside. The local vertical on Earth had become aligned with the long open direction from the MDA to the bottom of the OWS. An instantaneous reaction surfaced: I'm going to fall! After I clutched a handhold, I laughed at myself and realized I hadn't forgotten gravity completely.

"For several years after our return, every time I looked out a round porthole in the galley of a commercial airliner, part of me felt I was floating back in Skylab looking out a round porthole in the MDA except that from the airliner the horizon was flat and my vision covered half a city, not half a continent.

"Another but stronger feeling of height crept up on me during our spacewalks. I have found it difficult to step out the door of an airplane when skydiving. It was considerably easier to step out of the airlock even though we were 270 miles up. As long as I was close to structure, I still felt a part of it. But it felt different when I moved out and away.

"I have equated it in my mind to going to the top of a tall building and looking out. It's pleasant, relaxing. But now, what happens if I open the window and walk out to the end of a long springboard where a steel-fisted Hulk Hogan holds me by my ankles—head down. 'Intellectually,' I know I'll never fall. And even though I'm at the same height as I was inside, I'd have to admit . . . it feels a bit different.

"On an EVA I had that same feeling, just more of it. Head down, I'd glide over Earth at a very serene five miles a second. And the laws of Sir Isaac Newton gave me full *intellectual* confidence that I was up there to stay. But when I moved away from the main body of Skylab, like hanging off the sun end of the ATM, and looked straight down at Earth 270 miles below, I felt or saw nothing else around me. That's when that same little guy from lift-off whispered again from nowhere, 'Suppose that Newton guy was just a little bit wrong?'"

Though the possibility of an extended mission was already being explored well before their launch, officially, the target duration had remained at fifty-six days, as it had been for the second crew. By the time that duration was reached, the crew had a "Go" to stay. However, extensions were approved for a week at a time as the ground carefully monitored the status of the spacecraft, the crew, and the supplies.

NASA press releases issued at the time give the official view:

***Release No.:** 74-20*

Crew Cleared for Another Week in Space

The three Skylab astronauts, now in their 56th day in orbit, today were given a go-ahead for seven additional days. For the remainder of the mission, weekly evaluations of the hardware, consumables, and crew will be made by NASA officials. The first such weekly review was completed this afternoon. William C. Schneider, Skylab Program Director, said, the crew members "are in good spirits and excellent physical condition and the spacecraft is in good shape to continue." Originally, the three Skylab manned missions were planned, successively, for one of 28 days and two of 56 days. The first mission lasted 28 days, the second was extended to 59 days, and the third was then planned as an open-ended 60-day mission with consumables aboard to provide for as many as 85 days.

Release No.: 74-31

Crew Given Go for Another Week in Space

Astronauts Carr, Gibson, and Pogue now in their 63rd day in space were given the go for another seven days. For the remainder of the mission, weekly evaluations of crew, consumables, and hardware will be made by NASA officials. The second weekly review was completed this afternoon. Following the review of inflight medical data and the recommendation of Dr. Charles A. Berry, NASA Director for Life Sciences, William C. Schneider, Skylab Program Director, gave approval for the mission to continue until at least January 24.

Concerned about what effect the extended duration would have on the health of the astronauts when they returned to Earth, NASA doctors were getting more ambitious with the postflight medical protocols for the crew. NASA medical management had been persuaded, over the strenuous objections of some astronauts, that direct measurement of cardiac output by means of a catheter inserted into an artery in the arm and extended into the heart would give valuable data about the possible effect of weightlessness on cardiac function. A dry run was scheduled with the deputy crew flight surgeon as the test subject. A Johnson Space Center press release reported:

A volunteer subject fainted and suffered a brief loss of heart beat but was immediately revived during a cardiac output evaluation test conducted under controlled conditions for the Skylab medical program. The subject, Lt. Col. Edouard Burchard, required no hospitalization and was back on duty a short time later. The incident occurred after a needle had been placed in Lt. Col. Burchard's artery during the test. He responded immediately to the normal therapy that includes an injection of atropine and external heart massage.

The test, conducted at the Space Center Hospital near the Johnson Space Center, was a simulation of one of the post flight medical analysis checks considered for the Skylab III astronauts after their return to Houston. The purpose of the test is to get a precise measurement of cardiac output by introducing a dye into the blood system. Such dye dilution tests are routinely used in cardiac research diagnosis and medical officials said Lt. Col. Burchard's reaction was very unusual. As a result of the incident, however, Skylab program officials have decided that the test will not be performed on any of the returning [astronauts]. Lt. Col. Burchard is a West German Air Force medical officer detailed to NASA. He serves as deputy flight surgeon for the Skylab III crew.

So that was another worry off the minds of the crew. "Upon our return, we presented Dr. Burchard with a bottle of scotch with a note thanking him for 'his willingness to protect us with his life,' said Carr.

Another area of concern was the dwindling supplies available on Skylab. From the outset the food supply had presented a challenge—namely, there wasn't going to be enough of it to extend the mission as much as mission planners hoped. Further, the "gold-rush" attitude of scientists who wanted to get their experiments on the manifest after the success of Skylab II further limited the amount of space available for other cargo on the Command Module, which carried the Skylab III crew to the station. With a lack of food on the station and a lack of space to carry more food, there was a need for an innovative solution.

And an innovative solution was found: food bars. When the crew launched, they carried with them a supply of nutritional bars developed jointly by NASA, the U.S. Air Force, and the Pillsbury Company. "The difficulty with staying up that long was that we had only had enough food for fifty-six days and too many experiments to take up in the Command Module, which was already overloaded," Gibson said. "So we volunteered, actually agreed, that every third day we would eat nothing but food bars. That was probably one of the most supreme sacrifices anyone has ever made for the space program by a crewperson—food bars! Every third day we each consumed four of these little guys. Breakfast, which lasted about thirty seconds, consisted of four or five crunches, and that was it. There was no more. Meal's over. I still have a tough time looking a food bar in the face. But the bars worked, and we stayed. They had all the minerals and calories that we needed. It's not an ideal way to live, but they did work."

However, the lack of one item really bothered Gibson. "You just can't overestimate the value of a good butter cookie," he said. "We had an economic system on Skylab whose basic monetary unit was the butter cookie. But when we got up there, most of our money had been consumed previously by both the hungry Marine [Jack Lousma] and the Skylab I commander, Pete Conrad. It caused runaway deflation in the Skylab III economy."

As if eighty days of food bars every third day were not enough, mission protocols required that the crew follow their Skylab diet regimen for twenty-one days before the mission. Even after returning to Earth at the end of the mission there was no reprieve; postflight required another eighteen days on the Skylab III food-bar diet plan.

Other supplies also became of concern. "In mid-January 1973, when we were enjoying one of our 'days off,'" Bill Pogue said, "I was looking down at the Earth, Ed was at the ATM, and Jerry was doing an inventory of our remaining supplies. He floated down to his sleep compartment and left a message on the B Channel tape recorder for the ground control folks. Jerry was telling them that he had discovered a shortage of approximately ten urine sample containers, which we each used every morning to replace our individual containers that we filled the day before. A part of this task was to draw off and put 122 milliliters (about the size of a large ice cube) into the sample receptacle, place the urine sample into a freezer, and put the used urine storage bag into a trash bag for later dumping into the Skylab dumpster.

"The next day Capcom called with a solution. We were to change out the urine bags every thirty-six hours instead of every twenty-four hours; using this procedure would insure the remaining sample containers would last to the end of the mission. We followed this makeshift procedure, and everything worked out fine. Still we couldn't understand how the shortage had occurred because the people who prepared the mission equipment were highly competent. The waste sampling was to support a mineral-balance study conducted by the National Institutes of Health, and the principal investigator, Don Whedon, was most meticulous and careful. Once we got back to Earth we forgot the whole thing.

"Two months after our return the Astronaut Office had a 'Pin Party' for those Skylab astronauts who had made their first flight into space. The party is essentially a shindig where the backup crews roast the prime crews for many of their goofs and screw-ups during training and flight. The prime crews swallow hard, thank the backup crews for all their hard work, respond with good-natured humility and perhaps a few light-hearted jests of their own, and then make special individual presentations to their backup crews.

"Jerry, Ed, and I just about fell out of our chairs when Al Bean presented the missing urine sample containers mounted, on a plaque with a personal dedication plate, to each backup crewman. We looked at each other and burst out laughing. The 'Mystery of the Purloined Pee Bags' had been solved. They had been taken mistakenly by Al when the second crew returned to Earth. Of course we all quizzed Al and his crew about why they had developed a personal attachment to our pee bags."

Of course, the crew and spacecraft were not the only ones affected by the duration of the Skylab program. Neil Hutchinson recalled that Mission

Control was also feeling the effects of the passing months. "It wasn't like a prize fight where you train, fight, and it's over," he said. "In Apollo and now the early Shuttle flights, you train and train and train, then the mission goes, you work your tail off for a number of days, and the mission's over. Skylab was never over.

"Chuck Lewis, another Skylab flight director, got very ill. I flew the last flight with a kidney problem that ended up in a very serious surgery. It's not serious anymore, but in those days it was. It really, really took a lot out of people because you never got loose from it.

"We did all kinds of crazy stuff. We had our families in the control center for affairs to try and change the pace of things. I held a big dinner. Maybe all the flight directors did. It was a big sit-down catered dinner in the control center while the spacecraft was still up but during one of those times between manned missions. We were just trying to keep people's focus and attention. Still we had guys drop out of teams, and we had to change players. It wasn't that the control center was wilting on [Skylab III], it really was the sum of the three missions; we were all on duty for nine months.

"But by this latter part of the mission, both the crew and Mission Control were feeling really good because it all was going dramatically better. It became obvious we would get everything done and then some, and everyone could see the light at the end of the tunnel. As the last weeks of Skylab III went by, we all felt better and better."

Finally, the end of the mission neared. "I recall the last six weeks of the flight were very pleasant for me for two reasons," Pogue said. "One, we'd achieved the skill level sufficient to do the job quickly and accurately, and second, I no longer suffered from the head congestion that had plagued me for about the first six weeks of the flight. Midway through the mission it didn't seem to bother me much but became more like a low-grade headache that doesn't really hurt very much even though it still slightly decreases your efficiency.

"We all had a much better feeling about the whole flight toward the end. In fact they asked us if we would stay up for another ten days. James Fletcher, the administrator, had suggested it. Mentally we were prepared to come back, but more important, we didn't have any food left even though we probably could have scraped together enough for a few more days. But we came back on schedule after eighty-four days."

"Medically," said Gibson, "there were at least two reasons for our feeling so good: after our bone marrow greatly slowed its production of red blood

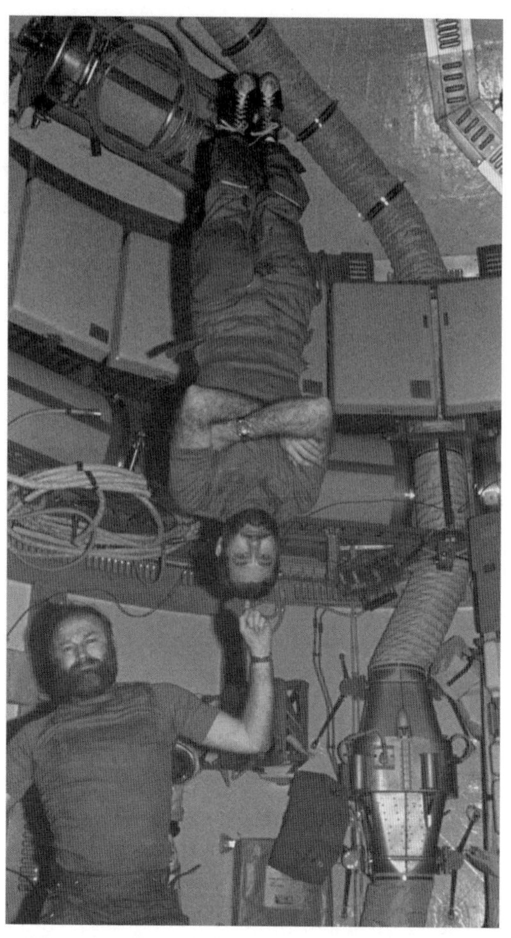

43. Carr and Pogue have fun with the possibilities presented by weightlessness.

cells, because our hemoglobin concentration had gone up in the first few days of the mission when we lost about three pounds of plasma from our circulating blood volume, it took a while for the hemoglobin concentration to drop low enough to trigger red blood cell production again. That production brought our circulating red cell mass back to normal, if not higher, toward the end of our flight. Also the tone of our cardiovascular systems had improved as measured by our response to the LBNP, which saw us reach presyncope [nearly pass out] about midway through our mission, before we significantly improved.

"From a personal standpoint, I would have liked to stay longer. I had come to think of our space station as an average, three-bedroom home, just 270 miles high and whistling over the ground at five miles a second. It felt so solid, so secure, that it didn't really feel like flying at all until we left it in our reentry vehicle. Then it felt just like leaving my home down here, sliding into a sports car, and accelerating back onto the road again. It was a comfortable home for sure, and I would've been content to live there for *many years*, if I had friends and family along . . . and maybe a good pizza delivery."

There were things about the Earth that the crew missed, though. In his book of humorous space anecdotes, *The Light Stuff,* Bob Ward, a newspaper editor at *The Huntsville Times* reported:

As the final Skylab flight approached the end of its nearly three months in orbit, Houston used the onboard teleprinter to send up changes in the plans for closing down the workshop and preparing for the trip back home.

Astronaut Carr noted that these teleprinted instructions stretched almost from one end of the space station to the other—about fifty feet. That evening, when the new team of flight controllers came on duty, Carr couldn't resist remarking that he fully expected Houston next to transmit the Old Testament.

Later Carr notified Capcom Bruce McCandless that the Skylab 3 crew wanted a book sent up via teleprinter that evening.

"War and Peace?" asked McCandless.

"No," replied Carr, "Little Women."

Then there was a brief pause and the astronaut added: "Bill (Pogue) says send him up a big one."

That the spacemen manning the last, and longest, Skylab mission may have had the opposite sex on their minds had been suggested a few days before the Little Women episode.

From orbit the astronauts had held a live press conference. NASA intended to include a set of questions from a sixth-grade class in a small town in New York, but time ran out. During the next orbit, flight controller Dick Truly went ahead and asked the children's questions, anyway. He saved for last the question of one sixth-grader who wanted to know, Did the astronauts miss female companionship after so long a time?

Ed Gibson, taken aback by the frankness of the query, responded: "What grade did you say that was, Dick?"

Then the astronaut answered the precocious child's question with a frankness of his own:

"Obviously, yes.'"

Though the lumps given the Skylab III crew by the media early in the mission would affect their reputation for years to come, coverage of their successes by the same media was not as forthcoming. As the mission neared its close, it was one of the first missions ever since the early Gemini missions that the flight itself was not extensively covered by the media. In fact they didn't even cover the return after the crew had set a time-in-space world record.

"At the time it happened, I didn't realize it," Gibson said, "because I was not looking at it from the outside. People were making a big deal out of it being an exception, but once I thought about it after a couple of months, I realized that in a way it was good. We're trying to make space to be more commonplace and space operations to be more accepted because they were being done repetitively and routinely. People can't be sitting on the edge of their chairs all the time, especially during long space station operations. So it's only natural that people's attention would drop off. I thought, 'Well, maybe we've reached a point in the space program where it's become more mature and lack of day-to-day interest it's only natural. So, let's accept it and move on.'"

And so, for Skylab's final crew, preparations for the trip home began. Before they left, this final crew of Skylab made sure to leave the welcome mat out for any future visitors. Although there were no plans for another Skylab mission, there were hopes that a crew of the Space Shuttle, which it was then believed would be in operation well before Skylab deorbited, might come up, check on America's first space station, and even boost it to a higher orbit to extend its lifetime.

A time capsule was even prepared for a visiting Shuttle crew to return to Earth. In it were a variety of materials, which would allow scientists to study the effects of long-term exposure to the spacecraft's environment. Although the time capsule was left inside Skylab, the venting of the atmosphere after the crew left meant that the materials were exposed to vacuum.

A few hiccups arose in getting ready to button up Skylab, close the hatch, and deorbit in the Command Module. "The frozen urine samples had to be put into an insulated container for their trip home," Ed Gibson said. "Each of these frozen samples, about the size of a very large ice cube and often called 'urinesicles' by the crew, had expanded just slightly beyond the size allotted for them in the return containers. Thus I had a problem. Reentry was a few short hours away, and the whole sample return for a major experiment was in jeopardy.

"As beads of sweat seeped out, clung to me, and soaked my suit, out to the rescue came the old trusty Swiss Army knife with its coarse file! The sharp plastic edges on the entry lips of the containers were all then filed down to a bull nose so that the urinesicles could be forced into each container with only minimal damage. To say the least, I was elated that the knife was on board. Because of the concern for inhaling particles, we were not allowed to have files in the tool kit, but the one in the Swiss Army knife had slipped by detection.

"In the midst of these busy preparations to leave Skylab, the back of our minds began reflecting on its future. I thought that Skylab was a great office, lab, and home that had set the bar high for all space stations to come. And I also thought that in another three to six years, our current home would be replaced by either Skylab B, which is now sliced up and residing in the Smithsonian Air and Space Museum in Washington DC, or another space station, which would be much easier and cheaper to build since we had the Skylab experience. 'We only needed a few large tanks, a couple of docking ports, a door for spacewalks, some first-class experiments, three or four CMGs to stabilize it all, and a few large solar panels hung on the outside for electricity. Nope, it's not hard. All of it can be off the shelf. Let's go do it.' But that was not to be when it was decided to throw away the booster capabilities we dearly paid for in Apollo and hang our future on only one access to space, a shuttle.

"It all had seemed like it would be so simple, yet it's come out so hard. The history of pioneering tells us that we shouldn't expect progress to take place in a straight line. Thus, I have confidence that in the future we will have fully completed space stations in Earth orbit, each manned with six to eight highly competent personnel and that their scientific and technologic productivity will be judged far worth the effort by all but the most ardent critics."

"Just as we were leaving Skylab, I almost had one last task to complete," said Pogue. "We had lost a coolant loop between the second and the third missions, so one of the first things I had to do when we arrived was to replenish and recharge the glycol solution in the failed coolant loop. It was that loop that we used for our water-cooled long johns [liquid cooled garment]) that we wore under our space suits on EVA. So I was really interested that it worked. We got it fixed real quickly. But just as I closed the hatch as we were leaving, the other loop failed. They asked if I wanted to go back in to fix it. I asked, 'Why?'

"After we got in the Command Module, we went through a long series of involved procedures. We were almost euphoric all during this period. Of course, we did a fly-around, and I took about seventy-five pictures of Skylab as we went around for the last time."

"When we undocked and made one trip around Skylab to photograph its condition," said Gibson, "it was obvious that the sun's ultraviolet light had greatly discolored all surfaces. What was white preflight was now tan. Even the white sunshade sail erected by the second crew had turned a golden tan with one notable exception. As we maneuvered over the surface that faced toward the sun, both sunshades rippled and waved in the gas stream from our reaction control thrusters. The sail erected by the second crew still displayed the creases from when it had been tightly folded in its stowage container before Jack and Owen pulled it out and hoisted it up the twin pole supports. Jack had done a great job of unsticking and unfolding the sail, an unanticipated chore, except for one fold that now opened up under the wind gust of our thrusters. Like light from a cracked door, the material inside the fold beamed back a stark white in contrast to its surroundings, a feature readily apparent in pictures today."

"Pretty soon after we separated," recalled Bill Pogue, "we could see Skylab going away. After we did the first deorbit burn, which brought us down to about 125 miles, I remember thinking that, after looking at the Earth from 270 miles for several months, it was almost like hedgehopping at 125 miles where you perceive the ground going by a lot faster.

"Almost everything worked out quite well except that we did have a problem with the reaction control system in the Command Module. One of our two rings [sets of attitude control jets] system had already lost pressure and had to be deactivated. The official record says that they told us to put on oxygen masks at this point, but we never heard the transmission so we never had them on.

"The problem came after we had separated from the Service Module. I looked over at Jerry as he was moving the hand controller to get the right entry attitude, which we absolutely had to be at for reentry to avoid landing in the wrong location or being cremated before our time, and nothing was happening! I yelled, 'Go direct.' Direct is a mode that is entered by going to the hard stops on the hand controller, which bypasses all the black boxes and puts the juice directly to the solenoids controlling the propellants in the reaction control jets. It worked. We got close to the right entry attitude and threw it into autopilot, which steered us during reentry. No problem.

"When we got down on the deck, we were hoisted aboard the aircraft carrier, and everybody was in pretty good shape. We later found out that Jerry had inadvertently pulled all of the circuit breakers to the Command Module reaction control thrusters instead of those for the Service Module, which were to be unpowered to prevent arcing when the guillotine cut all those wires between the modules before they were separated. The Command Module breakers were right above those for the Service Module. Since Jerry was floating a little higher in zero gravity than in the simulations on Earth three months before and it was dark, it was an easy mistake to make. Human factors should dictate that you don't put these sets of breakers adjacent to one another if you require that kind of a time-critical safety-of-flight procedure. He just pulled the wrong ones, which was a real easy error to make. But it turned out fine. That was our biggest excitement during reentry: Jerry moved that hand controller and nothing happened."

"At first," Ed Gibson said, "reentry was like living inside a purple neon tube whose brightness gradually increased when we began colliding with air molecules in the upper atmosphere at mach 25. About the time we got the .05-G light [reached a deceleration of one-twentieth of gravity], I felt myself start to tumble but in no specific direction. 'Strange,' I thought, but then my vestibular system hadn't felt any linear acceleration for eighty-four days, and my brain was trying to figure out how to interpret these faint murmurs coming from my inner ears. As the Gs increased, this feeling of tumbling was replaced by the strong sensation of deceleration that eventually hit over four Gs. The violet glow had progressed to a white-hot flame, the Gs and turbulence continued to build, and it was now more like living inside a vibrating blast furnace. The flames from the heat shield streamed by my window and out behind us. Sitting in the center seat, I could watch the roll thrusters fire as the computer rolled the spacecraft to bring us down precisely on target, exactly three miles from the USS *New Orleans*, the aircraft carrier that waited to pick us up.

"Eventually, the light and turbulence subsided, a firm explosion above our heads told us the nose-cone ring had departed, and small drogue chutes streamed out to stabilize us. At ten thousand feet the drogues also departed, and the mains appeared. At first they were held partially closed or reefed, and then they billowed out to three good fully deployed chutes, which we were all happy to see. But I felt confused. Once on the mains we were obviously pulling only one G. But then why did it feel like we were still pulling three Gs?

"We splashed down onto a calm sea with no wind. However, we still ended up in what NASA called Stable 2. Translated that means that we were hanging upside-down in the straps, bobbing up and down on the water in a closed damp cabin with the heat of reentry soaking back in—for me the most uncomfortable part of the whole flight or recovery!

"Before we got the balloons inflated that would right us, my mind flashed back to our training when we practiced what we would do if we remained in Stable 2 and had to exit the spacecraft by ourselves. We did the training in a Command Module mockup, very much like the real one, in a water tank in Houston. A lightning and thunderstorm was in full bloom as we began the exercise.

"As we got out of our straps, prepared to dive down into the tunnel to open the hatch, continue further down and out the tunnel, and swim to the surface, we noticed that the mockup was actually sinking! A relief valve had not closed properly and water was pouring into our habitable volume. No longer was this a casual training exercise; this was for real. With very little breathing air left, the last of us made it out and to the surface, using a procedure we had never practiced before. The technicians outside had a crane that they could have used to pull the spacecraft out of the water—except that its use was not allowed when there was lightning in the area.

"I looked around our Command Module for signs of water. There were none. The bags inflated, we popped over to Stable 1, and we gained access to the warm ocean air outside.

"We felt elated. We knew we had gone through an ordeal on this mission yet made many major accomplishments that contributed to the space effort. It was a mission of which we would always be proud perhaps even more so because we had worked through some early and very difficult situations before we turned it around and reached full stride.

"Nothing was left but medical tests, speeches, and a return to our families. Smiles were frozen onto our faces.

"Outside helicopters were hovering as frogmen jumped into the water and connected floatation devices and attachments to haul our Command Module aboard the deck of the USS *New Orleans* (LPH-11), an Iwo Jima–class amphibious assault ship (helicopter). It was clear these folks really knew what they were doing, since they had previously completed several Apollo recoveries."

The USS *New Orleans* had been commissioned on 16 November 1968, exactly five years before the launch of the Skylab III crew and wouldn't be decommissioned until 1 October 1997. In addition to supporting real space missions, it also supported the filming of *Apollo* 13, the movie. It could accommodate twenty-five helicopters on its 592-foot-long deck and reach the scene of the recovery at a speed of twenty-five knots. It was the third of four ships in the U.S. Navy to proudly bear the name of USS *New Orleans;* the fourth ship, a San Antonio–class amphibious transport dock, was launched in 2004 and is still in service today.

Gibson continued, "Once back on the carrier deck, a part of me was depressed. No matter how hard I pushed off, I could no longer float. And no matter where I went, I was painfully aware that once again I had to haul along massive amounts of meat and bone. Later rolling over at night became a real engineering challenge. But the exercises that we did during our easy, lazy days of zero gravity, paid off. Unlike some other crews and after months in space, we could walk as soon as we landed and suffered no lasting effects.

"Yet even with the G-suit squeezing my legs and the switch on forehead thrown to one-G, I felt just a bit wobbly. We were all glad we had those G-suits because climbing out of the spacecraft, crumpling into a ball, and rolling off the platform into the crowd would not have been good public relations. After about two hours I could maneuver pretty well but with my feet spread wide apart. I suffered no nausea just as I hadn't when entering zero gravity. Training, hard work, or fortitude had nothing to do with it—I was just lucky.

"After about two days, I could walk without any noticeable difference from my preflight gait, but it took about two weeks to hit my preflight performance in the balance tests.

"There was another disappointment to deal with when we came back. Without gravity in flight, each of our vertebrae had expanded a bit, and we each became about two inches taller. What a great deal! But our new height was short-lived as soon as gravity got us back into its clutches again.

"Upon return to Ellington Field in Houston, we had to stand on a platform for almost an hour trying to say something historic. The wobbly legs returned, and I again feared tumbling off the platform into the crowd. But the legs held and the wobble abated.

"Then I did a dumb thing. About four days after landing, I felt better than I thought I might, so I figured I ought to stop lollygagging around and get back to my standard exercise, a relaxed five-mile run. Wrong thing to do! Muscles and joints that had little stress on them for three months screamed their pain at me for the next two weeks.

"At least in addition to our pride and personal satisfaction, we were handsomely compensated by NASA. After all, we had traveled thirty-five million miles. Each and every day we received a government travel allowance, which because our meals, quarters, and transportation were government provided, came up to $2.38. Over our eighty-four-day flight, that came up to a whopping $199.92 for each one of us!"

There was great rejoicing when Skylab III—and the whole Skylab program—ended in success. A lot of tired people got to know their families again. And the participants busied themselves with documenting their lessons learned for the future, hoping that the future would soon include a permanent space station.

Skylab had clearly demonstrated the value of human intelligence applied in a hands-on way onboard the nation's first space station. It had also shown that humans retain all their abilities and needs even when they are several hundred miles up. Proper work scheduling, positive motivation, and meaningful communication are just as essential in flight as they on the ground, if not even more so.

Those directly involved began to reflect on what had been accomplished and contemplate our future in space. "But shoot," said Pogue, "talk about something that was successful, Skylab was highly successful! It was our first space station and focused attention on long-term reliability of systems and proper integration and support of the crews. Apollo flights lasted eight days or ten days. That's one thing. But when you stay up there for months, your systems and crews are going to see a much greater exposure to all the problems that are waiting for you in zero gravity and the space environment. Now we see how difficult it will be to design reliable systems and support crews for something as long as a Mars mission, which is nominally about two to three years."

Ed Gibson said, "It was a great sense of accomplishment; we had met all our mission objectives, averaged as many accomplishments per unit time as previous crews, despite our slow start, and had set a world record for time in space. But in only a few short years the Russians eclipsed our mark.

"However, our mission did set an American record that lasted for twenty-one years. Actually we expected and wanted to have our American record broken within four to six years—on an American station! But that was not to be. Norm Thagard, a very capable guy, broke our endurance record, but he did it on the Russian Mir space station.

"We all recognize Skylab was the beneficiary of the program that came before it: Apollo. Skylab itself was constructed in large part with hardware that became available when the last three missions to the moon were canceled. Unlike today it was launched all in one shot using the Saturn v, which was the greatest rocket system the United States has ever developed. It could put 250,000 pounds into low Earth orbit, about seven times what the Shuttle can do today. It launched many flights in the early years and got the whole space program off to a fast start including Skylab.

"My own view of this fast start began back in 1957, when I, my parents, and Julie, my girlfriend then and wife now, stood out in the backyard of our home in Kenmore, which is just north of Buffalo. We watched mankind's first satellite go over, the Soviet Sputnik. Back then, I'd never heard the word astronaut. But just fifteen years later, my parents stood out in the same backyard and watched me go over.

"But the rapid pace and success of the early years was not because of hardware alone; people and leadership made it happen. We in the astronaut office closely experienced one of the very best: Deke Slayton. He was one of the Original Seven astronauts but was medically grounded before he could fly. Rather than quit, he was driven to contribute wherever he could, and he was appointed the head of Flight Crew Operations, which included the challenge of keeping over forty headstrong astronauts under control, a daunting task. But he was tough and very mission focused. If you were also there, like him, to advance the mission, he gave you his full support. If you were to advance yourself, he'd rip out the flamethrower and turn you to a crisp in nothing flat. He was the right guy for the job.

"In Deke's demands for an overwhelming focus on the mission, which he applied unselfishly to himself as well, he was tough but fair, harsh but kind, someone I respected, trusted, liked, and feared all at the same time. I've seen many leaders in my career, some very sophisticated, but I regard Deke as one of the best I've ever encountered. He and many others like him made Apollo and Skylab happen. To the cheers of everyone around him, Deke finally did get to fly on Apollo-Soyuz, the joint U.S.–Russian mission after Skylab.

"Future space stations will have a hard time matching Skylab's high scientific accomplishments for its relatively low cost. Based on Skylab, we should expect that future stations will discover new materials of engineering and biological importance, as well as new knowledge of how our bodies function without gravity, important for better understanding how they function right down here on Earth, as well as on future long-term missions far from Earth. But with the more complex systems that must last for many years, not just months, we would expect that proper manning of a station should reach at least six crewpersons, preferably seven or eight, to keep the station in full operation, properly perform really top-notch science and technology experiments, and realize the real potential of a space station. There is NO GOOD REASON that we cannot perform Nobel Prize–quality science up there, just as we do down here!

"In the long run, despite tragedies and budget droughts the prospects for America in space remain bright. Most important, we have within our people, still, a spirit and will that wants nothing more than ever-deeper explorations of space, and its profitable use. We have physical facilities in America second to none. And charging in the front door, we have our youth, equally motivated and far better trained than those young engineers who took us to the moon and into Skylab. Lastly, we have our graybeards, engineers and managers with the knowledge and wisdom from decades of experience.

"Certainly, in time, we will complete and properly staff the International Space Station, return to the moon, land on Mars, and eventually explore the rest of our solar system. But beyond that? In the back of my mind I speculated when I was on Earth's dark side during EVAs. The stars were clear, steady, and not a twinkle to be seen in any of them. A dense hemisphere of stars swelled into existence as my eyes got dark-adapted. There's got to be life out there!

"As we find more and better ways to visualize planets around other stars, we just might visualize a blue planet, one with an oxygen atmosphere. Then the pull would be irresistible. We'd have a crash program for near–light-speed flight, then a mission that'd fire our imagination far more than any fantasy from *Star Trek*. But we all understand that the distances are immense. On our Skylab flight, we traveled thirty-five million miles, which is the distance that light goes in just three minutes. Yet it takes light over four years just to reach our closest neighboring star. Clearly when it comes to deep space travel, we've just barely put a few layers of skin on our big toe out the front door.

"But it's also clear that we're on the front end of something much larger than any of us can imagine, travels and adventures far greater than anything we can now picture. And it's also clear that we'll never stop exploring, never stop reaching outward—it's hard-wired into our psyche. I believe that if you scratch deep enough into the tough hide of even the most cynical, hard-boiled, space engineer, like a few of those you've encountered in this book, lurking at their core you'll find a Trekkie; that's someone who realizes that space probes and all their data are interesting, often exciting, but ultimately, it's we who have to go there, in person, to see and feel new turf up close before it truly becomes a real part of our own world.

"One day, certainly in the long-term, driven by the human spirit, we will travel in vehicles that are derivatives of Skylab and subsequent space stations out to the rest of our solar system and, eventually, beyond."

11. Science on Skylab

Several books could be—and have been—written to summarize all of the scientific experiments performed on Skylab. Almost one hundred different pieces of experiment equipment were manifested for the original launch. Thousands of hours were spent on science. Tens of thousands of Earth observation images were taken as well as over a hundred thousand solar astronomy images.

The two fields that were Skylab's greatest scientific legacy, as well as the ones requiring the largest time investments from its crews, were solar astronomy and life sciences in weightlessness. Research performed on Skylab would revolutionize both of these fields and would lay the groundwork for all that would come later.

Life Sciences

The Prologue: Early Spaceflight

Skylab was medicine's first, best chance to unravel the mysteries of weightlessness. Man's ability to fly into space and to withstand the effects of being weightless had been matters of controversy since the very beginning of NASA. Opinions were all over the map. Some believed the experience would be pleasant and of no medical significance; the original astronauts were in this group. Others speculated that disruption would occur in many body systems. Balance would go haywire without gravity to guide the inner ear; the heart would weaken; the passage of food "down" the digestive tract would suffer; urination might be impossible; and the isolation would induce a state of sensory deprivation, the "breakoff phenomenon." A lot of these hypotheses were published in medical journals, promoting the impression that space was dangerous and unknown.

The U.S. Air Force had begun to prepare itself to manage America's manned space efforts, and this preparation included medical support. It was the unquestioned leader in the field of aerospace medicine with three times the personnel and four times the budget of its closest competitor, the Navy. A distinguished German physician and physiologist, Hubertus Strughold, established in 1950 the first department of Space Medicine, at the U.S. Air Force School of Aviation Medicine. Other German scientists also did research for the Air Force.

NASA's predecessor organization, the National Advisory Committee for Aeronautics, or NACA, had no medical staff or expertise; all of its original experts were borrowed from the military. To provide medical support to Project Mercury, the Air Force contributed William Douglas, Stanley White, and Charles Berry; the Army, William Augerson; and the Navy, Robert Voas. These men brought with them the military method of qualifying humans for the stresses of flight. As aircraft flew faster and higher, pilot tolerance to and protection from acceleration, hypoxia, and disorientation had become major problems. The approach to solving them emphasized testing and monitoring both in laboratories and in flight, an incremental increase in human exposure with healthy skilled test pilots, and very close liaison between medicine and engineering.

The academic community's advice was quite different. It emphasized peer-reviewed scientific experiments by National Institutes of Health or university scientists and a great deal of animal research before exposing humans. The rationale was that the effects of spaceflight must be characterized and proven safe before people flew. Throughout the 1960s a continuous stream of criticism was heaped upon NASA by scientists: its programs were too ambitious, too rushed, not safe. One group insisted that NASA fly forty animals into space before committing to human flight.

Animals were the first to be sent into space. In December 1958 a squirrel monkey named Old Reliable was launched in the nose cone of a Jupiter missile to an apogee of three hundred miles; it survived the launch, but the nose cone was lost upon reentry. In May 1959 a rhesus and a squirrel monkey, Able and Baker, made the same trip and survived. Two chimpanzees, Ham and Enos, became the first animals to ride in Project Mercury capsules—Ham on a suborbital flight and Enos for two orbits. Both did fine. Mercury's medical support group believed that these flights, plus the reports from the Soviet Union of a successful six-day Soviet flight of the dog Laika on Sputnik 2 in 1958 (though recent information indicates that Laika's flight was far less successful that early reports would have had the world believe), showed that weightlessness was survivable, at least for short periods.

But biological scientists wanted still more. Led by Ames Research Center, they achieved NASA approval and funding for the Biosatellite project, which would launch and study various life forms on dedicated satellites. The first Biosatellite mission failed on launch. The second successfully flew plants and insects into space in September 1967. The third was to fly a fully instrumented monkey, Bonnie, on a thirty-day mission to pave the way for the Skylab program.

Biosatellite 3 launched on 29 June 1969. The spacecraft was built by General Electric Reentry Systems Division at Philadelphia, weighed 1,550 pounds, and was launched on a Delta into a 240 nautical-mile circular orbit. Reentry was commanded on the ninth day of the flight, on 7 July (just nine days before Apollo 11 launched to the moon). Bonnie was recovered but died less than half a day later.

Here is a letter on University of California, Los Angeles letterhead, to the editor of *Science* magazine, dated 14 August 1969. It's a copy of a copy of the original. At top left someone has written, "This has been submitted to SCIENCE for publication although NASA objected to certain portions thereof." And on the right the person wrote "SLAYTON."

It's a lengthy letter. Here are a few excerpts:

The recent flight of Biosatellite III with a male macaque monkey (Macaca nemestrina) was the culmination of more than five years' intense collaborative scientific effort. . . . The flight lasted only 8.5 of a planned 30 days. . . . The physiologic deterioration of the monkey . . . is mainly attributed to the effects of weightlessness. . . .

The monkey was in excellent physical condition at the time of launch. . . . All physiological sensors functioned perfectly throughout the flight and after recovery. There were 33 channels of physiological information. . . . The range of these measurements in different body systems and their detailed character are without parallel in any single previous experiment on Earth or in space.

The last sentence had been underlined, and in Deke's unmistakable hand was the comment, "That's what killed him." (Garriott joked that if the surgical preparations the monkey underwent, among the least violative of which included incisor tooth extraction and tail amputation, had been required for potential Skylab astronauts, NASA would have lost nine out of nine crewmembers.)

The letter goes on to describe the monkey's gradual loss of responsiveness to tests; the drop in body temperature, heart rate, and blood pressure; the emergency reentry, and the death in Hawaii from a sudden heart arrhythmia after hours of emergency treatment. Now the investigators sum up:

The well-documented sequence of events leading to collapse in this monkey suggest the need for a guarded approach to design of missions for man that might involve extreme effort after a considerable exposure to weightlessness . . . the important findings listed above characterize this mission as highly successful. They also indicate the great value of carefully designed animal experiments . . . especially where the physiological sensors and required experimental control are difficult or impossible to secure in manned flight. Sincerely, five scientists.

In the Q&A session at a press conference, held 22 October 1969, Dr. W. Ross Adey, the principal investigator, was asked, "Why believe one bad result in a monkey instead of seventeen good ones in astronauts?" He replied that the astronauts didn't do all that well, really—Dick Gordon never did get the tether attached (on Gemini 11), and he sweated profusely. Then the following exchange occurred.

Question: To follow up Bill Hines's question: might your experiments with this monkey indicate that monkeys are less adapted to spaceflight than men are?

Adey: Well, I think the question of individual susceptibility cannot be ruled out. After all, this is one monkey, and there are seventeen men. And here I would like to take off my hat as an experimenter and put on another one, as a member of the President's Science Advisory Committee, which has had a medical group looking into the question of the biomedical foundations of manned spaceflight. And their advisory document to the President has been released and is in the course of being published. And I would submit that the best considered opinion is that we do not now have the biomedical basis for going ahead with the very elaborate programs proposed in the way of space platforms and space stations which involve major new engineering developments, and that the biomedical competence—or rather, the body of knowledge, as the report says—and I think I quote it correctly; it says that the necessary biomedical basis does not exist in NASA nor in the scientific community generally, and that it is not realistic to go ahead with the planning of major new space systems and exclude from almost any consideration the question of the biomedical capability of man to not merely survive in space, which has been his requirement to this point, in essence, but to perform at a high level on a continuing basis."

Kerwin recalled, "Well. That truly threw down the gauntlet to us Skylab types. None of us were thrilled to hear that the Apollo 11 mission had demonstrated mere survival. But clearly it was up to us to show that we could perform at a high level—on a continuing basis—while floating around. That we could brush our teeth, go to the bathroom, take spacewalks, yes, even (gasp) DO SCIENCE—just like Bonnie had.

"Chuck Berry's response to the Biosatellite business was courageous and correct. He publicly and accurately identified the differences between the monkey's circumstances and ours and judged Skylab to be safe. Given a lot of bedside hovering, of course. He'd put himself on the spot, and when things didn't go perfectly early on in our mission, some medical pessimism returned."

Meanwhile, those seventeen astronauts had flown into space for durations that ranged from four and a half hours to fourteen days, and they'd all performed well and recovered quickly from any effects of the flights. There had been changes. There was weight loss, ranging upwards of ten pounds. There was loss of appetite and in some cases motion sickness. There was some muscle weakness after flight. Blood volume decreased, and a few astronauts had a tendency to be light-headed immediately after recovery. On one flight (Apollo 15, well after Biosatellite) there were disturbing heartbeat patterns in two crew members. Calcium excretion increased, and there was just a hint that bones might be losing structural strength. These gave rise to questions about the feasibility of long-duration spaceflight. Skylab was the place to answer them.

The Skylab Medical Plan

On Skylab, for the very first time, life sciences were not just along for the ride—they were going to have top priority as a mission goal. And the NASA life sciences people—despite their organizational fragmentation, differences of opinion, and constant criticism from outside—responded to the challenge with a well-planned, ambitious set of experiments.

One group would test the cardiovascular system, studying heart function during exercise and simulated gravity (using lower body negative pressure). Another would be a very careful metabolic balance experiment with exact

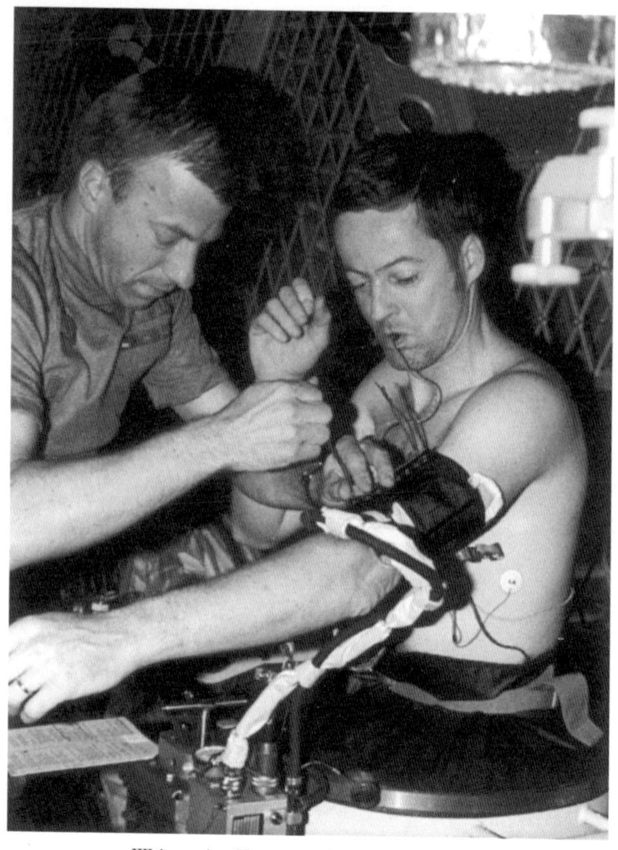

44. Weitz assists Kerwin with a blood-pressure cuff.

measurement of all intake and output combined with pre-and postflight measurement of bone loss using gamma ray densitometry, more accurate than the x-rays used in Gemini. Yet another would measure the body's responses to the stress of flight by measuring hormone levels in blood samples collected and frozen in flight—and observe whether the trend to a reduction in blood volume and red blood cell mass was continuing. And one would intensively evaluate the vestibular balance system in the inner ear, suspected to be the culprit in space motion sickness. There were cleverly designed "scales" capable of measuring the "weight" (the mass actually) of the astronauts, of any food and drink they were supposed to eat but didn't, and of their feces. There was even a special cap to measure and record brain waves during sleep, looking at duration and depth. The scientist members of the crews got to wear that one.

All these experiments would be carried out during flight, not just before and after, thanks to the ample size and weight of Skylab. But designing them to be carried out successfully was still an enormous challenge. The food system would have to accommodate the metabolic balance experiment. Feeding the crews was a pretty big part of getting ready for Skylab. Astronauts have to eat, and on this mission they'd have to eat for a long time. So there were a lot of requirements and considerations jostling each other for priority:

1. Learning how to package foods to be consumed in zero gravity.
2. Launching all the food for all three missions aboard the Skylab workshop because the Apollo spacecraft used as the crew's taxicab wasn't big enough to hold it. That meant selecting food treated and packaged to have a year-long shelf life in space—not the best setup for tasty meals.
3. Giving the crews food they'd like—making mealtime a positive experience on these long and isolated missions.
4. Keeping them well nourished, which is not the same thing as giving them food they'd like.
5. And last but definitely not least, discovering what happens to nutritional needs during long periods in weightless spaceflight. This would be one of the most important medical experiments.

Storage on orbit for up to a year at "pantry" temperatures was the most severe environment yet for space food. It ruled out fresh food, food that required refrigeration—any food you'd throw away at home if it hadn't been eaten after a month or two. Both weight and spoilage considerations dictated that the food should not be stowed mixed with water. If soup was wanted, dried soup was stowed, and the water was added just before eating. So a lot of rehydratable food, from orange juice to spaghetti, was on the menu. Adding water doesn't work for certain foods—for example, bread. The solution here was to irradiate it for preservation, then vacuum-pack it. Unfortunately, vacuum-packing sucks most of the air out of bread, making it an unpalatable paste. Bread was not a hit. But the sugar cookies—food system specialist Rita Rapp's own recipe—were delicious.

Foods that were to be served hot were packed in plastic bags, and the bags packed snugly in little flat round cans. The routine was as follows: open the can, add water to the food through a nozzle, smush it around to mix the food and water, put it back in the can, put the can in a fitted receptacle in an airline-style tray, and turn on the electric strip heater. An hour or so later, the item would be hot. This system worked well. It was a little time-consuming; one crewman would usually prepare three meals an hour or so ahead of time. And it did generate a lot of trash. Hot coffee was achieved a different way; the crew just added hot water to the instant coffee and shook instead of smushing.

Once the Skylab food system, or galley, was developed, the big question was, how many different kinds of food would be provided, and how much of each? And that's where the scientists came in. Their working hypothesis was that flying in space was like resting in bed, immobilized by illness or perhaps multiple fractures. You were in "negative metabolic balance." You lost appetite and lost weight, and muscles not used began to atrophy. It was intuitively obvious that in space many muscles weren't used much (those used for climbing stairs, for example). The energy needed for normal body activity must therefore decrease, and the need for food would decrease in proportion.

Along came Dr. G. Donald Whedon, an experienced and prestigious researcher, with a diet plan for Skylab. He proposed that all Skylab crewmen consume a diet of 2,400 calories per day, below their Earth-bound needs. The diet would contain precise amounts of calcium, phosphorus, and other electrolytes and specified amounts of protein and fat with very little variance allowed. Some additional carbohydrates—"empty calories" such as lemon drops—were allowed if the men were still hungry. Menus would be made up with enough variety to provide a six-day cycle, which would then repeat.

The crews would eat this diet for eighteen days both before and after flight. And—here's the key—both before, during, and after flight, every gram of matter that entered or left their bodies would be weighed and analyzed. Thus, whether the men were gaining or losing calcium from the bones or nitrogen from the muscles would be known with precision. It was a lovely experiment. But it gave rise to some practical problems.

First was standardizing the diet. The crew violently objected to the assumption that all of them would have to consume the same amount of food. Alan Bean weighed 150 pounds and had consumed less than 2,000 calories daily on his Apollo 12 flight. Jack Lousma weighed a fit 195 pounds and ate more than 3,000 calories a day on Earth. There was no way they could both be constrained to 2,400. The second problem was that the 2,400 number had been calculated based on the assumption that during spaceflight metabolic demands decreased and so did calorie consumption. This might happen, the crew argued, but it was unproven; and even if it did happen, it was wrong to put the men on the in-flight diet for nearly three weeks before and after flight.

On 1 March 1971 Deke Slayton wrote a memo stating in part, "We are not raising goose livers, and it is unreasonable and unrealistic to force-feed astronauts." Finally, the investigators agreed to tailor each crewman's diet to his usual intake. A week-long test using prototype flight food was organized, and the results used to construct the in-flight diets. Instead of merging all nine men into one data set, each one would serve as his own control. Feeding the flight diet before launch was retained, however. Sure enough, eight of the nine crewmen lost weight during the eighteen days before launch.

Another problem was, how do you weigh things in weightlessness? It's true; Justice's scales are useless in space. The little weights would just float away. But objects in space still have mass—they just don't have gravity pulling that mass against the scales. So Dr. Bill Thornton invented an ingenious device to measure the mass without using gravity. His theory was this: if you attach an object to the free end of a strip of spring steel, clamp the other end, and give the object a push, the steel will oscillate back and forth. And the heavier the object—or rather, the greater its mass—the more slowly will it oscillate. So you have only to attach whatever you want to measure, start it oscillating, and measure the time it takes to complete three back-and-forth movements—the "period" of the spring and mass. Skylab adopted Bill's principle, and the Air Force built one large device for measuring the mass of the astronauts, and two small ones, for measuring food residue, feces, and other small amounts of substances and small items. Given the opportunity, Dr. Whedon's team wanted to measure the mass of everything to the greatest accuracy possible:

1. The bags used to capture feces came secured with green tape; the crews were instructed to "weigh" the tape, separately, each time they used a bag.
2. They were then asked to mass measure each used fecal bag "wet" before putting it into a vacuum oven, where it was dried for return to Earth.

3. Both large and small masses were requested to be "weighed" to six significant figures—less than a hundredth of a pound for people, and a thousandth of a gram for food residue. That called for averaging many repeated weighings.

Whedon's team explored several methods for sampling sweat, but finally gave it up as impractical. They knew there would be considerable sweating but estimated that only a small percentage of the controlled minerals would be lost by this route. "The problem with these procedures was that we'd be spending an inordinate amount of time in flight doing them," Kerwin recalled. "With only three people aloft, eighty experiments to conduct, and a hotel to run, we needed everything streamlined and every nonessential task deleted. We compromised. The investigators would use the average weight of the green tape. We agreed to the many repetitions necessary to calibrate the mass-measurement devices in-flight to maximum accuracy, and they agreed not to require that accuracy in daily use.

"We also made another promise to ourselves. The rule was, if you didn't eat all of a food item, you had to weigh the residue to keep accurate track of your intake. We vowed that if we started a food item, we'd finish it, and avoid having to weigh it."

Many people in the medical field were involved in these issues, but the crew's principal point of contact for the experiment was Dr. Paul C. Rambaut, who functioned as the principal investigator's principal coordinating scientist. Crewmembers argued, agreed, and compromised with Rambaut for several years. In January 1970 he wrote, "The proposed in-flight procedures do indeed involve excessive and unproductive use of crew time for manual manipulation of food, water and waste. This situation is unfortunate and its correction has so far eluded the most vigorous protests of the Medical Directorate."

What Rambaut meant was that the Medical Directorate had fought hard for fully automated systems for collection and measurement of food and waste but had been spurned by the program manager because it would take too long, cost too much, and no one knew how to do it. The Medical Directorate was willing to make things as easy as possible for the crews. But they were absolutely not willing to compromise the validity of their experiments. In Apollo they had had to stand aside for operations. Skylab was their mission.

Progress was made during 1970. While the engineers were figuring out how to package soup in peel-top cans, NASA's nutritionists were working on the menu. By August a list of seventy-two items was given to the crew for evaluation, and shortly thereafter their deletions and additions were taken into account. Out went the strawberry wafers, the lobster bisque, and the cheese soup; in went the German potato salad, peanut butter, and—in a move that would prove lucky later on—the Carnation Instant Breakfast. There were five soups, ten drinks, twenty-seven meat-and-fish items including chili with no beans, eight veggies, seventeen desserts and snacks, and five breakfast foods. The contract was issued (to Whirlpool, a washing machine manufacturer!) and food production planning began.

One of the new items accepted was frozen prime rib. Frozen? Yes! The program office had agreed to provide food freezers with enough capacity for about four hundred of the little food cans—almost enough to provide one frozen item per crewman per day. (The freezer, though, did not include a refrigerator, so fresh foods that would have to be kept cool were still not an option.) Besides prime rib the choices included filet mignon, buttered rolls, and coffee cake. The investigator objected to ice cream at first, fearing that its high fat content would make it too difficult to fit into the straitjacket of his dietary requirements. But the nutritionists showed that they could do the job, and ice cream was added to the list. This decision and these items were major contributors to crew satisfaction with the food, and the mission.

And then there was item number ten on the beverage list: "Wine (rosé or sherry)." As the crewmembers discussed palatability and variety with the experimenters and attempted to make the diet as pleasant as practicable given the constraints, someone said, "Wine is empty calories too! Let's have some on the menu." Surely, wine is empty calories by the standards of the experiment—it contains little or no protein or controlled electrolytes. But getting it by the doctors wasn't so easy. Early in 1971 Deke Slayton wrote a memo requesting several changes to the food system. One of them was the addition of wine. This suggestion was indignantly rejected by the experimenters. The reply stated, "We disagree with the assertion that the provision of wine is mandatory to make the Skylab Food System flyable. Wine is not a necessary component of any nutritional regimen in any environment to which human beings are exposed. . . . The principal investigators of the M071 and M110 series of experiments are adamantly opposed to its use."

The formal objection by the investigators' representative, Dr. Leo Lutwak, stated, "Alcohol has effect on renal function via inhibition of anti-diuretic hormone. This introduces an additional variable even if consumed in the same amount daily by each man. Possible changes in retention and excretion of fluids and of hormones in flight (changes in kidney function with respect to water balance) are an important concern. . . ." But crew representatives argued that if it were backed off to once a week, any effects would be transitory and self-correcting, and the investigators reluctantly concurred. This is how they put it:

Recommendation:

1. *Delete all alcoholic beverages from menu.*
2. *Will accept: a) No wine first two weeks in flight or in first 48 hours post flight. b) 4 oz. of sherry (or equivalent stability wine) once per week thereafter."*

The crew accepted this compromise philosophically. The nose of the camel was under the tent. Now they had only to select the wine. There were a few requirements. Dr. Lutwak was right; they needed to pick a sherry or other fortified wine that could tolerate storage in plastic for a year or more. Ordinary table wine, red or white, was likely to go bad. The other requirement was to select an American wine.

A wine-tasting party was set up at Dr. Kerwin's house on 20 November 1971. Wives were invited but didn't get to vote. Kerwin had had the pleasure of narrowing down the list to six wines from Taylor, Paul Masson, Inglebrook, Wente Brothers, Almaden, and Louis Martini. The evaluation sheet outlined the rating code (a modification of the Cooper-Harper scale used by military test pilots to describe the flying qualities of fighter planes) and added these comments:

There are six entries, three dry and three sweet. All are domestic. A couple of imported sherries are at the end of the table if you care to taste them for reference.

Recommend about ½ ounce for tasting purposes, as medical science cannot cure a wine hangover.

Plastic cups are provided to simulate flight hardware. It is permissible (though not mandatory) to re-use your cup. Rinsing facilities are not available—wipe with napkin or shirtsleeve if desired.

To help you fill out the "comment" line a list of adjectives follows: unpretentious, robust, dulcet, uncompromising, reminiscent, ethereal, insouciant, devil-may-care, cynical, Earthy. A more complete list is being compiled for the flight checklist.

The Taylor cream sherry was selected in a close contest, and Rita Rapp set about her packaging duties. But frustration lay ahead.

All of the crewmembers worked in a few public appearances during training. One crewmember ("We won't tell on you, Jerry," Kerwin jokes) gave a talk in a southern state in which he mentioned that wine would be served on Skylab in the interest of gracious living and crew morale. Several of the listeners took umbrage at this, and letters began to arrive at NASA and congressional offices objecting to government-funded alcohol in space. NASA chose not to argue. Wine was quietly withdrawn from the menu, and the crews' kidneys were spared.

Despite this setback, the food system came together nicely as launch day neared. Procedures were devised for stowing most of the food in large overcans in the ring lockers in the upper workshop, each can carefully labeled with crewman, day, and meal. These would be brought down to the wardroom about a week's worth at a time and arranged, ready for each meal. There was "overage" also stored—extra food in case of spills, substitutions (discouraged), or mission extensions. Much of the overage was devoted to items that wouldn't affect mineral balance—lemon drops, butter cookies, black coffee. It was a little complicated in the days before bar codes, but everyone tried hard to make it work.

Also on the topic of mineral balance, there were the pills. It was very important to the investigators that the intake of protein, calcium, phosphorus, and magnesium be held constant. Protein consumption had to be imbedded in the food items themselves, but the minerals could be consumed as supplements. So the following routine was devised: all items not eaten by each crewman were logged and reported to Houston during an evening status report. If an item was partially eaten, the residue was "weighed" and the weight reported. Overnight, the medical team calculated how much of these minerals had not been consumed, and in the morning a teleprinter message told each man how many calcium, phosphorus, or magnesium pills to take. Munching the morning pills quickly became routine.

As a final gesture of solidarity, the dieticians managed to squeeze a number of fresh items into the crews' diets during the pre-and postflight quarantine periods. It was really nice to have a fresh salad with dinner amidst the cans and bags. The meals were pleasant and memorable and contributed to a team spirit that made the hard work of experiment compliance in flight manageable.

The other major intersection between research and operations was exercise. It was a design challenge and battleground between the crews, the researchers and, often, the managers. When the Mercury astronauts were selected, there was an enormous emphasis on physical conditioning and toughness based on a complete ignorance of the effects of weightlessness on humans. So the Original Seven, having been exposed to every stress the doctors could think up, concluded that staying in shape was their responsibility, and nobody was going to tell them how to do it.

As Mercury and Gemini flights took place during the years 1961 through 1966, all in small capsules with little or no opportunity for exercise and for durations extending to Gemini 7's fourteen days, a pattern began to emerge. Astronauts had eaten less during flight and returned having lost weight and (subjectively) some strength. There was evidence of a decrease in blood volume and a suspicion that bone density might be decreasing. Normal bodily functions were accomplished with no trouble, and the astronauts did not suffer psychologically—quite the reverse. They loved the weightlessness of space and declared their readiness to go to the moon.

More of the same was seen during Project Apollo. The crews accomplished their lunar surface excursions with enthusiasm and success, but they definitely paid a price, coming home tired and needing several days to recover their preflight weight and strength. Space motion sickness, first reported in the Soviet space program, began to occur in the larger Apollo spacecraft; and there was a bit of a scare on Apollo 15 when the two lunar surface crewmen developed cardiac arrhythmias during the return flight. This was attributed to a loss of fluid and electrolytes, especially potassium, during their extensive lunar surface activities. "No big deal," said the astronauts, and the missions continued with potassium added to the orange juice. But a case could be made that their strength and endurance, and thus their ability to perform challenging physical tasks such as spacewalks, would be compromised on very long flights.

The doctors tried their best to organize an exercise program during Apollo. These efforts were rejected. Here is a quotation from a memo from Deke Slayton to Chuck Berry, dated 27 March 1968:

Your recent offer to assist in development of an in-flight exercise program for Apollo is appreciated. . . . I believe it is clearly understood that crew physical conditioning is the responsibility of this Directorate. . . . Our intention is to provide each crew with the means and protocol to maintain a reasonable level of physical well-being. We have no intention of complicating the procedure by keying to station passes, data collection points, or dictated work levels. You will be provided the crew's best qualitative evaluation of their exercise program in the post-flight report.

That was the background for Skylab, which was to be the first opportunity for medical researchers to gain extensive in-flight data on human physiology in weightlessness. One of the centerpiece medical experiments was to be an exercise tolerance test. The astronauts would exercise on a bicycle ergometer to 75 percent of their maximum preflight capacity, while extensive measurements were made of heart rate, blood pressure, a twelve-lead electrocardiogram, oxygen consumption, and carbon dioxide production. The tests would be repeated every four days. The ergometer, without the measurements, would be available for exercise on the other days. It maintained good cardio-respiratory conditioning, but did little for strength.

Nobody raised any objection to the test. But there were several problems associated with the use of the ergometer for crew exercise. These ranged from whether and how a bicycle could be ridden in zero-G, and whether the data from zero-G would be comparable to that from pre-and postflight runs, to the question of how much daily exercise was the right amount, and whether the ergometer alone was enough equipment. There was another device onboard, the Exergym—a small rope-and-capstan device that allowed a certain amount of "isokinetic" exercise—leg and arm pulling and pushing at a constant velocity against a load. It was difficult to use and was used very little.

At the heart of the daily exercise debate was a fundamental issue. In order to understand the effects of long-duration spaceflight on humans, was it best to prescribe and constrain exercise or to let it vary freely and measure and observe what happened? The research community was in favor of prescription. They argued that unless all possible variables could be controlled, the changes observed would be difficult, maybe impossible, to interpret. They had the science of statistical significance behind them.

The operations community (astronauts and most flight surgeons) was in favor of "measure and observe." They argued that there was insufficient knowledge to write a good prescription; that there were too many variables whose control would have to be attempted—especially individual variations in exercise tolerance and preflight conditioning; and that more would be learned by allowing the nine crewmembers to react to the environment. Having a spread of in-flight exercise intensity was good, they said; it would provide a chance to see whether a dose-response curve existed. And of course, the crew still had that strong distaste for being regimented.

The Skylab 1 crew worked out a compromise agreement. They would devise and document both a preflight and an in-flight exercise plan and would carefully record all in-flight exercise. About six months before launch, the first crew performed their baseline exercise runs on the training version of the ergometer. The ergometer was a good aerobic device; it had been designed to accommodate loads of up to 300 watts for thirty minutes. But Bill Thornton had ridden the training version at 300 watts for nearly an hour and destroyed the motor; henceforth, it was "de-rated" to 250 watts. That turned out to be enough for the Skylab astronauts.

The baselines determined were the watts at which three minutes of pedaling would stress each crewman to 25 percent, 50 percent and 75 percent of the maximum heart rate of which he was capable; that would be the in-flight protocol. Conrad's baseline was 50, 80, and 120 watts; Kerwin's, 50, 100, and 150 watts; Weitz's, 100, 150, and 175. That's when Conrad decided it was time to get in shape. He exercised his command authority to require a session of paddleball daily with one or the other of his crewmates. All three improved their conditioning noticeably. But the researchers decided it was too late to change the baseline; the crew ought to have an easy time of it on orbit.

How It All Turned Out: Results

How did it all turn out? All nine astronauts returned to Earth safely, and all but one are alive and well, over three decades later. The exception is Pete Conrad, who was healthy and vigorous until his tragic death in a motorcycle accident in 1999. But did the Skylab team accomplish their goals? Did they measure the right things and measure them accurately? Were they able to draw conclusions? And did they discover what's needed to keep people well and effective in space?

First, all the equipment worked. Food was consumed, uneaten items logged, supplement pills taken. Urine was sampled and frozen; feces were properly processed and dried; both were correctly returned. Exercise was accomplished once it was learned how; oxygen consumption and vital signs were duly recorded. Lower body negative pressure and rotating chair devices did not fail. Nine men were measured as never before. And it's a good thing they were because NASA remained conservative and skeptical till the end. During the final flight, the NASA administrator required a weekly report—on Wednesdays, and in writing—that the crew was in good condition and certified to go another week.

In spaceflight, the first symptom to show up is usually space motion sickness. The term was bowdlerized to "space adaptation syndrome" by some thoughtful researchers, but some irreverent crewmen called it dreaded space motion sickness, or "DSMS" for short. It's a lot like seasickness, and a severe case can make you miserable. A problem with treating motion sickness is that if you are already sick and take a pill, you're likely just to throw it up. Shuttle crewmembers who get spacesick are now given injections to control their symptoms. The drug, promethazine, is quite effective but makes you sleepy.

The odd thing is that a person's susceptibility to motion sickness on Earth has no predictive value in space. You can be quite sensitive in boats or cars and never have a quiver while weightless, or the reverse. But repeated flights tend to diminish the severity of the illness. And after it goes away, usually in three or four days, zero-G is a lot of fun.

The Skylab I crew had no space adaptation syndrome during flight. Kerwin got seasick after splashdown and threw up in sickbay on the carrier, but seasickness was not unusual for him. The most surprising finding was that the rotating chair experiment showed that the crew was immune to motion sickness in flight once they had adapted. Kerwin said, "I think life on a rotating spacecraft will be easier than I thought."

Once they learned how to ride the bicycle ergometer in zero-G, their performance matched their preflight levels. In between medical runs, Conrad exercised the most, Kerwin the least. In contrast with exercise, the lower body negative pressure experiment was much more stressful right from the first run. Heart rates were higher, blood pressures dropped, and several runs were terminated early to make sure the subject didn't pass out. Kerwin was the most susceptible. This was a concern early in the flight, but subsequent runs showed no further deterioration in performance.

Those three experiments were the main source of in-flight data on adaptation—or deterioration—and the overall picture looked good. Also the crew ate well, slept well, and felt well. The physical work involved in their spacewalks gave them no trouble. Twenty-eight days in space seemed quite feasible. How would they react to gravity after landing?

Here are the basic findings. The crew had lost an average of 7.5 pounds of weight. They were unsteady and walked with feet wide apart. They were most comfortable lying down. Standing heart rate was 30 percent higher than preflight. Ability to exercise on the treadmill was correspondingly lower; each of those heartbeats was moving less blood through their arteries. Treadmill performance on landing day did correlate with in-flight exercise; Conrad did best, while Kerwin was too tired to use the ergometer at all until the next morning. Response to the first lower body negative pressure test was just about the same as to the last in-flight test—worse than preflight. Blood tests showed that blood volume was down, and the number of circulating red cells was smaller by 14 percent—the bone marrow was not producing them at all. Finally, postflight strength measurements on the arm and leg muscles showed big decrements.

Return to normal was relatively rapid. The crewmen ate and drank vigorously after their first good night's sleep, and much of the weight loss had been restored by the fifth day back. Performance on the LBNP and ergometer was close to normal preflight limits in a week, and completely normal in three weeks. Full leg strength returned more slowly, depending on what exercise the men chose to perform.

Some questions remained. Their bone marrow didn't start to produce red cells again until about three weeks after landing. Imaging wasn't sensitive enough to show visible bone loss, but examination of the returned urine and feces showed that they were losing calcium steadily with no sign of leveling off. The in-flight cardiovascular data from LBNP could be interpreted as "not yet steady state." And the large losses in arm and especially leg strength were a sign that the crew hadn't exercised enough. The next crew was strongly urged to do more.

Skylab II, unlike their predecessors, faced space adaptation syndrome on Mission Day 1. Lousma was most affected and vomited that evening. Bean and Garriott felt nausea; all three took Scop-Dex capsules on Day 2 to suppress the symptoms, and Garriott and Lousma continued this medication on Day 3. Garriott and Bean tried making head movements to hasten their adaptation but decided they were not helpful. All were well enough for full duty by Day 4.

Initial adaptation issues aside, this crew's in-flight experience was similar to that of the first crew. The big change was that they performed more exercise and did not allow exercise time to be preempted by other tasks as the first crew sometimes had. They launched with two additional exercise devices. After seeing the first crew's strength losses, Bill Thornton got permission from Deke Slayton to add capability, with strict weight and volume limitations. (Exercise was still an operational responsibility.) Bill looked everywhere for candidates, and found the Mini-Gym in John Rummel's back room. He rebuilt it with help from center engineering and flight-qualified it. It was designed to produce "isotonic" loads on the back and legs, and was used faithfully by the crew during longer daily exercise periods. The other device was an "expander" from Sears, consisting of up to four springs with handles that could be used in various combinations. (The Russians copied this device and used it in their Soyuz program for many years.) These devices largely solved the upper-body problem. The second crew also had a device to measure blood hemoglobin levels, to reassure the doctors that loss of red cells was not continuing.

Skylab II returned to Earth on 25 September 1973 at nineteen past one in the afternoon, and the Command Module was picked up by the USS *New Orleans* forty-four minutes later. Here's some conversation between the crewmembers and the crew Flight Surgeon, Dr. Paul Buchanan, as they prepared to egress:

Bean: "We're going to have to be careful; as I said, I was dizzy. Better stick with me. But I feel like if I could move around, just sit up, and maybe the dizziness would go away."

Garriott (who had already left his couch and taken the pulse of his crewmates): "I don't think it will, Al. You ought to watch for it to increase slightly. It's likely to increase, I think."

Buchanan (to Garriott): "Are you doing all right standing?"

Garriott: "I'm doing fine, but I can tell that any sort of motion induces the sensation of pitching or rolling, a little dizziness."

Bean: "That's what I thought when I just leaned up. I thought that I didn't feel like I was going to pass out, but I did feel like—that I would be unstable. . . . Let's not leave these guys standing out here too long."

Bean (ten minutes later, in the medical lab): "I didn't think we'd get back feeling this good."

They did feel good, and they recovered rapidly from the effects of their flight—Garriott fastest. Their weight losses had been for Bean, 8.6 lbs. (5.5 percent); Garriott, 7.7 lbs. (5.7 percent); and Lousma, 9.25 lbs. (4.75 percent). By "R+5," the fifth day after landing, their lower body negative pressure and exercise tolerance tests were "within normal limits"—that is, back within two standard deviations of their preflight results. That's not 100 percent recovery, but it's impressive. After R+9, Garriott and Lousma were returned to flight status in the T-38. Bean had strained his back on R+1 (showing that muscles were a little more susceptible to injury after a long flight) and didn't get back to flying for another week or so.

Loss of strength and endurance in the legs was virtually the same for this crew as for the first crew. That was a very encouraging finding and a worthwhile payoff for the additional time this crew spent on exercise (about an hour a day each). But the decrements were still on the order of 20 percent, and Bean's crew recommended to Carr's that they increase exercise even more on their flight, then being extended to eighty-four days.

Although legs showed a decrement with their diminished use in weightlessness, the arms were different. They were not only used for translation around the workshop, but there were two useful exercise machines available for heavy exercise. Jack was a consistent user of this hardware and his early postflight data show an average 15 percent increase in arm strength, while his two crewmates showed little change.

And that is when Dr. Bill Thornton, who was making the strength measurements, invented his "Thornton treadmill." Faced with even more stringent limitations on what could be launched in the Command Module, Bill Thornton and engineering built the poor man's treadmill from a strip of slippery Teflon, which could be fastened to the Skylab floor with fasteners already on board. The crew placed the Teflon near a side wall with a handrail to hold on to. Wearing cotton socks and the previously discarded ergometer waist restraint, fastened to the floor with 180 pound-force bungees, they canted their bodies about thirty degrees forward and "ran," feet slipping on the Teflon surface. It was much harder work than running, but it did the trick.

The Skylab III crew lived longer in space than anyone had before—a lot longer—and spent more time and concentration thinking about the lack of gravity and its effect on their body and their performance. For the first month they felt rushed and behind and didn't get enough sleep. After that they settled into a productive routine. How did they feel about it? Here is what they shared during the medical debrief:

Question: "When did you think you were in control of things and everything was OK?"

Ed: "Somewhere between four and six weeks. At four weeks we were over the hill and I think had started to settle down and by six weeks I felt locked in solid."

Jerry: "If you drew a graph from launch day, you would have a fully steep slope and the graph would bottom out the day the three of us bombed out on M092 (for Jerry, that was on Mission Day 16, 1 December). And from then on, you can chart a gradual increase with a fairly reasonable slope until probably Day 45, and then you could throw a gentle knee into it. Again there's a positive slope, and we were on the increase all the way to the end."

How stressful was the lower body negative pressure experiment? All three men came near to fainting on tests run during the tenth to sixteenth flight day. For these tests another astronaut was always assigned as an observer, able to watch the subject's signs and terminate the tests if warranted. The test involved subjecting the subject's lower body to a partial vacuum. There were three steps of increasing vacuum: minus thirty, forty, and fifty millimeters of mercury, the latter estimated to pull blood into the legs about as much as standing quietly upright in normal gravity. These tests were terminated early, before the end of the minus fifty exposure—probably making both crew and experimenters a little doubtful that they'd go the whole three months. But things got better from there. Gibson had only one more test stopped, on Day 71; Carr and Pogue had none. Here are some of their comments:

Ed: "I would say that thirty up there was like fifty down here . . . the amount of sleep, amount of water intake and amount of salt intake were significant [variables] for me. If I felt dehydrated or overtired, I knew I would have a problem with LBNP. . . . I've always felt a craving for salt on this diet."

Bill: "And related to that, perhaps equivalent, was the time of day . . . earlier in the day, you felt better."

These men could feel the fluids shift within their bodies. During LBNP they would feel less fullness in the head as their legs became engorged—and sometimes that would be followed by cold sweat and faintness. Exercise also shifted fluid into muscles, but it felt much better. So did EVA.

Jerry: "But the times when I could feel the fluid shifts the most were during exercise. As soon as you got on the bike and started pedaling, your head immediately began to clear up. And by the time you finished pedaling and the blood had shifted to your muscles, you felt great. Your nose was clear, the pressure was off your sinuses, and you felt good; tired from the work, but you felt clear.

Ed: "I recall that on the first EVA it felt good to get outside and do some physical work for a long period of time. It was comfortable to have a heart rate above nominal for the duration of the EVA."

In contrast, they felt periods of dullness or sleepiness during times of less activity:

Jerry: "I think that during a period of adjustment my stamina, reserve, or whatever you want to call it, was gone. I found myself essentially living from meal to meal, what we call a lowering of blood sugar, when you're getting really hungry and you start feeling fatigued and tired and dull. I don't know what it is. I call it a lowering of the blood sugar. That's what we felt up there, but as soon as we got a meal in us we would feel much better. You feel a little drowsy after the meal, but shortly after that you feel better. But then it only takes about two to three hours for you to become tired again. The old reserve and the stamina is gone and by 3:30 in the afternoon we were hungry again, feeling bad."

Voice: "Did all three of you feel that?"

Bill: "I feel that way, but I would have liked to snack all the time I was up there; candy bars to eat periodically. I could have used those throughout the mission."

Jerry: "I wish I could tell you where the calories go when you're up there."

Ed: "I felt the same effect as Jerry mentioned. I would have liked snacks. I would have liked three meals and then snacks in between, some high-energy food like Bill mentioned."

Question: "Listening to you, I get the impression that one might suggest shorter days or naps, recharging in that manner as well as eating."

Ed: "A siesta right about noon time would have been great."

Question: "Could you appreciate that your legs had diminished in size?"

Ed: "You could see it and I think I could feel it in my calves, after I had worked the Thornton device, for example. That really was an excellent device. I'm glad we had it along. When did we start using it, around two weeks into the mission?"

Jerry: "Yes, somewhere between Day Ten and Fourteen is when we finally got time to set it up and start using it."

Ed: "And from then on, we started to turn around."

Jerry: "I felt pre-flight that the LBNP could be used . . . to maintain condition. I always hoped that you would take one of us as a control and put us in it every day as part of exercise . . . use it as an exercise tool and not require . . . all the instrumentation. Just get in it every day for five or ten minutes then see what you ended up with, compared to the other two guys."

Question: "Which of you would have volunteered to be the no-exercise, no-LBNP control?"

Jerry: "Oh, I didn't say no exercise (laughter). We had already decided that there would be no volunteers for no exercise."

To summarize the crew's opinion, the LBNP didn't do much for them, but exercise was essential. And the postflight findings certainly bore them out.

They all adapted well to the topsy-turvy world of weightlessness. As Carr related: "The first two-thirds of the mission, my dreams were one-G dreams, and I shifted to zero-G dreams about the last third. It really surprised me that I had some zero-G dreams." They struggled a little bit to describe the feeling of carrying one's personal "up" and "down" around in space:

Jerry: "Looking out the window at the Earth, sometimes the horizon would be upside-down. It was convenient just to flip (my own body) upside down and look out the window . . . [but then] every once in a while you'd look inside at one of the other guys 'upside-down with his feet locked in the ceiling' and he looked funny that way. And the table was upside down. It was very strange."

Jerry: "The normal mode of moving was just to go head first, whistling right down through the dome hatch all the way to the trash airlock. [But] if I ever decided to go through the other way, feet first, I had a very distinctive impression that I was very high and sure didn't want to fall and hurt myself. It's just because I had my feet out in front of me and all of a sudden there was a reference that connoted high."

The following table compares data from the three Skylab crews.

Mission Duration	Food / Day (CAL / KG.)	Exercise / Day (WATT-MIN)	Weight Loss Percentage *	Leg Strength % decrease	Arm Strength % decrease
28 days	18.1	2,150	4.2% (1.7%)	20%	8%
59 days	19.7	4,686	5.3% (1.0%)	20%	0%
84 days	21.0	4,836	1.6% (0.0%)	14%	10%

Data Comparisons of Three Skylab Crews

* *Figure in parentheses is additional loss during prelaunch quarantine.*

Those averages hide some sizeable individual variations. On Skylab I Conrad exercised almost twice as much as Kerwin and munched many more calories per kilogram than either of his crewmates. On Skylab II Lousma was a tiger for exercise, but Garriott ate the most. And on the final mission Carr set a record by losing essentially no weight (only two tenths of a pound) in nearly three months aloft. He was only the second astronaut ever not to lose weight in space. The first was Alan Shepard, not on his fifteen-minute Mercury flight but on his Apollo 14 journey to the moon.

But the averages do tell the story. The story is, "Feed them, exercise them hard, and they'll do well in space." Or, to quote Dr. Bill Thornton: "The conclusion is that muscle in space is no different from muscle on Earth; if it is properly nourished and exercised at reasonable load levels, it will maintain its function."

Based on the Skylab findings, here's a combined narrative of what you could expect to happen to your body if you took a big "grand tour" flight to Mars.

When the rattle of the booster stops and you're in orbit, that strange feeling of floating will occupy your attention fully. Your head feels full, as if you had a cold. You've been briefed to be very slow and careful with head and body movements. You will take this good advice and feel OK—but the person in the next seat will suddenly feel nauseated and reach for the sick bag. The flight attendant (all are trained nurses, as in the early days of commercial aviation) will give your neighbor an injection, and he'll feel a little better in the morning and be back to normal in three days or so.

Your vestibular system adapts slowly to the absence of "down"—that constant acceleration of gravity. The same absence causes body fluids to migrate from the lower extremities up into your chest and abdomen; the body responds by decreasing thirst and effectively losing fluid. Your blood becomes a little more concentrated, with a higher red cell count. Your bone marrow notices this and stops producing new red cells.

You won't get space motion sickness, but your appetite won't be very good for the first several days, and learning to move, eat, go to the bathroom, and sleep while floating may seem hard. On the morning of day eight you'll wake up thinking, "Wow! I feel great today!" But you'll have lost six pounds since the day before launch. Your legs look ridiculously skinny. The chief flight attendant now puts you on a strict exercise program for the duration of the flight.

Unused muscles atrophy; exercise is needed to keep them fit, and you have to keep doing it faithfully. A treadmill keeps your heart and lungs in order and provides some leg exercise; other devices simulate weight lifting and tone the other big muscle groups. The compression also helps your bone retain calcium, slowing the gradual thinning of bone structure that is another feature of zero-G.

Your flight goes on for several months. Your workload is light. The novelty of looking out the window at the stars slowly wears off; you become bored. Small habits and mannerisms of one of your fellow passengers, "Herb," begin to irritate you. Psychological stress—"cabin fever"—is definitely a risk of long spaceflights as it is of any long isolation with a small group of people. Skylab didn't experience much of that, because the crews had been so carefully selected and extensively trained. We hope that your Mars flight has a compatible crew and good leadership. Talk to the flight surgeon about your problems with "Herb."

You make it through the flight with flying colors. As reentry day gets closer, everyone becomes excited and upbeat. You and "Herb" exchange e-mail addresses. You've almost forgotten what gravity feels like. Now you are about to be reminded! You survive the parachute landing at the Space Recovery Field and are helped out of the spacecraft. You don't feel very good. Every time you move your head the world moves too, a severe vertigo. You feel as if you weigh three hundred pounds, and you walk unsteadily, with feet wide apart. Your pulse races. You feel dizzy, tired, light-headed, and thirsty.

First your vestibular system needs to readapt to the long-absent gravity vector. Your weight is only a few pounds less than when you left, but you are low on body fluids and red cells in the blood—almost as though you had bled a pint or two from a wound. And your muscles, despite the exercise (did you slack off a little the last few weeks?) are weak. A very light landing day and a very long night's sleep start you on the road to recovery. Next morning you feel like someone getting up after a bout with the flu—a little light-headed still, but good. A period of rehabilitation, with careful, supervised exercises, will be prescribed for three weeks to get you back to normal.

The red cell count will be normal in a month or so. Most of the calcium lost from your bones will eventually return. You'll have been exposed to radiation during your trip, including cosmic rays; the amount will determine whether any future threats to health exist. The lifetime limit is probably one Mars trip per person.

The Skylab medical data set rigorously quantified all these changes and allowed the causes of many of them to be understood. It demonstrated the safety of three-month spaceflights sufficiently to allow NASA to plan and build the International Space Station—and incidentally to allow the Soviet Union to plan and build Mir. Both the United States and Russia have built on this data and now routinely fly for up to six months. But the strict quantitative metabolic balance study Skylab accomplished has never been repeated. It did the job. And repeating it would drive the astronauts nuts.

Solar Observations

For solar observations it is essential to note that our atmosphere is opaque to portions of the sun's radiation. Only the so-called visible wavelengths, from about 400 to 800 nanometers (4,000 to 8,000 Angstroms) can pass through the atmosphere without significant absorption. It is presumably no coincidence that this just happens to be the wavelengths at which our eyes are sensitive to light; while it is presumably a coincidence that the sun's major radiation band is passed through the atmosphere to warm our planet. But the sun also radiates substantially at even shorter wavelengths in the ultraviolet, extreme ultraviolet, and x-ray ranges. It is again fortunate that the atmosphere does absorb these rays as they would otherwise be lethal to all forms of life here on the Earth's surface. Without the thick atmosphere (and the Earth's magnetic field for protection against arriving charged particles) we might all be living in caves, underground, or in the water.

So the Skylab was equipped with eight large telescopes in a large canister called the Apollo Telescope Mount, most of which had to be above almost all of the atmosphere to function properly. Two of the telescopes looked at the sun at visible wavelengths called "hydrogen-alpha," radiation coming from excited hydrogen atoms in the solar atmosphere. As this wavelength can be seen with ground-based telescopes, it provided a means of coordinating observations with terrestrial observatories all around the globe to look at a particular feature and compare observations. Three more telescopes were sensitive to the extreme ultraviolet and two to the x-ray portion of the spectrum. The sun looks quite different at each of these wavelengths, and solar astronomers use these differences to deduce many things about the sun. For example, each wavelength of emission is directly related to a different temperature of the emitting atom and therefore to a different altitude in the solar atmosphere, which permits a much better picture of the sun's structure to be constructed.

The eighth telescope was a White Light Coronagraph, which could have worked on the ground except for the fact that the sun's disk is about one million times brighter than the faint corona. Light scattering in our atmosphere from the bright disk completely swamps the faint coronal light here on Earth. The one exception to this occurs at times of a total solar eclipse, when the moon passes directly in front of the sun and blocks out its bright light. So only for a few minutes every year or so, at some places in the world (usually remote, it seems), it is possible to see the solar corona. But on Skylab an occulting disk was placed in front of the telescope to replace the moon, and the crew could then see the corona continuously whenever they were not on the dark side of the orbit.

So the ground astronomers and the crew remained in close contact every day. The ground teams worked up plans for how the crews could best optimize their limited observation time. There could be at most periods of about fifty minutes of sunlight, coming about ten times each twenty-four hours. Although there were almost sixteen orbits each Earth day, the crew was sleeping or otherwise occupied on some of them. This still permitted substantially more observational time than in the original flight plan. The ground team noted what the likely interesting solar features might be, argued within their own ranks about which instruments should have priority on each occasion, then had to argue with other research disciplines when special circumstances arose, sometimes taking their case all the way up to the flight director who controlled the whole mission for a decision. Then when their plan was sent up to the crew on a teleprinter, the crews would plan the orbit's activity and usually follow the ground's suggestions. But the final pointing of instruments always had to be done onboard, and whenever solar conditions changed or the crew saw evidence of even more interesting phenomena, they might well change the ground plan and proceed with an alternative observational program.

With the extreme ultraviolet telescopes the view presented was of gases at tens of thousands of degrees Celsius, coming from the chromosphere above the photosphere, which is the part of the sun the naked eye sees. With the x-ray telescopes the crew was viewing gases at millions of degrees Celsius, where the emission comes from even higher in the corona. All of these features differ when the sun is very active as compared with a cooler and very quiet sun. The sun's activity varies over a cycle of about eleven years. When Skylab was first planned, it was hoped to have a very active sun, which would be especially interesting to most of the solar astronomers. But flight delays of several years pushed the launch date out into the quiet portion of the cycle. Then nature came to the rescue. As it happened during the second mission, the sun became very active, at least on one side, and the sunspot number (a measure of the number of visible black dots and groups on the surface of the sun) varied all the way from 10 to over 150 at other times. So very good observations could be made from quiet to very active solar conditions.

The White Light Coronagraph stood apart from the other experiments in that it looked at the very faint light coming from far above the disk of the sun. Of particular interest were phenomena then called "coronal transients," but now called "coronal mass ejections," or CMEs. In these cases a long magnetic strand lifts off the lower atmosphere and expands into an enormous loop far out into the corona like a stretched rubber band. Occasionally, the band is stretched so far that it breaks apart and the confined gases escape into interplanetary space.

When this happens a coronal mass ejection occurs and if it is pointed in the correct general direction, it will eventually arrive in the vicinity of the Earth. This usually takes two to three days. When it gets here, it frequently produces not only the beautiful aurora but also not-so-desirable power line fluctuations or outages and sometimes damage to sensitive satellites in Earth orbit. Even though satellites in Earth orbit may be damaged, it could be far worse for space probes or manned spacecraft far outside the magnetic envelope (the magnetosphere) of the Earth. In this case very energetic solar protons could be more dangerous to space probes and people, whereas within the magnetosphere, the magnetic field largely turns away the charged particles streaming outward from the sun.

Garriott reported one of the most exciting observations of his mission was the day the first major CME was observed. "As I recall, the ground called us and alerted us to the possibility that one of the low magnetic loops appeared to be rising. When we first looked at the White Light Coronograph, indeed, the loop was already extended roughly a solar radius or half a million miles above the photosphere. So I promptly took a Polaroid picture of the screen while we also activated others of the telescopes to record relevant data as well. Then when we came back around the Earth in about an hour and a half, I took another Polaroid photo, and sure enough the loop had expanded by perhaps another two solar radii into the corona. A quick calculation told us that the outward velocity of the CME was at least 500 kilometers per second, a phenomenal speed it seemed to us! As far as I know, this was the first visual observation of a CME and the first real-time measurement of its minimum speed. When analyzed more carefully on the ground, the same number was calculated.

"Also of interest to us were small spots visible in the extreme ultraviolet, but not on the hydrogen-alpha visible images. For lack of better terminology, we called them 'XUV bright points.' We didn't really know what they were or their relevance to other solar events. We hoped they might even be precursors to a solar flare, for which a full study was a very high priority. They didn't seem to live too long, maybe thirty minutes, then faded away. We spent quite a lot of time studying them in space, and subsequent analysis on the ground revealed that they are the source of much of the solar magnetism. In addition we felt that they could be the location of new solar flares, so we watched them carefully. Indeed, Ed used this clue to provide images of the very early stages of a flare by following the development of one of the 'bright points.'"

Another feature studied at considerable length was coronal holes, which are areas of very low emission. They can be discerned all across the spectrum, but especially in hydrogen-alpha emission and at XUV wavelengths. Both the ground and the space crew could see them, permitting close cooperation in selecting targets and timing. What was less expected was to see the close association between areas on the sun's face where there was very little emission (the "holes") and the solar wind, which permeates all of interplanetary space. It appears that the "holes" are areas where any local magnetic fields do not form loops, but instead are essentially open. For this reason, any hot gases from the sun, which would otherwise be trapped by the solar magnetic field, can now flow straight out into space.

While the results of Skylab solar observations seemed to exceed even the ground investigators' expectations, one of its most valuable benefits was the instruction it provided to later experimenters who wanted even longer continuous observations of particular features. The corona and CMEs? Now the researchers knew roughly their frequency and speed. They knew to connect them to appearance of coronal holes visible on the ground. Bright points? Solar physicists have now been studying them for decades building from the pioneering work on Skylab. Much of the next phases of solar research have been done with automated satellites based on the results found on Skylab.

The Challenge of the ATM

Designing and operating the manned solar observatory, the Apollo Telescope Mount, presented a challenge of epic proportions for the ATM team composed of principal investigators who remained on the ground, the instrument designers, ground control personnel, and the nine Skylab crewmen who operated the observatory in flight. Like other such challenges, it proved to be highly demanding, rewarding, and just plain enjoyable, especially for those fortunate enough to operate the observatory in flight.

Skylab, and ATM in particular, served as an illuminating milestone to both the test pilots and scientists in the astronaut corps: "Hey, even scientists have major contributions to make," and "Even test pilots can actually learn science." Enough lively kidding and chain pulling by both groups took place in training to field a cohesive, capable, and dedicated team of observers on each of the three missions.

Unlike most stellar observations, the detail visible on our nearest star demanded that crews be prepared to continuously make operational decisions with each instrument in space (instrument pointing), time (times of operation and film exposures), and wavelength (use of different instrument operational modes that covered the spectrum from visible light all the way down to x-rays). Our mother star is rich in a variety of features including granulation, supergranulation, sunspots, active regions, filaments, bright points, and coronal structure. Solar events also cover a wide range of the required time resolution from its eleven-year activity cycle down to flares, which are solar explosions that can rise within seconds. It was the challenge of the ATM team to put instruments in orbit with cutting-edge observational capabilities and operate them in the way that maximized further understanding of known solar features and events and discoveries of new phenomena.

But just how does one meet this challenge? Basically, there were four essential components to the ATM team's approach:

Place in orbit state-of-the-art instruments with cutting-edge capabilities.

Provide the in-flight observers with real-time, appropriate feedback on relevant features and events on the sun to enable them to make the best choices in space, time, and wavelength for observation.

Ensure each operator has the background and training to effectively operate the ATM observatory.

Ensure the flight plan has the flexibility to accommodate optimum ground scheduling and observer real-time modifications.

Meeting the Challenge

1. State-of-the-Art Instruments

The principal investigators designed and had constructed instruments with cutting-edge capabilities. When designed, the instruments that made up the observatory had the highest feasible resolution in space, time, and wavelength and best combination of capabilities to meet the observing challenge.

In the design phase the potential observers, the astronauts assigned to work on the ATM, had little impact on the design except in two important areas: the instrument controls and displays and real-time data fed back to

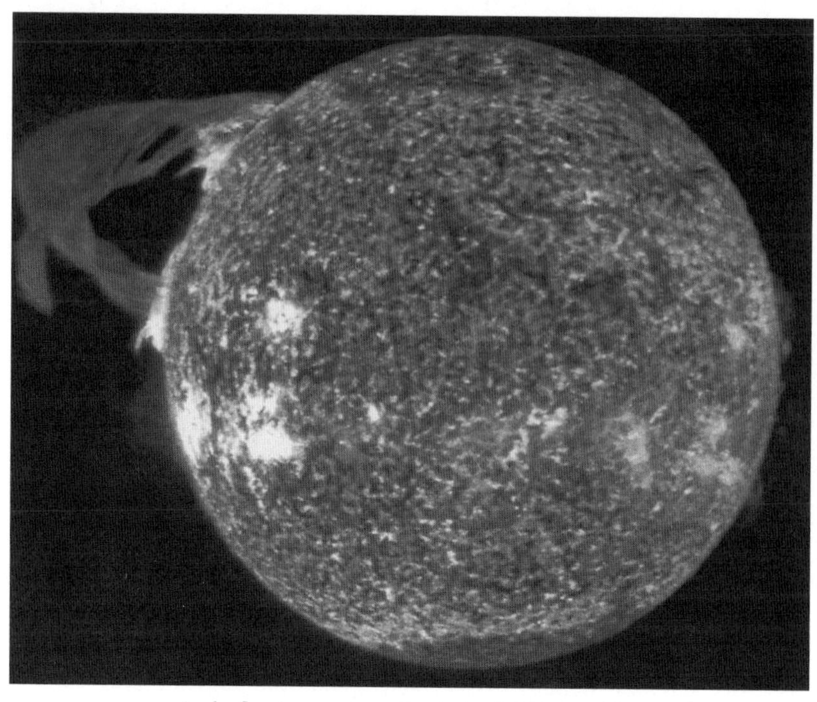

45. A solar flare image captured by the Apollo Telescope Mount.

the observer. As the integration of the observer into the design matured, the mandate for an integrated, simultaneous operation of the individual instruments was manifested in the instrument control and display (C&D) design. Design engineers each wanted their own C&D real estate for their instrument to do with as they saw fit. "We, the crew, wanted to stress the functional similarities in the way each instrument was operated as well as the sequence of each control operation," Ed Gibson explained. "After 'spirited discussions' and hard work on both sides, we arrived at a 'functional (vertically aligned)—sequential (horizontal order of operations)' layout. For both training and in-flight operation, this layout proved to be highly effective."

With today's computer capabilities it is hard to imagine that many of the operational modes and exposure times for the instruments had to be frozen into the design many years before launch and could not be changed as knowledge increased before and during flight. However, the PIs did an exceptional job of incorporating a wide range of possibilities into their design.

Despite the thought, sophistication, and expense that went into the observatory, the crew still managed to demonstrate the adage, "When the going gets tough, the tough use duct tape." Crib sheets, Polaroid cameras, and food containers were among the many items that were accommodated at the C&D panel by the astronauts' universal tool—duct tape. However, when it was taken outside on an EVA, the glue on the tape was either frozen solid by the cold blackness of night or turned to soup by the heat of direct sun.

2. Observer Feedback

"The flight crewmen were provided an exceptional array of highly useful data," Gibson explained. "We had displays in white light and at a wavelength called hydrogen-alpha, both very useful and also visible on the ground. We were provided individual displays of the sun in extreme ultraviolet and X-rays as well as a display of the corona. Lastly we had a highly versatile display in the ultraviolet that allowed us to read the intensity of radiation across the ultraviolet spectrum at a point or over a region of the sun at a single wavelength."

Although all PIs firmly supported the central role of the in-flight observer in the data acquisition, the astronauts had an insatiable appetite for display of far more real-time data than the instrument designers could afford or provide without degrading reliability. In the design phase a classic chicken-and-egg controversy ensued: "Why do you astronauts need to see the atmosphere of the sun in real-time? We've never seen any rapid changes from down here!" versus "But we don't know for sure what the sun's atmosphere is doing unless we look at it. If we only take data at preprogrammed times, we're likely to miss some very important observations." Over many months the spirited discussions oscillated between lofty observational philosophy and detailed nut-and-bolts design. Fortunately, the instrument designers were able to make a few accommodations, which, knowing what we know now, produced some very important real-time understanding of the sun for both ground and in-flight observers and the capture of data on events that otherwise would have been missed.

3. Optimum Observer Response

Given the exceptional instrument array and real-time feedback on solar structure and events, it was then up to the flight crew to respond in ways that maximized the new and important information in the returned data. Thus, the task before flight was to provide the observers with the best possible training. But the training program had a few challenges: the time available was limited by the approaching launch dates and demands of other mandatory training and by the backgrounds of the observers ranging from test pilots with little science expertise to scientists who had a good understanding of physics and some of solar physics. In the end each crewman had to possess a working knowledge of solar physics and an expertise in performing the ATM tasks. The first challenge was a given, and all that could be done was done to carve out as much time as humanly possible. In meeting the second challenge, each and every potential observer demonstrated the nature and mental disposition to maximize their learning. Some understood that they would be expected to operate the instruments with negligible errors as a highly skilled technician. Others understood that they would be required to be not only highly skilled technicians but to also alter the in-flight operations in real time as their scientific judgment dictated. All nine crewmen who flew responded to the very best of their individual abilities.

Along with the excellent procedures trainer that was provided, a major amount of solar physics classroom instruction was accomplished. "Our classroom trainer was a godsend: Dr. Frank Orrall, a practicing solar physicist and observer, a highly dedicated instructor, and a man of exceptional humor and patience," Gibson recalled. "His knowledge and enthusiasm left their marks on every one of us. Upon the conclusion of his instruction, he was presented a picture signed by each of the crewmen he instructed. The photo was one of the whole sun that clearly displayed its supergranulation, the large, nearly circular cells that crowd together on the surface. Some of the crewmen who had previously made lunar flights labeled a few of the cells with the names of craters they had studied on the moon—a way to pull Frank's chain about how much they had learned about the sun. He loved it."

4. Flexible Observatory Operation

At the beginning of every orbit, the crews had a planned set of observations but also the freedom to deviate if they saw a more information-bearing feature or event on the sun's surface; that is, they approached the observatory operations with an open mind but not an empty one. The planned observations, which were usually sent up from the ground on a teleprinter pad, were organized into Joint Observing Programs that defined how each instrument was to be operated as a particular feature was observed, such as a solar filament, bright point, or active region. They also had a JOP to cover the occurrence of a flare, a very time critical and film-consuming set of joint observations. Very useful background data on the solar state was also provided from NOAA (National Oceanic and Atmospheric Administration) and the PIs. Periodically they also had voice conferences with the PIs, which turned out to be useful but rather stiff because of the many restrictions on communication with the crew. Fortunately these restrictions have softened since then, and Shuttle crews are now able to discuss joint experiment concerns and procedures with ground-bound observers in a less formal and restraining way.

The rapid rearrangement of magnetic field structures on the sun often leads to large explosive flares that are accompanied by large coronal mass ejections. Large masses of gas are hurled high into the solar atmosphere sometimes with enough energy to escape the sun entirely. Occasionally some of that mass in the form of high-energy particles enters the solar wind and subsequently rains down on Earth, causing the northern and southern lights (the aurorae) and major disturbances in electrical distribution grids. The ejection of these masses upward through the corona was dynamic, majestic, and very rewarding if captured in the data from beginning to end. On the third Skylab manned mission, a major CME was recorded from its inception because of a real-time tip from an observatory in Hawaii that saw a large prominence start to lift off. Later in the mission, the largest such event was observed by Skylab. "This liftoff of a major arch of gas, which covered one-eighth of the solar circumference, has become an icon of the Skylab solar observations," Gibson explained. "Much to our embarrassment, it was all recorded by the ground's remote operation of our instruments as we floated in our sleep."

The grand prize for any Skylab observer was to record the birth and life of a flare. All the clues on how and why a flare occurs are revealed by the details of its inception. "Our flare warning systems told us when a flare was occurring, but not when and where one was about to occur," Gibson said. "Also, since the JOP for a flare demanded a high burn rate of the limited film housed in most instruments, a shotgun approach was unacceptable; we had to find a way to pick off one with a rifle. The answer lay in patience and close inspection of the most energetic active region as seen in X-rays and the XUV. As the magnetic field structure of an active region became more unstable, one or more bright points would surge and pulsate in intensity. It gave one the impression of a pot of water just starting to boil. The trick was to pick the right bright point, then the right time to call a surge in brightness the early eruption of a flare. This technique practiced on the third mission rewarded us with the capture of a flare rise just as were about to conclude our observations and come back home. It was rewarding yet frustrating—why didn't we fully develop this technique earlier?"

Though certainly not designed for it, the ATM took advantage of a real target of opportunity: Comet Kohoutek. The comet was much fainter than anticipated and certainly very much fainter than the bright solar features that the ATM was designed to observe. "Nonetheless, we did get some interesting yet faint pictures, some with the coronagraph as we pointed at the center of the sun and some as we pointed the whole Skylab cluster at the comet before and after it swept around the sun," Gibson said. "These later maneuvers were cumbersome (twenty keystrokes were required for a single maneuver) yet a testimony to the ingenuity of the ground control team that we could make any off-sun observations with the ATM at all!

"In addition to what we could capture on film, we recorded on paper what our most sensitive and versatile optical instruments onboard could detect—the human eye," Gibson continued. "The comet came out from behind the sun on the day we had scheduled a spacewalk to replace the ATM film. Even with the strong filtering of our space helmet visors, the spike of brightness that pointed at the sun and away from the tail was evident. Over the next week we monitored Kohoutek, especially its sunward spike, and made sketches of what we saw, which are now on display in the Smithsonian in Washington DC."

Ed Gibson said, "The operation of the ATM observatory was complex, exhilarating, frustrating, rewarding, tiring, and totally absorbing. All reference to where the space station was over the ground would be lost; only the time remaining in daylight was of importance. The C&D panel was complex, demanded one's full attention, and invited errors—even after all the effort that went into its design. I sometimes forgot some of its nomenclature, even though I was central in the design process. Some of the readouts were in decimal (base ten) and some were in octal (base eight), which could also cause confusion. The combined procedures were sometimes very complex or required alerts and timers to remind the observer of actions to take or interlocks to make sure the some actions were not taken. During design we all tried our best to put these alerts, timers, and interlocks in place, but we fell short of optimum. Also the more an observer knew about solar physics, the larger the dilemma he faced: 'Do I use some of our valuable time in daylight to search the sun for potentially more rewarding targets than sent up from the ground, or do I just punch the buttons on cue as requested?' The compromises made were often followed by many could-of's and should-of's.

"And yet, as the weeks went by a simplicity of operations emerged for me: If one fully understood the capabilities of each instrument, the physics of the sun's surface, and the needs of each PI, the JOPs could be pushed into the background. The task really became one of matching the state of the sun's surface with the capabilities and needs of each instrument; that is, the sheet music (JOPs) were put away and the ATM played by ear (full utilization of one's knowledge and intellect). Of course, I never took this extreme latitude on those days that the ATM was scheduled to be operated. However, on Sundays, our day off, I chose to give the PIs some bonus data and operate the observatory in this way. I felt somewhat like a piano player in the silent movie theater; the instrument was played to match the visible action. After six to seven continuous orbits of observation, I felt exhilarated yet drained, rewarded yet frustrated by what was left undone.

"Of course, each of us could have performed better with more in-flight time, more training, additional displays and interlocks, and more direct communication with the ground scientists. Even so, the ATM observatory operations were a milestone that far surpassed the contributions that a scientific operator in orbit had so far demonstrated and set the bar high for future human utilization on space missions."

The Future

Despite the extraordinary effort, sizeable investment, and success of the ATM solar observatory, the question remains: For future solar observatories in Earth orbit, should not the observers and instrument operators remain on the ground? After all, several unmanned solar observatories have flown since the ATM and made exceptional scientific contributions. Gibson believes the answer depends on the nature of the flight opportunity, the seriousness with which a manned solar observatory mission is approached, and several other factors.

Certainly electronic data collection capabilities and air-ground telemetry rates have seen explosive growth in the past few decades. Also, except for repair and instrument upgrades, such as utilized on the Hubble Space

Telescope, the expense, extra complexity, use of less-than-fulltime and best-qualified solar physicists as observers, and other restrictions of manned missions argue in favor of the observer remaining on the ground. However, if a manned mission will be in orbit and solar observations can be accommodated, the lessons of ATM are applicable. The International Space Station might present this type of opportunity—if it can be continuously manned by six to eight crewpersons in total with at least three of them full-time, best-qualified solar physicists who are devoted 99 percent to observations. This situation will not likely become a reality unless cheap, frequent, and dependable transportation to and from ISS becomes a reality.

The extrapolation of the lessons of ATM suggests the inclusion of:

A routine observations program with a prioritized shopping list of targets of opportunity and the freedom to modify operations as judged best by the operator.

A dedicated observatory with round-the-clock operations and stable solar pointing on the day side of the orbit.

Full-time dedicated observers who are best qualified to operate the observatory whether in flight or on the ground.

Dedicated continuous communication loops with ground scientists for two-way, free exchange of data and commands.

Stability, instrument resolution, and display resolution that matches the best available capability (currently approaching 0.1 arc second).

Instruments that synergistically cover the visible down to the x-ray range of wavelengths.

At least one instrument that observes the sun's magnetic field, which drives all solar phenomena in and above its surface.

At least one instrument that observes the Doppler shift of several wavelengths to detect line-of-sight velocities at various heights in the solar atmosphere.

Onboard quick-look capability for most data sent to the ground.

"Unfortunately, considering our current manned spaceflight programs and proficiency, it is not likely that the opportunities and capabilities for a manned solar observatory are likely to materialize in the near future," Gibson said. "Thus, the ATM mode of operation should be viewed as a rare milestone that will not be soon duplicated or surpassed.

"However, two general conclusions can be drawn from the ATM experience. First, mental challenges of the type offered by the ATM operations are essential on long-duration flights if for no other reason than for intelligent and motivated crewmembers to retain their mental sharpness and positive outlook. Second, there is no good reason that Nobel Prize–quality science, utilizing the space environment, cannot be accomplished in an orbiting laboratory just as we realize in our best laboratories here on Earth."

Garriott would prefer a more modest (and perhaps realistic) goal for the scientifically trained crewmember. From his perspective, and thinking in terms of the next fifty years or so, spaceflight is still expected to be a marvelous, but seldom encountered, personal opportunity. It seems more likely to him that scientifically trained people will be most valued as generalists and not specialists in one (or even two) disciplines. They will be needed as observers working in close cooperation with the best researchers around the world, helping them in conducting their specialized activity. This is not unlike the roles of the Skylab science pilots but extended as hardware capabilities and knowledge expands. While Nobel-competent astronauts are not to be excluded, he believes their "Ah-ha" insights leading to new scientific discoveries and even a Nobel nomination are more likely to arrive in quiet contemplation near their home office or in team meetings with their fellow specialists in interdisciplinary discussions on the ground.

Habitability

The first two major priorities of Skylab were space medicine and solar physics. The third was habitability. How was a space station to be designed so that humans could live and work in it effectively?

Engineers had been designing spacecraft for human operability since the Mercury program, but they hadn't yet made a methodical study of the subject. Skylab was their opportunity because of its generous weight and volume available and its long missions. The opportunity was recognized and taken.

Two experiments were submitted and approved. The first was M487, habitability/crew quarters, the purpose of which was "to measure, evaluate and report habitability features of the crew quarters and work areas of Skylab in engineering terms useful to the design of future manned spacecraft." Its little brother was M516, crew activities/maintenance study, designed "to evaluate Skylab man-machine relationships by gathering data concerning the crew's capability to perform work in the zero-g environment on long duration missions."

Plans for gathering data were made, prominently including crew voice reports and questionnaires, film and video, and measurements of how long tasks took to accomplish. Tools were provided to measure light, sound, air movement, temperatures, and forces. Procedures were tested during the SMEAT simulation and improved. The engineering approach suited the crews, who recognized the importance of the effort and were glad to furnish their evaluations and opinions. The experiment was continued into design. Many different types of restraints, handholds, equipment tethers, and even door openings were provided, so that the crews could show and tell which worked and which didn't.

It's been said by some observers that the astronauts were constantly complaining about Skylab systems and accommodations. Actually, they were doing their jobs. They thought it more important to describe inefficiencies and suggest solutions than to praise successes, although there was plenty of the latter, especially postflight. And the human-factors engineers behind the experiments, ably led by Bob Bond, knew how quickly memories fade and wanted the evaluations during the missions, not after. So there was a lot of air-to-ground communication on how things worked and how to make them work better.

The results, gathered over the ensuing two years into "Skylab Experience Bulletins," filled seventeen volumes.

Two very important, unanticipated observations were made about the crews' bodies: first, they became taller by one to two inches; second, they adopted a characteristic "zero-G posture," flexed at knees, hips, and neck. These two facts greatly affected the way the people fit into space suits and at workstations of all sorts and led to some important recommendations for future spacecraft. One was "No Chairs!" The body doesn't want to flex into a chair; it's uncomfortable and unnecessary. (On the space shuttle, chairs are used for launch and entry and stowed in orbit.) Don't make an individual crouch at a workstation by putting key instruments below his or her eye level; it's not feasible to slump. Without gravity to help, bending over to tie your shoes is harder. And size the space suits and clothing with some extra length. Future engineers would look with interest and amusement at the photo of the fifth percentile girl beside the ninety-fifth percentile guy, and spend some extra time designing one workstation that will fit either.

If chairs are no good, how can you restrain a person to do a task? Foot restraints are the answer, and the best ones give a firm purchase that allows both hands to be occupied with the task. More casual restraints were OK for one-handed tasks.

There was quite a bit of debate before flight about whether people would feel more comfortable in a work space that looked like an Earthly room with floors and ceilings and all the signs oriented the same way, or whether in zero-G they could operate nicely on walls and ceilings, allowing space to be more effectively utilized. The answer to both questions was "yes." The report says: "The Skylab crewmen were able to operate equipment easily from any orientation. They quickly established their own coordinate system in which the location of their feet signified 'down.'" But the one-gravity architecture of the OWS crew compartment was preferred to the put-it-anywhere radial arrangement in the MDA. The latter was a bit disorienting when you entered it; it was OK once you'd reached your workstation. Within a workstation everything had to be oriented to the same "up."

Improvements were suggested in the overall layout of Skylab too. "Don't ever put the airlock in the middle again," was a good example. Having the airlock right on the main pathway between the systems center in the MDA and the work and living center in the workshop meant that nothing could be stowed in the airlock lest it impede traffic. It also meant that if the airlock hatch wouldn't close after a spacewalk, the workshop became uninhabitable and the mission was over.

This work, combined with systems evaluation, resulted in a very thorough set of design criteria for future spacecraft, especially permanent space stations. It had some impact on Shuttle design, although Shuttle was already well into its design process when the Skylab results were promulgated in 1975. It had considerable impact on the European Spacelab module in whose development several Skylab astronauts participated. And it significantly assisted the Russians in the design of Mir, although no Americans were invited to take part directly in that space station's initial design. Later, of course, NASA astronauts would live and work with their Russian counterparts aboard Mir.

In the eighties the Skylab human-factors lessons and several other sources were combined into one massive habitability design book, NASA *Standard 3000*. It was used in the design of Space Station Freedom and its successor, ISS. What other lessons from Skylab were not learned by the designers of ISS is material for another book.

Student Experiments

"The Skylab Student Program came into being because some of us involved in the space program were concerned over the decline in interest of our youth in science and engineering fields in 'post-Apollo' days, the first moon landings in 1969 having recently been made," said Jack Waite, who served as the student experiment project coordinator while at Marshall Space Flight Center. "A number of NASA headquarters and field center personnel (MSFC and MSC, now JSC) discussed ways to stimulate the American youth interest in these fields. It became apparent that they could, even should, be an integral part of the Skylab Experiment Program. At headquarters Joe Lundholm was a key player, along with the program directors Bill Schneider and John Disher, Reg Machell at MSC, and myself from MSFC."

Waite's position as head of the Experiment Development and Integration Office at Marshall and also as a representative on the Manned Space Flight Experiment Board allowed him to facilitate this work over the next several years and to follow up the student careers for decades to follow.

The program developed was a nationwide contest for seventh- through twelfth-grade students and patterned very much like formal university spaceflight projects are selected. A "Request for Proposals" was sent to all fifty states and nine overseas high schools. These proposals were to be related to research experiments in seven basic areas of study: astronomy, botany, Earth observations, microbiology, physics, physiology, and zoology. The Marshall Skylab Experiments Office, which Waite headed, was designated to manage the student program. A total of 3,409 proposals were submitted and evaluated by the National Science Teachers Association, with participation from National Council for the Social Studies, NASA, and even Skylab flight crewmembers. From that total, twenty-five were selected as winners and twenty-two ended up being incorporated into the Skylab flight program, some carried out on each of the three missions. Most involved some modest hardware elements that were designed by the student principal investigators working with engineers at Marshall and structured as would be any professional program. Crew safety was always considered as well as

46. Students chosen to participate in the Skylab science program gather on the steps of Marshall Space Flight Center's headquarters building.

compatibility with all other experiments and systems. Design reviews and mission operations were reviewed, as were all other experiments by the Marshall Experiments Office.

The student program was and is considered a tremendous success. A yardstick for success should be the degree to which student interest and enthusiasm has carried over into career activities. While it is not possible to objectively measure the degree to which high-school students nationwide may have been moved to a better science and engineering awareness and career involvement, it is possible to follow most of these particular students, which Waite has done with great perseverance for decades. Among the twenty-two projects which ended up with approved and flown activities, the student PIs achieved careers as follows: six became science teachers (elementary, secondary, and university levels), seven were engineers and/or scientists, three became medical doctors, four had business careers, one was a military officer, and one became a monk. Many university advanced degrees are present in their biographies as well.

"Likely the most publicized experiment performed was the one proposed by Ms. Judith Miles from Lexington, Massachusetts, to observe how 'Cross' spiders spin their webs in space," Waite said. "At our Skylab thirtieth anniversary celebration in 2003, we were pleased to have Judy and some of her family in attendance. The widespread publicity associated with the two spiders' adaptation to weightlessness apparently so fascinated the general public that many adults still remember the experiments conducted over thirty years ago and have related their stories to new elementary students, so that they also now ask questions about the experiment."

But some youngsters well before high-school age were already thinking about the mysteries of space and the sun. One of the youngest was Amy Eddy, the then-seven-year-old daughter of Jack Eddy who was a coinvestigator and the crews' teacher in solar physics. His daughter's poem was published in the NASA science publication *A New Sun: The Solar Results from Skylab* along with a drawing:

The Sun

how did the sun get in her place
with her round and shiny happy face
who cast the shadows high and low.
I do not know, I do not know.

Science on the Cheap

One example of the diversity of the scientific experiments on Skylab was a contribution by rescue and backup crewmember Don Lind. During Apollo, Lind had worked with Dr. Johannes Geiss of the University of Bern in Switzerland on an experiment that would use thin metal foils to capture solar wind particles on the lunar surface. That relationship would carry over into the genesis of a similar experiment on Skylab.

"When Dr. Geiss came over for the first Apollo launch," Don Lind said, "he stayed at our home. And he and I were over in the simulator building one day, standing in the Skylab mock-up that was getting ready for this next mission series. And one of the two of us said, 'Can we do any good science on Skylab?'

"We said, 'Hey, we could put some of those foils on the outside struts that hold up the ATM' and we could pick what were called precipitating magnetospheric particles. These are particles that were coming down the magnetic tail of the Earth, headed toward the Earth. When they struck the Earth's atmosphere, that's what caused the aurora. Now the assumption was those may be solar particles, because the energy that they had was exactly what the dynamo effect would produce from the solar wind.

"We proposed that this be added to the list of experiments, and it was called S230. Every NASA center had to evaluate all of the experiments, so they sent out this proposal to all the different space centers. Every space center said, 'We recommend that it not be approved, because all the other experiments were approved a year and a half ago, and it's simply too late.'

"But I knew exactly what had to happen. It was to be mounted on a fabric background, so I went over to the guy at Marshall, and said which fabric is the best?' He said 'armalon,' so we proposed armalon.

"I went down to the Cape and said, 'What is the best installation sequence that will not give you guys any problems?' He said if we would install the sleeve at one point in the countdown and the foil panels at a different time, there would be no stowage problem. That is exactly what we requested.

"So nobody had any reason not to approve it, except that it was, quote 'too late.' But there was really no technical objection to it, so it was proposed, and it was the last experiment that came in.

"All the guys had to do on the flight was, every time they went out to pick up the ATM film canisters, they had to pass this spot on one of the trusses, and they'd just take off the next foil and bring it in and bring it home and that would expose the next foil below it.

"A typical space experiment costs a million dollars or ten million or something, and it usually weighs a couple hundred pounds, and it takes several hours of astronaut time. Well, our experiment cost $3,500, and it weighed less than ten pounds, and all it did was require an astronaut to take literally thirty seconds to pick up one of these things on a traverse that he was going to do anyway. And we made a significant scientific discovery for $3,500.

"Kenny Kleinknecht, who was the head of the Skylab program at that time, said to me one time, 'Lind, this is my favorite experiment.' 'Well, why's that?' 'It's the cheapest.'"

12. What Goes Up . . .

"In April of 1982 I was lucky enough to be assigned the job of NASA senior science representative to Australia—'NASA Rep,' the Aussies called it," Joe Kerwin said. "So I got on the plane in Houston, and some twenty-two hours and three stops later dragged my weary body into the Canberra airport terminal.

"My new secretary met me in the official Ford, and we decided to drive to the office before she deposited me at the motel. 'This is your desk, mate.' she said—and the phone rang. 'Hello,' I yawned, 'NASA representative.'

"'G'day, mate,' said a voice. 'I've got a bit of Skylab here and was curious; were you still buying them back?' Incredible! Not in the country two hours, and Skylab, my triumph and embarrassment, had followed me there like the bottle imp.

"By the way, the chap, from a small town near Perth, didn't have a bit of Skylab. He had a great chunk of it, an intact oxygen tank about eight feet long and well-charred. We informally certified it for him, but NASA had enough samples and didn't want this one. As far as I know, it's still adorning the entrance to his pub. Skylab had returned to Earth, sure enough."

Although the Skylab program had officially been designed to include only three manned missions, there had been discussions of other possibilities, from reboosting the station to launching the second Skylab workshop. Even NASA's space-race rivals were intrigued by the possibilities the facility offered. If representatives of the Soviet space program had had their way, a much larger-scale version of what eventually became the Apollo-Soyuz Test Project would have involved both nations' space station programs as well.

Marshall's George Hardy recalled that before the launch of Skylab, James Fletcher, the NASA administrator, met with the head of the Soviet space agency to discuss the possibility of cooperation in the event of an emergency in spaceflight, among other topics. "So we got an invitation to go to Russia in 1969, and Bob Gilruth was selected to head that delegation, and then my name was on there," Hardy said. "I got selected to go because nobody really knew what the Russians wanted to talk about, and there was some possibility that maybe they wanted to talk a little about Skylab, although our management was not terribly interested in talking about Skylab."

The meeting was cordial but unproductive, so it was agreed that a Soviet delegation would come back to the United States the next year for a meeting at MSC to further discuss rescue mission cooperation. "The discussion had been very general about rescue capability—what would we need: common docking systems and things like that and rendezvous capabilities and all that sort of thing," he said. "And it was very generic discussion.

"About the third or fourth day, they came in and the head of their delegation opened that meeting that day and said, 'We'd like to make a proposal,'" Hardy said. "And he just laid out flat a proposal where there'd be two missions: one where Soyuz would come to the Skylab, and one where the Command and Service Module would go to the Salyut."

Under the plan, the Soviets offered to host the first mission on Salyut, and then their mission would be flown to Skylab, which at that point NASA believed was about a year from launch. While the delegations were in the next session, Gilruth called headquarters, and NASA deputy administrator George Low flew to Houston. "Basically the bottom line was that we didn't want to have a mission with them with Skylab," Hardy said. "We didn't want to complicate Skylab to that extent. We thought it would delay Skylab. Turns out Skylab was delayed anyway."

The NASA delegation returned to the meeting with the Soviets and explained that they didn't believe it would work out to add such a major new element to the Skylab program. There were already obligations, they explained, to principal investigators that wouldn't allow for such a substantial change to the program timetable. The Soviet delegation agreed that was understandable and suggested that the joint mission could be flown the with second Skylab station, which they knew had been constructed. The NASA delegation explained that there were neither plans nor funding for the launch of the second Skylab.

Savvy to the U.S. system of allocating budget funds on a year-to-year basis, the Soviet delegation said that they understood that the Congress hadn't appropriated the Skylab-B funds and that they could wait until that happened. "They were told no, that wouldn't happen," Hardy said. No matter how the NASA delegation tried to explain it, he said, the Soviets wouldn't believe that a space agency would build an entire space station with no intent to fly it.

"They couldn't believe it," he said. "And it was almost like—they didn't say this—but you kind of got the impression they were feeling like, 'These fellows just don't want a space station with us.' There has to be another reason, and that's the reason they would assume."

With a trip to Skylab off the table, the talks of any sort of joint space station operation fell apart. "It just wasn't the time for it. And once that didn't happen, then the trip to Salyut didn't happen either, and that's how we ended up with ASTP," Hardy said. "We didn't want to go to a Russian quote, 'space station,' and the Russians didn't want us to come to theirs if they couldn't come to ours." Though the talks about the joint space station operation ultimately proved unsuccessful, Hardy said he has many vivid memories of the process.

"My introduction to that, I can remember it well," George Hardy said, "I got a call from [Marshall deputy director Eberhard] Rees. I was here doing my job, and this was one morning at ten o' clock or something. He said, 'Can you be in Dr. Gilruth's office this afternoon, by three o'clock?' I said, 'I guess so, I'll see. What am I going for?' He said, 'Well, they'll tell you when you get there.' That was strange. So anyway, I did; I caught a plane. I got down there; I walked in that office and introduced myself, and [the secretary] said, 'Oh yes, Dr. Gilruth's expecting you soon. Have you got your tickets for Moscow yet?' And that was the first time I knew I was going to Moscow.

"I really had a good time with [Gilruth]. We were coming back from that trip over there in '69. The State Department had briefed us, and things were pretty contentious back in those days between the two countries. One of the things they told us was typical I guess—they told everybody that went over there, don't lock your suitcase, 'cause they're going to open it up anyway. Don't lock it; they'll break the lock on it. We all tried to adhere to that, but Gilruth evidently forgot it or something. They broke in, busted the lock on it.

"Here we were coming back through Heathrow Airport, and Bob Gilruth had on his cowboy boots and that ten-gallon hat. He was carrying a suitcase that was tied up with a rope around it, and he had a big bag under his arm; he'd bought a fur or something for his wife over there. It looked like Texas walking down through there.

"When they came over here there must have been twenty, thirty, forty, or fifty of them. I know they had a whole wing at the Holiday Inn down there. And they had buses to take them to the Galleria, shopping. They loved to go to the Galleria. They'd buy flashlights and flashlight batteries; that was their favorite thing to take home."

Hardy said that he has often wondered since then what would have happened if things had worked out differently. "I don't know what that would have cost," Hardy said of the Skylab-Salyut proposal. "I don't know how complicated that would have been. I don't know what the political payoff of something like that would have been, or the scientific payoff, but that would have been a real joint mission, a real joint mission."

In fact, he said, depending on how interested the Soviets had been in the program, it might have been possible to work out an arrangement that would have allowed the second Skylab to be flown. Funding issues aside, one of the biggest issues facing a second Skylab program was the limited number of U.S. Saturn launch vehicles and Apollo spacecraft remaining. Supplementing those with Soyuz rockets and capsules in a cooperative program, Hardy noted, could have opened up new possibilities. "For example, you could have used Soyuz preferentially—four of theirs to our last two Command Modules, all sorts of options," he said, adding, "They just seemed amazed that we built an entire vehicle and wouldn't fly it."

Scientist astronaut Phil Chapman was a member of a committee started after Apollo 14 to study space station possibilities. He was part of a group that advocated launching the Skylab-B workshop after modifying it to make it refurbishable. "The modifications to the workshop mostly involved provisions for replacing consumables," Chapman recalled. "I think the most costly change was mounting the CMGs in palettes so they could be replaced when necessary. As I recall the additional cost was about $50 million in 1970 dollars. We could have had a permanent space station in 1975 with more real utility than the ISS for a total cost twenty times less. Once that was up and running, our proposal was to build a reusable crew transfer vehicle, launched initially by a Saturn IB, and then to work on a reusable flyback booster."

Chapman said that the possibilities posed by the Space Shuttle were seen by those in charge as making this proposal unnecessary. He said that NASA's decision to pursue the Space Shuttle was one of the principal reasons he left the astronaut corps in 1972 rather than waiting for a chance to fly.

Plans for the future utilization of the Skylab hardware were not to be. When the Skylab III crew closed the hatch as they left, it was to be the end of the operational program. On 9 February 1974, just one day after the return of Jerry Carr's crew, Mission Control did some final systems testing, maneuvered Skylab to a gravity gradient attitude (perpendicular to Earth, small end up and workshop end down, an orientation in which it would wobble but remain pretty stable without the need for electrical power or propellants) and turned off the power.

Experts at Marshall forecast that Skylab would descend from its end-of-mission altitude (about 235 nautical miles) only very gradually. If nothing were done, drag from the very thin atmosphere at that altitude would inexorably pull it down, and it was estimated that reentry and burn-up would occur around March 1983. This estimate was based on the average density of the atmosphere at various levels of solar activity. It was known that between 1974 and 1980 solar activity would be increasing, approaching solar maximum—a time when sunspots, flares, and the ejection of solar particles to and beyond the Earth would be more frequent. Increased solar activity heats the upper atmosphere, causing it to expand. The slight increase in density at Skylab's altitude would increase drag, causing it to descend more quickly. All of this was factored in.

But the rise in solar activity during the 1970s was far from average. It was the most active solar cycle ever recorded with modern instruments. And through the years from 1974 to 1978, NASA and NOAA differed in their forecasts. NOAA was forecasting higher drag than NASA.

NASA's plan when Skylab was deactivated was to visit it again when the Space Shuttle was operational, and the first flight had been scheduled for early 1979—plenty of time. A remotely operated system would remove a propulsion module from the orbiter and place it in Skylab's docking hatch, either to boost it to a high, safe altitude while plans were made to somehow activate it or to deorbit it safely to a remote ocean. Neither component was yet being built because of NASA's very tight budget.

But as the sun gradually began to move Skylab's demise earlier, the Shuttle schedule began to move later. An early 1979 launch began to look risky. Skylab would have to be visited earlier than the fifth flight, during the so-called test phase of the program, and Shuttle management didn't want to risk that. Could anything be done to keep Skylab aloft longer? Maybe something could. In February 1977 a team of eight engineers—four from Marshall and four from Johnson—went to Bermuda to try to wake Skylab up. Bermuda was the only NASA ground station that still had command capability using the "old-fashioned" UHF radio band.

As Bill Chubb, at that time leader of the Support Team for Attitude Control at Marshall, explained, "Four years had passed, and we had no idea of the condition of any of the systems nor did we even know if they were commandable from the Skylab ground station network. It was critical that we establish communications, interrogate, and activate these systems to facilitate a controlled reentry.

"In order to evaluate what options were available to us, the state of the onboard systems had to be determined. Ground tracking told us when Skylab would be within communication range. Onboard batteries of the power system were most likely fully discharged. Power would be available on the vehicle only when the solar panels were pointing toward the sun. There was no way of knowing if its attitude would be such that the solar panels would be pointed toward the sun during the passes over Bermuda, making power available to the onboard telemetry system. Even with power available we did not know whether it was operable. On March 6, 1978, as Skylab passed within range of the Bermuda Ground Tracking Station, the onboard Skylab Airlock Module command and telemetry system was commanded 'on.' Numerous 'on' commands were sent until at last data from Skylab came into the Bermuda Station. It was a moment none of us would forget!"

Charlie Harlan was appointed by Chris Kraft, director of Johnson, in 1977 to create and head up the Skylab Reentry Flight Control Team. He recalled: "We'd send a command to charge the batteries, but there had to be juice on the bus for it to be received. So those guys just sat out there, sending command after command after command. Eventually one would get through." With persistence the remaining good batteries were finally recharged, and Skylab was ready to be commanded out of its passive gravity-gradient attitude into one that would enable control of drag. What attitudes could the aging control system sustain?

Now that control of Skylab attitude had been reestablished, what was the desired direction in space? The lowest drag would make the station travel like an arrow (end-on) and provide the longest lifetime. But to best control the point of reentry and final destruction, a higher drag and shorter lifetime would be better. During this control period, Associated Press space reporter Howard Benedict (later the executive director of the Astronaut Scholarship Foundation for many years) noted that "Jack Lousma, a member of the (second) crew to live aboard the station for fifty-nine days, came by the Control Center and asked if the station could be inhabited again. Chubb said 'Yes,' there was enough oxygen and nitrogen for perhaps a ninety-day mission. Lousma noted that there probably was still plenty of asparagus aboard, too—left by past crews." It was one of their least favorite dishes!

One of the three Control Moment Gyros used to control Skylab's attitude had failed during the third crew's mission, and another had developed increased bearing temperatures—a possible sign of impending failure. There was very little nitrogen left in the cold gas backup attitude control system. Two CMGs had to be enough to control Skylab and possibly even one. Was this possible?

Charlie Harlan again: "There were some heroes in this story, and maybe the biggest ones were Hans Kennel and his colleague John Glaese at MSFC. They came up with what many of us thought was impossible—new control laws for the CMGs to control Skylab even if two of them failed, and new attitudes we could use to control drag. They brought in four guys from IBM who had done the original control system. They completely rewrote the software in Skylab's IBM computer. They wrote over all the code that wasn't needed like crew displays and controls. We turned their stuff into commands and sent it up. I remember one day we tumbled Skylab doing that. They'd sent us a matrix, and somehow the rows and columns got transposed and we sent up a bunch of garbage. But they caught it right away and we corrected it."

Since the goal at that time was to stretch Skylab's lifetime, drag had to be minimized. The attitude invented for that purpose was called "end-on velocity vector (EOVV)."

Hans Kennel recalled: "To reduce the orbital decay the attitude control of the Skylab had to be regained and the drag had to be reduced by pointing the long vehicle axis along the orbital velocity vector. Leading up to reinstatement of active control of Skylab and activation of EOVV was the discussion about the health of one of the two remaining functional Control Moment Gyros. It had shown signs toward the end of the original Skylab mission similar to the one that failed, and many thought failure of this second CMG was imminent.

"That would have left one CMG remaining, whereas Skylab needed two good CMGs for control. We were told the reactivation mission was 'off' unless we could come up with a single CMG backup for EOVV. Along with the other things described, we had to develop such a thing and we did. It would have been a 'hairy' operation, but it looked like it would work. On the basis of that, the go-ahead was given, and we even adapted the same momentum management methodology developed for single CMG control for use with the two functional CMGs.

"As a side note it was learned after the reactivation phase that the ailing CMG worked better if it was exposed to the sun and not in shadow for extended periods and so we developed maneuver plans for flipping the vehicle from orbital workshop and habitation module forward to Apollo Telescope Mount forward, depending on which end was more favorable for heating the ailing CMG. From that point on it never again showed signs of problems and worked all throughout the remainder of EOVV and TEA [Torque Equilibrium Attitude]. We developed the necessary control methods, built a simulation to verify the operation, including what kind of data the ground controllers would see, and with the help of IBM the necessary algorithms were implemented on the onboard computer.

"All this was done in record time (we were made aware of the problem on 20 March 1978, IBM got the necessary equations for the onboard computer on 26 April 1978, and EOVV attitude was successfully entered 11 June 1978). And it worked very well reducing the orbital decay. This fast response was only possible because we were fluent in APL (a high level computer language); we had all the necessary simulation components due to previous experience; practically all red tape was cut; and the official documentation was done much later."

The low drag attitude was expected to increase Skylab's lifetime by about five months, into 1980. But early in 1978, the risk of a Skylab reentry was abruptly dramatized by the Soviet space program. A Soviet spacecraft, Cosmos

954, entered the atmosphere and broke up over northern Canada, spreading nearly a hundred pounds of nuclear fuel over a broad swath of forest. Critics began to question NASA's plans. NASA assured the public that Skylab contained no radioactive material.

There was talk of extending Skylab's life by moving it into a higher orbit. Engineers looked at the possibility of launching a booster that could be attached to Skylab. The Space Shuttle, it appeared, was the best answer. Launch on an unmanned rocket would mean figuring out how to automate the reboosting. With the Shuttle, the crews could carry the booster to Skylab and attach it.

However, the same program that offered hope for Skylab's salvation also brought that hope to an end. The first spaceworthy orbiter, Columbia, was plagued during testing with the loss of insulating ceramic tiles as well as trouble with its engines. A better system for applying the tiles was needed, and the first launch attempt slipped to 1981. Skylab was going to reenter.

The control team now knew what it had to do and prepared a plan for review by NASA headquarters. First they looked at how Skylab's orbit varied as it moved around the Earth. Some orbits passed over many densely populated land areas; others spent most of the time over water and desert. Using a population-density map prepared by the Department of Defense, they estimated the population under Skylab's path for each orbit.

The next step was to forecast how Skylab might break up, how much of its bulk would survive reentry and hit Earth's surface, and over what area. An analysis had been prepared by NASA before Skylab's launch and was used. The end-of-mission Skylab weighed about 173,000 pounds, and about 50,000 pounds of that was expected to survive reentry.

Putting the two analyses together, NASA estimated that there was an average chance of one in 152 that someone might be struck by debris. But on the very best orbits, it was much less than a tenth of that. These were the orbits that passed over southern Canada, then swept southeast over the Atlantic, skimmed just south of the tip of Africa, up the Pacific and Indian oceans to cross Australia, then across the Coral Sea and Pacific Ocean until reaching North America again. If reentry could be contrived to happen east of North America and west of Australia on one of these orbits, Skylab could be safely disposed of.

So the plan was to put Skylab back into its old standard solar inertial, high-drag attitude, then carefully track what effect that drag was having on its altitude and reentry point. As altitude decreased and drag increased, it would be impossible to maintain solar inertial; asymmetric drag would twist Skylab out of control. No problem; the unflappable Hans Kennel and his team had invented torque equilibrium altitude, a variation that perfectly balanced all the forces. Nominally the point of no return would be reached at seventy-five miles altitude. At that altitude, controllers would command Skylab to turn off its CMGS; it would immediately tumble. The known, lower drag of the tumbling configuration would result in a known entry location. And by varying the altitude at tumble time, the team believed it could stretch or shorten reentry to place it on one of the five "good" orbits for that day.

Charlie Harlan recalled, "Another hero was Richard Brown, a Rockwell contractor engineer. He figured out how much power we'd need to perform each maneuver and what attitudes would achieve it. Since our power margins were very small, we'd call Richard in whenever we were planning an attitude change." Headquarters agonized over the plan. There was a faction that didn't want to give the public the impression that NASA was in full control; if the scheme backfired, there would be much blame. "This was the 'God' faction—they basically didn't want to do anything," Harlan said, "so they could blame it on God. But I'm a Deist. I believe God put us on Earth with certain capabilities, and expects us to do our best. My team and I were ready, and pretty optimistic."

Finally, the call came from headquarters. John Yardley, acting as liaison between Johnson and the administrator, approved the plan. Harlan recalled telling Chris Kraft this and Chris saying, "Charlie, you got your answer. Hang up the phone and don't answer it again."

Headquarters had insisted on using predictions of the North American Aerospace Defense Command (NORAD) regarding reentry rather than NASA's, "so that there would be one official source." The NASA team was pretty sure their prediction was better, because they had a better knowledge of vehicle configuration and drag. And in June the NASA prediction was running about two days earlier than the NORAD one. "We knew the predictions would converge as we got close," Harlan said. "But the media really wanted to be here for the big event, so we told them unofficially, 'If you don't want to miss it, get set up a couple of days early.'"

But before the final demise of Skylab, the general public had lots of advice for NASA. Headlines in the 5 June 1979 edition of the *Huntsville Times* reported that "NASA Chief [Robert Frosch] Is Chided for Skylab's Fall." When asked where he would be at the time of Skylab's return, Frosch said that "if not at NASA Headquarters, he would probably be at a BBQ in his backyard."

Congressman Robert Walker, R, Pennsylvania, was "somewhat incredulous" that nobody had given any thought at all to tell the public what to do. NASA general counsel Neil Hosenball said, "Our people are the last in the world to know what to say or how to do it [alert the public]."

Other more specific advice came from the public: "Fill a robot plane with TNT and crash into it." Or "shoot a missile at it." All these many letters were sent to William O'Donnell, NASA's director of public information, who said they were all answered. One of those giving him "most pause to compose" suggested having "the astronauts attach balloons (filled with helium) so it will float into outer space."

A New York restaurant invited people to partake of Skylab cocktails—"two of these and you won't know what hit you." A large baseball mitt was erected at Cape Canaveral to catch the station. Another radio station (KMBZ, in Kansas) offered $9,800 for a piece of the station. Beanie hats with propellers and T-shirts were sold in San Francisco with a large "X" imprinted saying, "Hit me." (There were jokes that shirts like this would keep the wearer safe—there was no way the government could actually hit anything it aimed for.)

A psychic from California somehow got Harlan's home phone number. "She called me several times with predictions. I'd say, 'How do you know that?' And she'd say, 'Numerology.' But she predicted impact on Dover, Delaware, and she never called back afterwards."

Meanwhile the press was having a field day. Some people in Washington DC, had started a Chicken Little Society. There were bumper stickers ("Chicken Little was Right!" "Good to the Last Drop"), T-shirts, slogans, and contests—a kind of gallows humor. The *New York Times* chided NASA soberly. Officials in England offered advice: "Being inside a house would protect you from small pieces. . . ."

Garriott recalled: "I was greatly amused and annoyed by what I considered to be a gross overreaction by the press and criticism of NASA. I had an interview request from one of the Houston press about it. I noted that I/we didn't invite him to drive down into our community. And since he did, he was exposing our children to a greater risk of being hit by his car than we at NASA were exposing his family to by Skylab reentry. That was not too well received and didn't make it to print. The statistics were simple and required some estimation, but I believe they were true."

On 9 July headquarters opened the Skylab Coordination Center to keep everyone informed. On the tenth the forecast was made that entry would occur the next day between 7:00 a.m. and 5:00 p.m., Eastern daylight time. On the other side of the world in Australia, headlines warned of the impending reentry. Sydney's *Sun* newspaper ran a front-page headline, three lines deep in bold letters two inches tall, "SKYLAB ON AUST CRASH COURSE."

"Skylab is on a crash course that could bring debris down on southwestern Australia, American authorities said today," the article, dated 10 July, read. "But it could still re-enter Earth's atmosphere on any of 12 final orbits—including some over Sydney. The Western Australia State Emergency Service went on full alert this morning. 'All we've heard is rumours,' SES director Mr D.L. Hill said today."

Another Sydney paper, the *Daily Mirror*, announced the same day "SKYLAB ZERO HOUR NEAR." Stacked below it was the headline for another article, informing readers, "But here's some down-to-Earth good news—$10 a week tax cut plan."

Harlan and his team stood by to make their last decision. At midnight it appeared that Skylab would reenter on the very best orbit. But the predicted debris "footprint" was immense—nearly four thousand miles long by one hundred wide because the heavier pieces would be less affected by drag and would travel a lot farther. And it looked like the western edge of that footprint might just overlap the U.S. east coast. So the tumbling command was given early, with Skylab just under eighty miles high, and the impact footprint moved east as predicted.

But things rarely go exactly as predicted. Skylab's breakup altitude had been calculated from its design structural strength requirements. The actual vehicle was stronger than the specs required. It held together longer than was calculated, breaking up over the Indian Ocean. Most of the debris fell harmlessly into the water, but some chunks fell in western Australia along a line from south to northeast of Perth. (Nine days later, Perth hosted the Miss Universe pageant, and a piece of the fallen spacecraft was on display during the event.) "Thank God—and Charlie's team—no one was hurt," Kerwin said.

47. Johnson Space Center officials and flight controllers monitor the reentry of Skylab.

A ground track of Skylab in its last hour or so of existence on 11 July starts from mid-Canada and moves easterly out into the North Atlantic. It then moves southeast into the South Atlantic, just as planned. It passes south of the Cape of Good Hope and turns northeast across the Indian Ocean toward Australia, beginning to seriously break apart and the lighter pieces to burn up. Some smaller pieces scattered down on tin roofs in Esperance and other nearby cities, but a few of the larger chunks (such as the film vault and the oxygen and nitrogen tanks) presumably carried on overhead into the Outback. Some are doubtless still there, awaiting some adventuresome explorer to find them.

The raining of debris on Australia prompted legal action—the town of Esperance fined the U.S. State Department four hundred dollars for littering. Kerwin recalled that his primary emotion at seeing the end of his one-time home was simply relief that nobody was hurt. "I think we'd seen it coming for long enough not to be surprised or regretful," he said.

In Australia the reentering spacecraft put on a show for those who saw it. The Skylab control center actually got a phone call from the captain of a commercial aircraft flying along Australia's west coast. It was night, and he excitedly described the multiple streaks of flame blazing through the sky.

That was the clue for the team to turn off their consoles and go home. Airline pilot Bill Anderson gave an even more useful visual report when he noted that he and his passengers saw separate fireballs change from a "bright blue into an orangey-red" as the debris broke up and descended into the lower atmosphere.

A woman in the town of Esperance in southwest Australia was among those on the ground who saw Skylab fall. It seemed like "a shower of sparkling lights—like a rocket—passed overhead with no sound, until about a half a minute or so, [then] there was this loud boom," she said.

Once more, giant bold letters graced Sydney's *Sun.* Above the headline "SKYLAB HITS WA STATION," was followed by a distinctly local angle to the story: "The world stood in awe today as Skylab tore itself apart in a spectacular display and spattered the Earth of a remote West Australia sheep station," the 12 July article begins. "But Noondinia Station manager John Seiler's only complaint was: 'It scared my horses.'"

When Skylab fell, Stan Thornton was a seventeen-year-old truck driver's assistant living in Esperance, a remote coastal town set in the Bay of Isles, some 440 miles southeast of the capital Perth. He calls his resort hometown "a real paradise."

That momentous evening in a region known for incredibly clear night skies, Stan traveled with his sister and some friends to a local lookout and watched in fascination as a profusion of bright, colorful man-made meteors ripped across the starry heavens, indicating the end of Skylab. The following morning his mother Elsie "went out to [their] backyard, which had only been mowed and cleaned the previous day, and found charcoal pieces spread all over the grass." After Elsie had told her son about the burnt chunks of debris, he gathered up a few sizeable pieces and went to the local State Emergency Services (SES) office with his friend Ray Rose.

Local SES manager Phil Arlidge contacted a Perth radio station at 6:00 p.m., as he'd heard about a reward on offer for the first person to deliver an authenticated piece of Skylab to a newspaper office in San Francisco within seventy-two hours. The radio station was up to the challenge Arlidge's call presented and confirmed with him that the *San Francisco Examiner* was indeed prepared to pay a ten-thousand-dollar reward—on the proviso it reached their office in America within three days. Stan Thornton was about to be involved in the race of a lifetime.

48. Owen Garriott with an oxygen tank, one of the largest pieces of Skylab debris to be recovered, at the U.S. Space and Rocket Center in Huntsville AL.

Things began to happen in a hurry, according to Stan. “The radio people were in Esperance with the help of a Swan Brewery Lear Jet within two hours. They had already contacted Qantas to arrange my ticket and passport, and the next day I flew out of Perth for the United States. In San Francisco I was greeted at the airport by Qantas manager Gil Whelan, who had arranged a limousine, and I was taken straight to the downtown *Examiner* office.”

Stan had delivered the pieces with twenty-four hours to spare and was presented with his bounty. The newspaper’s reporters wanted to know everything about him and how he had found the Skylab debris. “There was a press conference at the *Examiner* office, and after this they locked the pieces into a briefcase to be sent to NASA,” he recalled. He suddenly found himself an instant celebrity and will never forget the experience. “For someone who had not been out of Esperance, it was pretty over the top,” he recently reflected. Within a couple of days, Ray Rose and Stan’s family had joined him in San Francisco to celebrate his good fortune. NASA examined the charred fragments and Stan said they later told him the pieces were "some type of balsa wood from the insulation."

Recently moving to a new home just south of Perth with his partner Kerry and their two children, his famous "dash for cash" still brings back vivid memories for the truck driver/laborer, and despite the passage of time Stan said he still has a degree of friendly notoriety among his family and friends. "The only part of my life that really changed over here," he reflects with a shy smile, "was the word 'Skylab' was placed in front of my name. Even today I am actually still greeted as 'Skylab.'"

People said a lot of nice things about Harlan and his team after it was over—including a headline in the *Toronto Star,* "How Charlie saved Canada from Skylab!" But the kudos he remembers most fondly came in a splashdown party skit put on by his neighbors. It featured a song called, "A Salute to Charles: He Couldn't Keep It Up." An excerpt (to the tune of "The Eyes of Texas Are upon You"):

The parts of Skylab are upon you
All the live long day
The parts of Skylab are upon you,
We hope they will decay;
Did you hear the Skylab coming,
Was it a big surprise?
Little ladies in Australia
Are saying, "Damn your eyes!"

13. The Legacy of Skylab

Perhaps part of Skylab's greatest legacy is the extent to which its legacy is often overlooked. Tucked between the glory of Apollo and the much longer Space Shuttle program, Skylab's importance to human space exploration can be lost in the shadows of its older and younger siblings. However, it is a testament to how effective Skylab was in breaking new ground in human spaceflight that the lessons it taught are now largely taken for granted.

Before Skylab, however, many of the things that today seem to have been part of spaceflight forever were still largely terra incognita. Nearly all of the transition from the early explorations of Mercury, Gemini, and Apollo to the living and working in space on the Space Shuttle and the International Space Station was a result of the successes of the three Skylab crews. They truly were able to homestead the space frontier and pave the way for those that would come after them.

So What Was Learned from the Skylab Experience?

It is possible to live and work in space

If Skylab taught the world nothing else, that one legacy alone would have made possible the future of human spaceflight. Three successive new records for spaceflight duration proved that people could live in space for far longer than the fourteen-day Gemini 7 mission that had set the bar for NASA previously. The depressurization tragedy at the end of the Soviet Union's twenty-three-day Soyuz 11 mission meant that much of the information from that flight was lost. And while some basic experiments had been conducted within the relatively limited confines of previous spacecraft, Skylab would prove that meaningful scientific research could be conducted in orbit—a fact that was vital to the success of the Shuttle's science missions.

An astronaut can spend months in space and live out a healthy life

The fact that no one had ever lived in space for the duration the Skylab astronauts did also meant that no one knew what would happen when someone returned to Earth after that long. It was one thing to know that a person could live in space for that long, but unless they were able to then return safely home, that knowledge meant very little. Extensive medical baselining before and during the flight gave confidence that the crews could come home in good health, and longitudinal data collected afterwards—to this very day—monitored their recovery after the flight.

Space motion sickness is a problem, but not an insurmountable one

Prior to Skylab there was only limited awareness of the problems that can result during adaptation to the microgravity environment. Skylab certainly demonstrated that space motion sickness could be a serious short-term problem for astronauts, but it also laid the groundwork for dealing with it.

Meaningful work can be conducted on spacewalks

Today's International Space Station has myriad translation aids on its exterior to allow astronauts to move around while conducting repairs and improvements to the spacecraft. Skylab, on the other hand, had next to none, outside of the path from the airlock to the ATM. The fact that Skylab was unprepared for the sort of repair work its crews would have to conduct shows just how little was understood at the time about the possibilities presented by extravehicular activity. Everything from the construction of the ISS to the repair of the Hubble Space Telescope owes a debt of gratitude to the precedent set by Skylab. For many years the patch on NASA's EVA space suits featured three stars in its background, each representing a key EVA—one for Ed White's first U.S. spacewalk, one for Apollo 11's first steps on the moon, and one representing Conrad and Kerwin's first Skylab EVA.

A space station is a viable platform for research

There had been numerous proposals for space stations over the years, but Skylab was the first time NASA implemented one of those proposals, and the first time anyone in the world manned one spacecraft with multiple successive crews.

Skylab also marked a turning point in spaceflight. Previously the focus of human spaceflight had been working toward exploring another world. Low Earth orbit had been only a steppingstone to the moon. Skylab proved that humans could make valuable scientific contributions in orbit, paving the way for spaceflight in the coming years.

Valuable astronomy and Earth observation can be conducted from space

Skylab revealed how complex and dynamic the sun is and marked the beginning of a new era in mankind's understanding of the active nature of the sun and its relationship to the Earth. While the idea of conducting astronomical research in space did not begin with Skylab, the success of the work done there lay the groundwork for future generations of space telescopes. Skylab also laid the groundwork for a field of Earth observations research that has continued since on both the Space Shuttle and the International Space Station.

Perspectives on the Legacy

Alan Bean, for whom life on Skylab was a follow-up to his experience of walking on the moon, said that the station was a vital step in the history of manned spaceflight. "My impression of it was that it was the best possible investment of NASA money," he said. "It's so much more valuable than sending one other mission to the moon, which they could have done with that Saturn V.

"It was so much more valuable, I felt, as far as understanding the future of spaceflight and taking the next step. Taking another step on the moon would have been nice, but it wasn't going to do much different. Different rocks, nothing new particularly. And so we made a breakthrough on another branch of spaceflight, and started out in a way that gave us a foundation."

In retrospect it's easy to forget how many unknown factors the Skylab crews dealt with, how many things astronauts today are able to take for granted because the Skylab crews proved that they could be done. "My biggest concern before we flew Skylab, or anybody flew Skylab, was, at the end of twenty-eight days, and then at the end of fifty-six days, would we be strong enough to go outside and recover the film and do the work in our space suits? Now it seems like a dumb idea. But I can remember thinking that is one of the real huge hurdles that I wondered if we'd be able to do. As it turned out to be, we stayed just as strong as we were at the beginning."

Bean said that he regrets the extent to which some of the knowledge gained on Skylab was lost during the decades before the United States once again became involved in long-duration spaceflight. While on orbit, he said, the crewmembers spent time each week answering habitability questions about life on Skylab—everything from how well the equipment worked to what they thought of the colors in the workshop. Their answers were recorded on tape, dumped to the ground, and then written up. "We talked about everything you could think of," Bean said. He said that, in talking to members of the astronaut corps about the International Space Station project, he's found that the habitability reports produced based on the experiences of the Skylab crews have gone largely unread.

Bob Crippen agreed: "The one thing that I really worried about was—we did it, then we let it sit up there until it was falling out of the sky, and now we're starting to finally get a space station up there again." Crippen said that the time gap between Skylab and the first International Space Station crew was very disappointing to him. "I thought we learned a lot of lessons, and it wasn't obvious when you get that big of a gap that you can transfer a lot of knowledge. That's the only disappointment that I felt." (It is worth noting that this is not due to a lack of trying on the part of the Skylab astronauts, many of whom were involved in the planning for the later space station programs either while still at NASA or as contractors.)

Overall, Crippen said that he was very proud to have been involved in the program. Learning of the problem with the micrometeoroid shield at launch, he said, was one of the lowest points of his career, and the recovery from that disaster amazed him. "I thought we'd lose the entire mission right there when that happened. The way the team worked to pull off getting the thing back flying again I thought was fantastic." The can-do attitude that began the program, he noted, continued throughout with the work that was done during the mission on such things as changing out the gyros.

"The Russians were over here about that time, and they were impressed at how we could do on-orbit repairs, some of those kinds of things," Crippen said. "In fact, we took some of those lessons into Space Shuttle, that we needed the capability to do in-flight maintenance on things we didn't even know we were going to have to work on. So I thought it was a great program. I think everybody that participated in it will tell you that."

For Bob Schwinghamer also, the effort to save Skylab after its disastrous launch was a defining moment. "I mean, that enormous recovery effort, and the ability to pull that off—in retrospect, you just wonder how people were able to do that in the short time that they did," he said. "And then the astronauts did their thing. It was something you can remember with pride, everybody that had anything to do with it. It was an uplifting experience."

The program, Schwinghamer said, lived up to its expectations. "I think it more than did."

That sentiment was echoed by George Hardy: "I think Skylab was in some ways, without question, the best space program this country has ever had. Now there are some space programs that had maybe higher, greater technical challenges, but I guess if you were to characterize it as 'bang for the buck,' I don't think there's been a program that's gotten the bang for the buck that Skylab got."

Hardy, who went on to be involved in the space station program, also said that he has been disappointed with NASA's failure to fully capitalize on the lessons learned during Skylab when moving forward. While he noted that many parts of the Skylab experience—such as the lessons learned about crew and ground operations for a long-term program—informally influenced later programs, lessons of other parts of that experience have been largely lost.

"I don't think there was a formal, structured program for carrying 'lessons learned' from Skylab into space station," he said. "I don't know exactly why that was. I think for the most part a lot of people working on space station thought to themselves that space station would be so far advanced from Skylab that there wouldn't be any useful lessons to be learned. And that was a big mistake."

George Mueller was even harsher in his views on the subject. He was very proud of the accomplishments of Skylab—"I thought it was great"—and very disappointed with what he saw as NASA failing to learn the lessons of that program in the space station project. "The design of the station really did not take advantage of what we learned on Skylab," he said. "That was my impression the first time I walked through that mock-up; I thought, 'We haven't really learned anything.'"

In particular, he said, the decision not to use a heavy lift vehicle, which would have allowed large volumes to be launched into orbit, was a major limitation in the space station program. "Volume is tremendously important in living conditions," Mueller said. "Engineers want to build everything into little boxes, but if you're going to live in it, you'd like to have some distance, you'd like to have some privacy, and you'd like to have some things that are pleasing to look at."

Jack Lousma said that he also is proud of the groundwork that Skylab laid for the future of spaceflight. "I think we demonstrated that you could live and work in space for a long time and do a good job," he said. "One of the most important legacies was the demonstration of long-duration flight, demonstrating that not only could you survive up there but you could do useful things, that you could work up there like you would in any laboratory back home. We demonstrated that we could do zero-gravity spacewalks. It gave us the confidence to go on to longer missions, having a sense that if you stayed there a year, you'd still be OK."

Where Are They Now?

Alan Bean

After resigning from NASA in 1981, Bean devoted himself full-time to his painting. He has found a devoted audience for his work, which is based on his experiences as an astronaut on the moon.

Bo Bobko

Bobko flew on his first space mission in 1983 as the pilot of STS-6, the first flight of the Space Shuttle Challenger, seventeen years after he was first selected as an astronaut by the Air Force. Bobko also went on to command missions of his own, Shuttle flights 51-D and 51-J, both in 1985. Bobko left NASA in 1988 to work as an aerospace contractor.

Vance Brand

Despite missing out on the Skylab rescue mission, Brand would fly into space four times—on the Apollo-Soyuz Test Project and as commander of the STS-4, 41-B and STS-35 Space Shuttle missions. After the retirement of John Young in 2004, Brand became the senior member of the astronaut corps still working at NASA, serving as the deputy director for Aerospace Projects at Dryden Flight Research Center until his own retirement in 2008.

Jerry Carr

After leaving NASA in 1977, Carr was heavily involved as a contractor in providing consulting input for the development of the International Space Station. He founded CAMUS, a family owned business that combines his consulting with wife Pat Musick's art and sculpture.

Pete Conrad

After leaving NASA months after his Skylab experience in 1973, Conrad remained involved in aeronautics and space exploration until his death in 1999 as a result of injuries sustained in a motorcycle accident.

49. At the Skylab Thirtieth Reunion (*from left*): Alan Bean, Jerry Carr, Joe Kerwin, Owen Garriott, Bill Pogue, Paul Weitz, Jack Lousma, and Ed Gibson.

Bob Crippen

Crippen, who served as director of Kennedy Space Center before leaving NASA in 1995, was awarded the Congressional Space Medal of Honor in 2006 in recognition of his role as pilot of STS-1, the first Space Shuttle mission. In addition to serving as pilot of that mission, Crippen commanded the STS-7 and 41C and 41G Shuttle missions.

Owen Garriott

After flying into space one more time on the STS-9 Spacelab mission of the Space Shuttle, Garriott resigned from NASA in 1986 and is now an adjunct professor at the University of Alabama in Huntsville. He spends additional time in charitable activities and as a founder of two new businesses.

Ed Gibson

In addition to working with aerospace contractors after leaving NASA in 1974, Gibson also authored two novels, *Reach* and *In the Wrong Hands.* He recently retired as senior vice president and contract manager with Science Applications International Corporation.

Joe Kerwin

Kerwin served as the NASA representative in Australia in 1982 and 1983, then as director of Space and Life Sciences at JSC from 1984 to 1987. He worked for Lockheed from 1987 to 1996 and finished his career as senior vice president with Wyle Laboratories, retiring in 2004.

Chris Kraft

Per Kraft: "Chris Kraft is now retired and trying to stay compos mentis. He is very proud of his part in the early days of manned spaceflight and especially the Space Shuttle, which has been unfairly maligned by the media, NASA and other ill-informed engineers and scientists."

Chuck Lewis (MSFC)

After Skylab, Lewis spent a few years in the propulsion division, working on the SRB booster separation motors among other things. He then worked in crew training and interface for Spacelab missions. At retirement in 1996, he was Chief of the Mission Training Division at MSFC, where he had the pleasure of watching young engineers continue to support the flight crew with great training and man-systems design for the remaining Spacelab flights and well into the ISS era.

Jack Lousma

Lousma made one more spaceflight after Skylab, as commander of Columbia's STS-3 mission, the third Space Shuttle flight, in March 1982. He resigned from NASA in 1983. In 1984 he won the Republican primary for one of Michigan's seats in the U.S. Senate but was defeated by the incumbent in the general election. Since that time, he has been involved in several technology-related businesses and still lives in Michigan with his wife Gratia.

George Mueller

Mueller continues to be involved in the advancement of space transportation, actively pursuing the fundamental physical requirement for a viable space society: a completely reusable space vehicle capable of delivering tons of payload to low Earth orbit at a cost of dollars per pound, a goal he describes as well within our reach.

Bill Pogue

After leaving NASA, Pogue turned his attention to the next generation of space stations, serving as a consultant for what became the International Space Station, and to the next generation of explorers, making spaceflight accessible through his books *How Do You Go to the Bathroom in Space* and *Space Trivia*.

Don Puddy

Puddy served as a Shuttle flight director and as the lead flight director on the Shuttle approach and landing tests. He subsequently served as a special assistant to the NASA administrator, as the deputy director of the Dryden Research Center, and as the director of Flight Crew Operations at JSC. After retirement he succumbed to cancer.

Rusty Schweickart

Schweickart left NASA in 1977 to join the staff of Gov. Jerry Brown of California. He served as commissioner of energy in California for nearly six years then as a senior executive in several space and telecommunications companies. He now chairs the board of the B612 Foundation, which champions plans to protect the earth from asteroid impacts. He is the founder and past president of the Association of Space Explorers, the international professional society of astronauts and cosmonauts.

Bob Schwinghamer

Schwinghamer retired from Marshall as assistant director, technical in January 1999. Since that time he has consulted for NASA on the Shuttle Columbia accident and also on the successful "return to flight" of the Space Shuttle. He still resides in Huntsville.

Phil Shaffer

After Skylab Shaffer served as a principal interface between JSC's flight operations organizations and Rockwell International for establishing operations requirements for the Space Shuttle. In the early 1980s after leaving NASA, he served as a consultant supporting several of the NASA contractors supporting the Shuttle and the Space Station. Phil Shaffer died in June 2007.

Jim Splawn

As a director of marketing and business development for Boeing, Splawn is responsible for acquiring new business through advanced technology applications for the Department of Defense. These technologies may be used in either an offense or defense weapon mode, including missiles. Applications include military operations, both local and foreign, and Homeland Security.

J. R. Thompson

Before leaving NASA, Thompson served as deputy administrator at NASA headquarters and as director of Marshall Space Flight Center. Today he serves as vice chairman, president, and chief operating officer of Orbital Sciences.

Bill Thornton

Scientist astronaut Bill Thornton waited sixteen years after his selection by NASA before he first flew on STS-8 in 1983. The mission further explored human adaptation to the microgravity environment, with much of the research being performed with equipment he had designed. Thornton flew once more, on the Spacelab-3 51-B mission in 1985, on which he was responsible for medical investigations. He left NASA in 1994 and returned to academia at the University of Texas Medical Branch and the University of Houston, Clear Lake.

Paul Weitz

Weitz commanded the sixth Shuttle mission in 1983. He served as deputy chief of the astronaut office until 1986, then was appointed deputy director of the Johnson Space Center. He retired from NASA in 1994 and moved to Flagstaff, Arizona, where he says he "took up birding, fly fishing, and loafing, in reverse order."

Nine men were fortunate to live the Skylab experience, and their names will forever be recorded as NASA's first to truly live and work in space.

But while the names of those nine are the best known of the Skylab program, their voyages were made possible by thousands of men and women who helped establish an outpost on the frontier of space.

"Their legacy is that it is possible for humans to live and work in space for extended periods—but only with a terrific 'Home Earth' team to support them," Joe Kerwin said. "We, the homesteaders, thank them for their gifts of problem solving, support, and cooperation.

"It was fun to learn how to fly."

Appendix

Alan Bean's In-Flight Diary

MD 1

[illegible]

LAUNCH DAY

I AM WRITING THIS EARLY IN THE
MORNING OF DAY 10 - [illegible] COULD NOT
SLEEP, EVEN TODAY, SO THOUGHT I MIGHT
CATCH UP. [illegible]

SLEPT WELL EARLY TONIGHT, TOOK
SECONAL AND HIT THE BED ABOUT 7PM SO
DID JANE & [illegible]. DAN LIND HAD
COME AROUND TO PICK UP MY THINGS, BRIEF
CASE, CLOTHES, GIFTS TO TAKE BACK TO
HOUSTON - SOME TO SUE AND SOME TO
HOME - SOME ARE IN THE ISOLATION TRAILER
IN HOUSTON WHICH WILL STAY LOCKED
UP TILL WE GET BACK.

AMAZING [illegible] TIME BY AL SHEP. HE
& DEKE KEPT TRACK OF US THE LAST
FEW WEEKS MORE THAN USUAL. THIS
WAS MIXED BLESSINGS.

EARLY MORNING URINATION IN BOTTLE &
WEIGHT IN GYM - SEEMED STRANGE TO
SEE PAUL BUCHANAN, EDWARD [illegible] &
DEE STANDING IN THE GYM WAITING. FIRST
THERE WAS THE MICROBIOLOGICAL SAMPLING.
THEN PHYSICAL THEN EAT - WE WORE
WHITE TERRY CLOTH ROBES.

NO TRADITIONAL SOUR BALLS FOR THE LAUNCH
CREW. LOOKING BACK NOW MAYBE WE
SHOULD HAVE.

50. The first page of Bean's diary exemplifies its author's "artistic" handwriting.

The following is the complete text of the diary that Skylab II commander Alan Bean kept during his time on the space station. It is presented unexpurgated and largely unmodified, with the primary exception of formatting. Bean's handwriting does not differentiate between capital and lowercase letters, and the diary included minimal use of punctuation. For the sake of readability, those issues have been addressed.

Bean numbered the entry with the day of the year on which it was written, so the first entry, 209, for example, refers to the 209th day of the year, 28 July 1973.

209

Launch Day

(I am writing this early in the morning of day 10—Could not sleep, EVA today, so thought I might catch up.)

Slept well early tonight, took Seconal and hit the bed about 7 p.m., so did Jack and Owen.

Don Lind had come around to pick up my things, brief case, clothes, gifts to take back to Houston—some to Sue and some to home—Some are in the isolation trailer in Houston which will stay locked up till we get back.

Awakened on time by Al Shepard. He and Deke kept track of us the last few weeks more than usual. This has mixed blessings.

Early morning urination in bottle and weighing in gym—seemed strange to see Paul Buchanan, Edward & Dee standing in the gym waiting. First there was the microbiological samples. Then physical—Then eat—We wore white terry cloth robes.

No traditional sour balls for the launch crew. Looking back now, maybe we should have.

Al Shepard rode in van as far as the Launch Control Center. I watched him because he held the RDZ book—when he got up to get off, he forgot he had it and I had to ask. On the way he told us he was the last minute back up—he then mentioned Glenn having his suit at the suit room prior to Al's first flight.

4, 3 (engine noise and damn the machine starts to shake), 2, 1, more and more violent shaking—it seems to want to go like a car spinning its wheels

0—God, and you feel it pull away from the launch site—vibration, rough jerking, much much feeling of an unleashed power house wanting to go skyward—almost the same feel in reverse when you step off the high diving board—You can hear and feel the beast start to accelerate. Jack and Owen are spellbound, so am I for that matter. Lift off, tower clear, whew, that's a big one, roll & pitch program. I call—my voice sounds OK, don't sound too nervous, that's good—Jack and Owen OK.

210

Slept in the OWS sleep compartments last night—Place not fully activated but better than CSM. Slept pretty good because I was so tired, and sorta sick to my stomach. Owen looked so-so but Jack looked real bad.

Stuffed up head feeling present and will probably be with us the rest of the flight—Nose gets lots of buggers in it & when I blow it it expels blood. Dry climates like Denver do the same.

May have had the straps too tight on the bunk last night.

Breakfast in CSM, no water yet in the workshop—Not a pleasant get together—Nobody wanted to eat but knew we had to. No one wanted to think of the things we had to do today—I was behind with stowage, putting the rate gyros together (a last minute add on) and trouble shooting the condensate leak. In fact spent most of today responding to ground request for troubleshooting the leak—seemed disorganized—kept doing things too fast, causing delays, lost 50% of my time today for things I should not have let loose.

211

We are farting a lot but not belching much—Joe Kerwin said we would have to learn to handle lots of gas.

Got to stop responding to ground so fast and just dropping what I am doing—causes us to run behind on the time line. Do not know just what to do about this.

I am feeling good in the morning and between meals. Meals themselves tough to get through.

Still losing a lot of things, too big a hurry. Wish the flight planners would let up. The time taken to trouble shoot the condensate system shoots the whole timeline. Got to stay on schedule.

Floated too much at a work station—wish my triangle shoes were adjusted correctly.

Guess we had the failure. Because except for a stuck thruster it's about the worst we can expect—We handled it well but not perfectly.

212

Jack was taking a cooked fecal out of the dryer—laughing—well, here is a real nice ripe one, I said, bet you are a good pitza cook. No, said Jack, pancakes.

We had too many fecals and vomitus to cook—

(Look at down voice and we can see what we did each day in ref. to Flt Plan.)

213

Without triangle shoes you can get a free return to our early ancestors; namely holding and swinging feet, legs, arms and wedging feet and arms, feet and butt or feet and back to hold position, you get good at it where you do not have to pick out specific place for each limb or which technique to use, but it comes naturally.

All were in extremely high spirits today, first day we all feel good.

Discuss location of items in particular pockets—left lower leg, trash.

Owen said that today we ought to ask for a reduction in our insurance rates because we were no longer running the risk of drowning or auto accident.

217

Left SAL vent open last night after water dump. Thought I was so good at it, did not use check list—fooled because this was first night without experiment in SAL.

Wonder what happens when we cross the international date line multiple times a day. Well, no matter.

Saw Cape of Good Hope.

Owen let Arabella out of the vial. She had been in there since ______ days prior to launch. She had not come out so Owen got the vial off the cage, opened the door, shook her out where she immediately bounced back and forth, front to back, four or five times, then locked onto screen panels at the box edge provided for visualization—there she sits clutching the screen. Owen and I talked of giving spider food because she has not moved one half day. Owen said "no" because when she gets hungry is when she spins her web. (More description.) She can live two-three weeks without if she has too.

First back-to-back EREP. Jack VTS sites Lake Michigan. But got Baltimore instead. Or Washington, his prime site.

Saw what we thought was a salt flat but turned out to be a glacier in Chile. We could see Cape Horn—Cape Horn and Good Hope all in one day, fantastic.

Owen wanted to know if we had tried to urinate upside in the head. He said it psychological tough. Jack said he tried it and he peed right in his eye.

Diving thru workshop different than in water—here the speed that you move (translate) is controlled entirely by your push off so for some spins you can have a ½, 1 ½, 2 ½, 3 ½, etc. Difficult to push off straight and to get speeds you want. You must watch your progress as you spin—it's tough to learn but to keep from hitting objects, it's a must.

It was a great day—first back to back EREP and it came off perfect. Jack and Owen good spirits for EVA tomorrow—we worked all afternoon and evening on prep, much more fun than on Earth in 1G.

Owen worked 22 hours today because he counted his sleep cap time. Talked with Sue over Guam tonight. She asked about RDZ from Dave Scott. He said 1 quad out at so some skill required—he is going to Flight Research Center, great! Every day is filled with memorable experiences—sites, sounds, emotions, hope, fear, courage, friendship. I just wish we could go home to our wives at night.

My urine volume lower than Owen and Jack. Been drinking a lot but must do better. Been concentrating on eating too much. Owen said meals were the high point of a day on Earth and here too. Only difference is there it's the start, up here it's when you finish. I got ahead today with a snack during EREP, don't ever fall behind.

I cut a hole in the bottom of my sleeping bag near the feet—too hot, had to tie a knot to keep from freezing in the early morning.

Heard about leak in AM primary and secondary cooling loop. Pri should last 17 days and secondary 60 days. Wonder what ingenious fix they will come up with.

No CSM master alarm today.

Almost a no mistake day. But just prior to sleep Crip calls and ask we turn the ESS (medical) off. I had just ridden the bike.

218

EVA day. I had a tough time sleeping. OK for first 6 hours or so then off and on—finally writing at normal wake up time, 1100Z (0600 Houston) because they let us sleep late. Bed is great. I am going to patent it when I get home. The bungee straps and netting for the head and the pillows were my idea. Might come in use someday because no other simple way to make 0G feel like Earth.

Jack sleeps next to me then Owen at end—the reason, his sleep cap equipment fits better.

Funny how good we feel now, I think we all would have said "to hell with this, let's go home." I—we were not old enough to know time would pass and we would feel better. No one ever said it in words but that was the way we all looked at each other around day 2 and 3.

Sleeping is different here because the "bed clothes" do not tend to rest or touch your body. This causes large air spaces about your body, that your body heat just doesn't heat. It's difficult to snuggle down. Have to put undershirts (long) and t-shirt on during the night. I cut feet out of the long handles then use them for pajamas. Also I mod'ed my bed by cutting a hole in the netting near the feet, too cold at night so close it up with a knot.

Little worried, Funny—Owen's PCU is #013 and his umbilical is #13. I'm not superstitious, but . . .

Started taking food pill supplements today. Kit is junkeys paradise.

Jack discovered new way to shake urine to minimize bubbles. I called ground and said, "we even have our professionals—Owen ATM checkout, me condensate dump, Jack urine shaking."

EVA thoughts

Owen was having trouble with the twin poles—the elastic was tight at either end of the pole plate and as he pulled the poles out the rubber grommet locking the lock nut would tend to roll off. We had not pulled out any poles at the pre check last night because we were afraid too, we might break the elastic. We could not even break it today as Owen tried. He used his head and stopped—thought of a new approach, right foot out of foot restraint to pull pole the right end and left out to take the poles.

We are going to have the twin pole sail configuration on our medallion when we get back.

Watching out the STS window as Jack worked in the dark; I could not look at him in the light as he was too close to the sun, it was fantastic to see the sunrise. It began as a light blue band which grew with a fine yellow rim near the Earth's limb—the blue gets larger then.

The line with the gentle curvature of the Earth and the fact it became dimmer as you looked off to either side of the sun's future position.

Just before sun up you could see dim flashes of light toward the horizon where thunderstorms were playing. This pin pointed the coming horizon which was not discernable against the dark of the Earth from the lighted cabin.

Gold grows at last 15 sec to cover much of dark blue then bright orange and a bright glint as the sun. As it rises the Earth's horizon moves slowly from head to toe on Jack as he is silhouetted against the blue line. It gives the feeling of going around a big planet, a big ball rather than just a disk moving from in front of the eye. The science fiction movie effect was fantastic.

Pole deploy was also difficult because line got tangled by 180 degrees—had to roll back the grommet and then unscrew the nut and remove it.

Poles nice and straight and not bobbing around—white tape at forward edge foil/nylon sail stuck together—did not want to unfold but Jack pulled it in and then out again as sun sets.

Jack said, being out on the sun end, was a little like Peter Pan—or that you were riding a big white horse—feet spread wide across the whole world—the Earth is visible on both sides, at the same times and you can see 360 degrees—riding backwards.

Jack kept teleprinter flight plan as he was going to bed. Owen ask why—Jack said "I want to keep the memorable and unique days"—I said "don't lose your day off flight plan then."

219

Passed the LBNP today for the first time. Think I was too far in it and squeeze around stomach cut off blood, will move saddle from 9 to 6.

Did a lot of flying about the workshop just before sleep tonight. Skill needed, but great relaxer. Wish Owen would move Arabella.

Arabella finished her web perfectly. When Owen told Jack at breakfast, Jack said "well that's good, I like to see a spider do something at least once in a while."

220

Lost our day 5 menu card today. Had to fake it at breakfast. I finally found it as I was looking in the toolbox for a Phillips head screw driver for the wardroom foot restraints. My green copy of Childhood's End floated by. If you wait long enough, everything lost will float by. A dynamic environment no one can be stranded in center of a space because small air currents have an effect.

Tried to fly (like swimming) last night. But air currents much more dominant.

Fire and rapid Delta P drill today. Owen needs them the most but hates them the worst. I try to stick with him and do this together, Jack goes alone—when I am distracted, Owen will be doing other things not drill related and I must get him back.

Slept better last night (upside down) because it was cooler from the twin boom sun shade.

Arabella ate her web last night and spun another perfect one.

222

Day off—we had mixed emotions. We were tired and needed rest yet our chance to do good work was almost one-fourth over. When each flight hour represents 13-14 Earth training hours then you can make worthwhile a lot of pre-flight effort with a little extra in flight effort. We did however do some ATM and some S019. We asked for extra plus housekeeping. Wipe, dry biocide wipe, the place is immaculate and not a predatory germ within miles, much less traveling at 18,000 mph.

Got a thrill today. Tried to put our a urine bag with the metal fine filter for the head in it [the Trash Airlock] in addition to three urine bags. It would not eject. I tried to close the doors and breathed a real sigh of relief as it came closed. I removed the filter from the now shortened bag and tried again, this time it moved 1" or so then stuck. I tried gingerly to close the door but this time it would not. My heart was beating fast. Can this be happening to us? I tried the ejection handle again, and no luck, the door was stuck. Finally the only way was to force it. I rapped it again and again at first no success, but finally a little at a time then she broke free. My heart still beat fast but maybe a lesson was learned. Why did they not build the lock as an inverted cone so whatever was in there could always be moved down the ever expanding diameter.

Owen did the spider TV three times. Once because he recorded it all on channel A, once because the TV select switch was in the ATM and not OWS position, the last time it was okay. He got behind and I did some of his housekeeping as he was still up when Jack and I were headed for bed. Jack said "Owen, do you have anything left I can help you with?" Owen said "no." But that's the way Jack is. Noticed Jack uses aftershave on his crotch. Old Spice after shave and skin conditioner complete with a NASA part number.

Notice we do not seem to reflex to catch something when we drop it as we did the first few days. It's enjoyable to just let a heavy object float nearby.

223

Jack replied to Hank Hartsfield, he felt real good today. Owen said, uh oh, Jack must be sick again. Go look for a full vomitus bag in his bunk room.

Had a PPO2 low caution alarm—ground said, we can inhibit as it was erratic recently.

We moved our suits up to the MDA getting them closer to the CSM and out of the way for the maneuvering experiment later in the week—we moved three silent friends and stashed two at the front end of the MDA and one under the ATM foot support.

Almost screwed up EREP pass today—put in switch command to go to ZLV then did not enter bias in the DAS. Time to get to attitude starts over which the DAS bias maneuver is entered and when have Hank reminded me from the ground I had not enter—I did—& immediately knew I should not have—Owen and Jack were up there in a minute to say I should not have. We ask the ground for a new maneuver time to still get there and they quickly gave us 18 minutes where I entered and turned out okay, although we were one minute late to local vertical attitude. This is where I believe all our training paid off—a foolish mistake but we caught it and recovered to make it go—got three sites and then could not get a volcano in because of the overcast clouds.

Owen called me up to the ATM and then took my picture with this Polaroid, out came a picture of a naked gal with big boobs. He took some others, turns out they were put there by Paul Patterson prior to launch. Owen said he said he was going to call Houston and going to tell Paul, but I said-careful, do not pique the curiosity of the newsmen because they will want to know what's up and the world has a few little old ladies that do not want pin up pictures in a U.S. space station.

Owen was discouraged today—the experiment

Jack mentioned there was only one requirement for peeing in OG, and that was keeping your pecker in the cup—Owen allowed, well, he found that he could minimize the urine drop at the end by being aware that the bladder was near empty then really press and increase the stream's speed, only a small drop remains.

224

Had a thriller, was writing in my book when caution tone then warning tone came on—Jack on toilet—Owen and I soared up and found cluster ATT warning lt and ACS (attitude control system light) on. We looked at ATM panel and found much TACS firing and x gyro single, y gyro okay, z gyro single. A quick look at the ATT power showed multiple TACS firings. Both Owen and I were excited, it had been some time since we practiced these failures, plus we are in a complicated rate gyro configuration—we both really were looking at all things at once—DAS commands, status words, RT gyro talk backs, momentum and CMG wheel position readouts. We elected to go ATT HOLD but TACS kept firing, so we then turned off the TACS, looked at each rate gyro and set the best of one set back on the line. We would have gone to the CSM but with our quad problems that would be a last resort. No, we had to solve it right then. We put the rate gyros back into configuration then enabled TACS, then did a nominal momentum cage—this seemed to make the system happy—namely TACS quit firing. Owen and I had settled down by then and were solving the problem again and again to insure we have not forgotten any step. We came into daylight—were two degrees or so in X and Y off so went to S.I.—maneuvered too slow so we set in a five sec maneuver time and selected S.I. again—Houston came up and I gave them a brief rundown—Owen, never giving up ATM time, started my run for me while I went down for dessert of peaches and ice cream.

EREP passed today, Jack got four targets, we then had an EREP CAL pass taking specific data on the full moon—all three of us working well together, we have trained a long time for this chance and we want to make the most of it.

Jack made a suggestion to walk on the ceiling as the floor for a few minutes—we did and in less than a minute it seemed like the floor although covered with lights, wiring runs and traps it seemed like a new place—cluttered but nice—the bicycle hung overhead was different as was the wardroom table but many lockers and stowage spaces were much easier to see and reach—I might use this technique to advantage when hunting a missing item or looking in a lower drawer.

Had to ask Capcom, Story Musgrave, to give us more work today and also tomorrow—we are getting in the swing—when you're hot, you're hot. We will have about 45 more days to do all the things we were trained to do for the last 2½–3 years—time is going fast and we must make the most use of it. Much of what we learned will have no application after Skylab—such as how to operate specific experiments, systems, where things are stored, experiment protocol, how to operate the ATM, EREP, etc.

225

Had bad experience today, sneezed while urinating—bad on Earth—disaster up here.

Did 10-15 minutes on dome lockers. Handsprings, dives, twists, can do things that no one on Earth can do—fantastic fun and I guess good limbering up exercises for riding the bike.

I went up and looked out of the MDA windows that face the sun, but at night. What an incredible sight, a full moon, Paris, Luxembourg, Prague, Bern, Milan, Turin all visible and beautiful wheels of light and sweeping under the white crossed solar panel of the ATM. Normally you cannot look out these windows because the sun glare, I could not watch Jack and Owen on their EVA. Now we are in the Bay of Bengal. In just 16 minutes we swept over Europe and Eastern Asia, Afghanistan, Pakistan, and finally over India. Too cloudy to see Ceylon (Sri Lanka). Sumatra and Java will be here soon. We repeat our ground track every 5 days but 5 days from now as we go over the same point of ground the local time there changes so that in 60 days we will have seen all points between 50N and 50S at 12 different times of the day and night. At least once we can watch Parisians (Paris residents) getting up having breakfast—

It's not so much that we are 270 miles up in space that isolates us from the rest of the world it's that we are going so fast. To come home we must most of all slow down, not too much or we would come in too steeply and aerodynamic forces would be too great. And not too shallow for the upper atmosphere is tenuous and might not slow you enough and you would enter off target and there is a lot of ocean.

Owen and I spent his first night in 17 days just looking out the window during a night pass. We came over France then down over Turkey, the Dardanelles were visible, then he pointed out the Dead Sea, the Sea of Galilee—I said I had been as high as anyone on Earth and had been to the lowest point on Earth, the Dead Sea last year. Owen talked of the night air glow—the fine white layer about a pencils width along the surface of the Earth.

We had looked last night for Perseid meteor shower with them burning up below us. Did not see any doing S019—funny to hit the atmosphere and make a shooting star, they all flying past us—with no meteoroid shield anymore, hope we do not contact any one.

Flew 509 today, disappointing because took too long to go anywhere, just jumping and diving more fun—strange we did not realize that prior to flight during training. Kicked up much dust and items that we had not seen for weeks-M092 subject cue card showed up, thank God because we needed it.

Went to bed wondering if we would have a master caution warning with the rate gyros but maybe the computer patch for course gains will work—we can't point as accurately but we should have less disagreement between gyros.

226

Fixed my sleeping bag today, safety pinned at two top blankets and took up slack in blankets—too much volume of air to warm at night. Have been waking chilly about 1 to 2 hours prior to 6 o'clock (normal wake up time) Houston time and having difficulty going back to sleep. Maybe this will help, sleeping upside down has helped, the cooler OWS as a result of the twin pole sun shade deployment is perhaps the greatest contributor.

Normal morning sequence is wake up call from Houston, I get up fast, take down water gun reading, then put on shirt and shorts for weighing in the BMMD. Take book up and weigh while Jack gets teleprinter pads and Owen reads PRD. I weigh, Owen weighs, then Jack. I fix breakfast after dressing, with Owen a little behind. Jack cleans up, shaves, does urine and fixes bag and sample for three of us, I finish eating as Jack comes in and I then clean, shave and sample urine, I'm off to work at first job as Owen goes to the waste compartment. Jack is eating and about 30 minutes later we all are at work.

A sudden realization hit me this afternoon—there is no more work for us to do—ATM is about it. Except for more medical or more student experiments what a sad state of affairs with this space station up here and not enough work to do—with S020, T025, S073 gone there just isn't much left.

We could think up some good TV productions getting 5000 watt/min of exercise per day and that should be enough.

Boy oh boy have I been farting today. You must learn to handle much gas up here and I wondered if we would forget when we went home. Owen said can't you just see Jack in his living room with all his family and friends around and he forgets.

Funny you want a flight so badly you work hard when you get it you can't wait to launch but once you are in orbit you're still thinking of entry. Doesn't make sense when examined closely. We are on the world's greatest adventure, experience, sight seeing and it's that desire to get home before something breaks mechanical thing do you know.

I am so glad that Owen and Jack and I are on the same crew. Our personalities fit one another well—Jack always working, always positive, always happy—Owen always serious, well may be not always.

Owen looks funny lately as he has not trimmed his mustache hair. Shaved under his neck too well—our little windup shaver and the poor bathroom light being the problem. I don't look too great either, my hair getting long, wonder if "O" or Jack will cut it on our day off.

Talking with Sue on phone VHF through Canary Islands and Madrid—she wanted me to call her in London and Zurich somehow and gave us the number. These calls are great, best entertainment—last one I had was on s-band and it was a bomb. Only talked last 3–4 min. Sue getting ready for school and for London trip. [?] in Las Vegas and in Los Angeles. Home sounds close but it's still 6 weeks away. Lots of living to do till then, lots to see, lots to accomplish. Amy sounded sweet. I love her more than my heart can understand.

I look forward to Jack and Owen's calls home too because it is a way of getting news without the censoring influence of open communication.

Took the crew pictures tonight—they will look like mug shots when we get home. Should have taken them earlier for a better comparison with later in the flight.

Our tape recorders are great entertainment but continually need drive wheel cleaning. Like mine best when peddling the ergometer.

Owen got his ego bent last night. He had been commenting about weight loss, wanting more food, and salt—peanuts a favorite, Dr. Paul Buchanan called on his weekly conference and told Owen, Jack and I we're doing okay but he needed to have a chat with him (Owen). Paul said, Owen we have been looking at your exercise data (over the last two days) and don't think you are doing enough, maybe your heart isn't in it—Owen about flipped because he takes great pride in his physical program, pound for pound he does more than Jack and I. He could hardly hold back, afterward he worked out till sweat was all over his body then called on the recorder to tell Paul and those other doctors the facts of the matter. Maddest I've seen him in months.

227

We have been trying to get the FLT planning changed. I especially have had a lot of free time, Owen and Jack to a lesser degree. Jack keeps on the move all the time, so my suggestion of cutting down some was okay. Owen thought otherwise, he has a long list of useful work that he brought along, things that other scientists have suggested, worthwhile. How do I accomplish this feat of us producing our maximum without infringing on Owen's time. He deserves some amount per day to do with as he chooses.

In a way space flight is rewarding but on a day to day it is awfully frustrating. Up here we are manipulating thousands of switches, controls, dials, etc. to accomplish some precise tasks/experiment and so with all the actions you make many mistakes, more than you would like to—each session on an experiment I say this time I will do it perfectly, but 90% of the time not so, it's a difficult assignment. Some miscues do not mean a thing, some ruin experiment data. Hopefully none ruin an experiment. It's hard to stay up for all these experiments day after day, but that is one of the challenges that is a real part of the job. Jack today spent whole night pass taking star/moon and star/horizon sightings on his own time to satisfy an experiment. When the pass was over, 20 marks made, he was debriefing and as he was talking he said, well, I did those sightings with the clear window protector still on. He had not noticed it in the dark. The data would be off .001 arc sec or so and that just didn't suit Jack. He told the experimenter on record that he would repeat them later.

Food stains get all over. Not yours so much but your crewmates' food gets on you—shirt, face, hair, you don't know it but at the end of the day your shirt has little spots where orange juice flew over, or steak juice. Once it starts to move it doesn't hit the floor, it keeps going in its initial direction till it contacts something or someone. You tend to shield yourself when you open your own food so the spray heads toward your crew mates who are not watching.

> Teleprinter message: To Bean, Garriott and Lousma
>
> We have been watching and listening as the three of you live and work in space. Your performance has been outstanding and the observations that you are making are of tremendous importance. Through your efforts Skylab 3 is a great mission.
>
> Keep up the good work.
>
> Signed,
>
> Jim Fletcher
>
> George Low

Received this today. Why do they not send something similar when you are not doing too well, like days 2–4, we appreciated this but just wondering not only them but about myself.

Went to bed on time, do not feel as energetic as usual so feel something was coming on. Sleep is the best thing to repair me, it always works on Earth.

228

Good day today, first part anyway. One hour after wake up had an ATM pass. This was the first of our new schedule. We are taking good ATM pics and lots of them. We must be ahead of our nominal mission plan, I hope so. I have not felt good the last day or two. Think it's my low water intake, I had hoped nature would take its course, but it did not. Better start forced drinking.

Failed my LBNP—had to cut off at 1 min 2 sec to go of a 15 min run. I was at 50mm and started to feel tingley—BP dropped and my heart rate started down, it took a while (5 min) to feel normal again. Pulse got as low as 47—don't think I'll go that far down again. This is not like Earth in that you cannot put your head low and gravity force the blood from your head and heart from your legs—it could get a bit sticky if your heart beat got too low.

Jack was observer for the run and hated to put the results on the recorder I could tell, I had to remind him—he has a fine heart. Jack opened up the 2nd tape recorder and it also had a failed drive motor belt. Two failure of a complicated expensive recorder because of a plastic circular belt one-half inch wide and 3 inches long.

Performed TO13 today, went well, all equipment worked. Hope the engineers can use the data to design control systems for future space stations and there is a need to know just how man moves the station because many experiments have delicate pointing requirements and the control system must maintain this accurately pointed during manned occupation. We had wondered if it would be difficult back in training to bounce back and forth between the force measuring panels but it was not. Hand-to-hand, feet-to-hand, hand-to-feet, all were simple. Jack feels a bit 2nd level, because for the last few days with 509 and TO13 he has observed and reported me, rather than do it himself.

We received a "heads down" report down from our Dr.—it means don't look out the windows at certain times because of possible nuclear explosions—this time it was near the French testing sites near Tahiti—French

Polynesia. When we awoke there was a teleprinter message for three consecutive days.

229

Just found out our day off was tomorrow—Boy, the week goes fast as I honestly thought it was 2 days off.

Jack made a boo boo today. (I was flying M509 suited, boy did the umbilical give problems, especially on the rate gyro and HHMU modes—initially the thruster impingement on the suit caused big disturbances.) I mentioned to Houston (Dick Truly) how the back pack seemed to be too far back and the hand controller seemed too short—he came back and asked if we had taken the back spacer out—I looked at Jack and he looked at me. If they asked us later, we were going to say we flew it both ways—we made sure we could say so by flying it for 2 min with the back spacer out.

You soon find out not to set items down free without a tie down strap or spring or in a crevice or in your pocket. Even if released with zero velocity, the air currents will soon take it away, and which way is anyone's guess. If the object is too big to tie down for just a minute you release it carefully and visually check it every 10 secs—after 2 or 3 checks you will have to reposition it like you put it originally. It's a habit easy to break but when you do, it then can result in many minutes of lost time hunting.

Owen and I stayed up and looked at the night pass over Mexico, Houston, New York, Nova Scotia and then Paris, the Mediterranean and east Africa. We looked so big and strong and steady as we glided over the Atlantic in 15 minutes. Our ATM solar panels were reflecting in the light of the near full moon behind us. We did not see Paris because of early morning haze but could Sardinia, Greece, Crete, Turkey, Israel, the Galilee, the Dead Sea and Africa where Montgomery and Rommel—Bet they would have given anything for a manned satellite to provide observation of the others troops and armor movements.

Owen became excited over the northern U.S. as he looked out and saw some aurora—this was a stranger yellower than the fog bank like aurora the other night but spectacular. There were vertical shafts of light almost like close spaced yellow green search lights.

230

Our first real day off. Best news was in the morning science report where it said we would catch up with all our ATM science as well as the corollary except for medical which was reduced by 25 hours the first half of the mission, we would do the rest—I called and discussed the additional blood work, hematology and urine analysis (specific gravity) that Owen had been doing and wanting them to count that.

We did housekeeping a bunch and had to plan two TV spectaculars. Since we have ATM all day we had to schedule it in the 30 min night time crew eat (first night pass for 487). Hair cut next, then acrobatics, then shower. Lots of planning for 3 ten min shows but think the folks in the old U.S.A. will enjoy.

17,112.8 mph is our orbital speed. Had used 18,000 MPH on TV and wanted to be sure—(look at TV tapes for details of what we did) it's fun but time consuming to do the TV. Just as we were to finish the shower scene my phone call to Sue came in—so Jack got out and waited—when I got back in 10 min or so he undressed and got in and lathered up—we took the TV scene, he rinsed off, we then took the last shot—the one where he floats out of the shower with his shorts on. Owen came back down in a minute and told us the TV switch was in ATM so we did not get it—back in the shower Jack went and we ran again—but before he could completely dry his call came in so as he dried I ran up and configured the comm in the CSM for him and talked to Gratia—when she heard what was holding Jack she laughed and said "you photograph him in the strangest places." How can we make him famous if we don't do it a little different. Anyway we finally got it done—I did more work on the day off than a normal work day. It was late but I peddled 5004 watt/min and took my first shower.

It was cooler than I like it—the biggest surprise was how the water clung to my body—a little like Jello in that it doesn't want to shake off. It built up around the eyes, in the nose and mouth (the crevices) and it gave a slight feeling of trying to breathe underwater—I would shake my head violently and the water would drop away (not down but in all directions) some to cling to other parts of my body, some to the shower curtain, some sort of distended the water where they were and snapped back. The soap on the face stayed and diluted with rinse water. Tasted sour when I opened my mouth—that little vacuum has sufficient pull but is rigid and will not conform to the body—so does not do too well there but okay on the inside walls, floor and ceiling. Jack had said it was better to slide my hands over the body and to scrape the water off and over to the shower wall. This worked for hair, arms, legs, but difficult for body especially back—two towels were required to dry off because water did not drain.

Owen was a little annoyed with the TV effort. He did not want to practice the acrobatics for the second time. He usually gets over it in an hour or so. It's part of our job and we should give it equal time.

231

Flew T 20 for the first time. Jack as usual had the dirty work but was trying harder because of his error yesterday. The work was slow and tedious because it was the first time around and because the strap design was poor—the whole thing was.

Jack said "I've done some pretty dumb things in my life but I never got killed doing it—in this business that is saying a lot"—

Owen said "Well the dumbest thing I can remember was flying out to the observatory near Holleman NM—short hop so I decided to do it at 18,000 ft—as I neared there I started letting down, called approach control-we talked and as I descended their communications faded out—I kept thinking why should they fade out—it suddenly dawned, shielded behind mountains—full power and a rapid climb in the dark saved my ass—I think of that incident several times every month over the last three years."

Jack was saying that when we got back he and Owen might be considered regular astronauts—Owen laughed—it was beyond his wildest dreams to be classed as a real astronaut.

Been wishing Owen and I had taken pictures of the Israel area the first time we stayed awake to see it—I want to give pictures of this region to some of my religious friends.

Sue told me last night that Boe Adams would be indicted for borrowing twice on the same bank stock—she told me about the 37 murders in Houston—we had not heard because on our nightly news they tend to eliminate the stressful news. Putting our 5,000 watt/min on the ergometer (or is it ergrometer like some say) in one 10 min and one 20 min run—have to stop at 15 min usually and let my 185 heart rate reduce to 150 or so, then finish.

Was supposed to photograph the Antipodes Islands southeast of New Zealand but was too cloudy—Antipode is the opposite place on Earth and the opposite from the islands is on the English Channel—so the original discoverer must have been from near there—these are on the 180th meridian.

We are crossing the date line every 93 minutes so we gain a day but of course as we go east we lose it an hour every time (look up some news on Glenn's flight, might discuss this). 24 hours, so it all averages out.

Jack's having his ice cream and strawberries. Jack's food shelves when we transfer a 6 day food supply are almost full of big cans plus a few small—Owen and I have half full shelves with more or less equal amounts of small and large cans—he really puts down the chow.

All are in a good mood, morale is high in spite of all the hard work, we are getting the job done.

I was supposed to take pictures of the plains of Nazca in the Andes in Peru (about 300 mi SE of Lima). Criss-cross patterns are the dominant features but I could not pick out even the 37 mi by 1 mi place called the airfield of the ancient astronauts. I was supposed to photograph with the 100 mm Hasselblad and the 300 mm Nikon and identify the plains and describe the geometric pattern. Also describe the general appearance of the plains and geographic relation to the adjacent features.—All in 5 min 10 sec of viewing—weather was broken clouds and I did not see the target. Could have several days ago many times—we must change our approach to ground targets.

232

A tough tough day. Worked almost all day on trying to find the leak in the condensate vacuum system—hundreds of high torque screws, stethoscope, soap bubbles, 35 psi nitrogen, reconfiguring several pieces of 20 equipment—we never found the leak—that effort must have cost $2 ½ million in flight time.

Jack feels down, the S019 stuck out and would not retract—we will keep using it but may have to jettison if it won't come in. He's hit the bed early, tired—the last ATM pass of the day he ran all the S055 in mechanical reference (102 steps less than optical) when it should have been optical—he was dragging—it was partly my fault as I left it there after my last run.

Owen got the word that the citizens of Enid would be putting their lights on for him to see—I went up with him—it was the clearest, prettiest night we've had—we could see Ft. Worth-Dallas perfectly—a twin city, one of few—then Oklahoma City then Enid then St. Louis then Chicago—Owen made a nice narration. He said he started to say he saw Tulsa up ahead and realized it was Chicago. Paul Weitz said that was the one thing he never became accustomed to on his flight—the speed which you cover the world, especially the U.S.

Passed my MO 92 today but used saddle position 5, may be too high out of the LBNP and not pull my blood so hard—Owen and I had a laugh by saying my delta P of 50 never gets lower, I just keep getting further out of the can—we visualized late in the mission sending TV to the doctors with just my feet in the unit but still pulling 50mm.

Heard tonight we may put in the rate gyro 6 pack—I told Owen I would do it because they did the twin pole and because that sort of work fits my skills better than Owen's—hope it did not hurt his feelings but that is the way I see it and that's my job—Owen even brought it up by saying "I think you want to put out the 6 pack and that's okay with me—I'm glad to do it but know you want to"—I said you're right we don't need this job but if it comes up we will pull it off.

233

Today was a special day—found out we were going to put in the rate gyro 6 pack—who to do it—Owen still wants to do it and so do I. Made up my mind that it would be Owen and I but after reading the procedures realized that I should stay in because of my CSM experience—Owen and Jack are just not up on it and it is the best decision—Jack will do the 6 pack as he is the most mechanical—Owen does not do those things as well as Jack, it will be tough to tell him tomorrow—I was awake for about two hours trying to put the pieces together and think Owen and Jack outside—me inside is the best way.

Got S019 in today—just decided to turn the retract crank hard—did so for 10-15 turns but the clutch slipped and it did not move. When I looked in the optics. I think turned it out (extend) then in (retract) then out then in and low and behold it came in.

I took it apart and looked okay as far as possible ice—looked as if a chain synchronizing the X screw drive was loose and may have stuck on a sprocket. I checked with Houston, adjusted a idler sprocket and it seemed to work okay. Jack looked at it too and noticed another chain that might be too tight so we loosened it. The mirror was so cold that it had condensation on it for several hours. Houston (Karl Henize) wanted it out in a vacuum so the moisture would sublimate and minimize residue deposited.

Had to laugh, Jack said he couldn't fart in the LBNP because he was being pulled down too hard on the saddle—needs a hole in it said Owen—we all pass gas up here—the nice way to say it is "watch out for my green cloud" if someone else comes near.

234

Told Owen and Jack about EVA crewman, they both seemed happy, told them what factors influenced me and who I felt most qualified for each position. Called Houston and told them later, they seemed happy. I started looking at the equipment for the job—all in good shape. Gingerly tested the tool for the interior plug removal and it looked promising. Will have to work on check lists later—need many questions answered from Houston.

Day went fast as it always does—only hard part was exercise 10 minutes (1 min 100 watt, 1 min @ 125, 2 @ 150, 2 @ 175, 2 @ 200, 2 @ 225) on the ergometer. Then some Mk I and Mk II work—then 25 min where I increase my 1700 watt-min to 5000 watt-min. It's a good workout and more than I did on Earth.

Owen seemed a little distant later today, don't know whether it is the hours or the EVA.

Worked on the coolant loop leak inspection. Jack made a break through by taking out screws which we could not turn using vice grips—did not finish procedure but found no leaks either, must be outside (I hope so).

Owen reported an arch on the UV monitor in the corona yesterday. We called it Garriott waves to the ground—he was in the LBNP and was embarrassed and told us to knock it off—we were happy for him. Today he heard the ground could not it see it in their taped TV display—he went back and checked and found it to be a sort of phantom or mirror image of the bright features of the sun except reflected in the camera by the instrument. He'll get over it (maybe that's why he was distant).

Crippen woke us this morning with Julie London singing "The Party's Over." Jack wanted to make this Julie London Day, so did Crippen so he could call her but Owen won out with Gene Cagle Day.

Got my exploration map up on the wall of my bunk room—Have to write propped up on the wall since my bunk is upside down.

235

Would you believe it we get better H-alpha pictures at sunset than we do at sunrise because our velocity relative to the sun is less and that effectively changes the freq of the filter in our H-alpha cameras and telescopes—not a small item either.

Owen's humor—I said "watch your head" as I pulled out the film drawer. Owen replied "I'll try but my eyeballs don't usually move that far up."

We were laughing about the malfunctions we had after we discovered the water glycol leak—I wanted to call Houston and say "Jack is working on the CBRM mal, Owen on the camera mal—tomorrow after we fix the door mals, the 9 rate/gyro mals and the 149 mal of the nylon swatch mal, I'll start work again on the coolant loop some more or the water glycol leak mal".

Everyone feels better about EVA—I worry too much and Jack will pull it off. Funny how easy it looks now that we are going to do it—did it get easier as we understood the plan or did we just want it to be doable? Morale is high—did perfect on my MO 92/171.

236

EVA day. I was talking to myself during EVA and Jack wondered what I was saying—I told him "I was just shooting the shit." Jack quipped, "get any?—what's the limit on those?"—Owen was saying "Come on, . . . hustle . . ." Give us some of that (positive mental attitude)—PMA (he doesn't believe in it. But knows I use it on me and them also.) "Go Earl (Earl Nightingale)." I said "you need it, it's worked on you whether you like it or not."

Jack had a difficult time with connector number two—it was difficult for me to keep from asking questions of Jack as I wondered if that would be the end of the show but he said don't talk for awhile and just let me work on it. He did for a very long 5 minutes and then reported connected.

Owen was elated with the view over the Andes—the 270 degree panorama with 3 solar panels in the field of view to form a perspective or frame work they were flying over all the world outside of the vehicle going 17,000 MPH (get transcript of EVAS). Lost three shims and one nut taking off the first ramp.

We have only 1 to 2 min of TV recorder time left so have to hold it for Owen's return to the FAS. Owen had to come out of the foot restraints to remove the ramps from the SO 56 and 82 A doors. Sun end EVA lights worked this time—Jack said he can see many orange lights, we were over mid Russia—not many cities—Orion came up, a beautiful constellation, Owen still working on bolts at Sun end.

Got a master alarm—CMG SAT—S/C going out of attitude—I put it in ATT Hold—TACS when was out 16 degrees.

Sometimes, like on a tall building, you get a controllable urge similar to jumping off which is to open a hatch to vacuum—or take off a glove or pop a helmet—fortunately these are passing impulses that you can control but it is interesting to know they take place.

Great EVA today—all happy tonight. Owen summed it up when he said "It's something that needed doing." Jack said he thought I made the right decision—so do I and we are all satisfied.

Owen bitched about the medical types that take care of our food because they told Paul Buchanan our food cue card were wrong for optimal salt and they had not bothered to update and had been making them up with supplements.—Owen flew off the handle because he has been needing salt.

It is comforting to know someone (many someones) on the ground are working your space craft problems faster and much better than we. We generally perform a holding action if we can. Till help and advice comes, then take the info or suggestions and do them, this is the only way you can free your mind to do the day to day tasks, the productive tasks while someone is trouble shooting your problems.

Heard yesterday that Fred Haise crashed in a Confederate Air Force T-6 and was badly burned but would survive. Seemed his engine out and he dead sticked it into a field, a wing tore off and the whole thing caught fire—burned him over his whole body except his face, so that's lucky—heard today that he was on oral foods so he must be improving.

Rearranged my bunk room—put a portable light on the floor near the head of my bed and turned my beta book bag upside down so I could grab the items inside easier. I use the door of the bottom locker (pulled out about 30 degrees) as a writing desk. Stole the power cable for the light from the spiders' cage—hope Owen doesn't get upset. He has been getting messages to feed them both filet and keep them watered. Will we bring them to the post flight press conference?

The teleprinter is a device about which you can have mixed feelings—it would be hell to get the information any other way so it must be cared for as an expenditure of effort. But at the same time every time you hear it printing you know it's more work for you to do. Wish we would get a non work related message sometime.

237

Felt good to have ATM film again. Operating the ATM telescopes and cameras is one of the most enjoyable tasks here. It is challenging, you can directly contribute to improved data acquisition—Owen has effectively changed the method of operating it in just ⅔ of a month. The Polaroid camera and the persistent image scope have made a significant difference.

I had Houston give all the ATM passes tomorrow, our day off, to Jack and I, so Owen could finish some things he is behind on and do some additional items that he has planned prior to flight—the flight is ½ over and he has little spare time—he needs some to be happy.

Using the head for sponge baths (bad name because sponges squirt water out when pushed on the skin).

(Tell story of MDAC test) has become more pleasant as I have been less careful about sprinkling water about. I tend to now splash it somewhat. And after the bath is complete wipe up the droplets on the walls. None on the floor like on Earth if you did the same.

We passed Pete, Joe and Paul's old space flight mark, in fact we now hold the world record for space flight—it feels good to be breaking new ground said Jack today. We will be ½ into our mission tomorrow night.

Our TV got too hot during EVA and quit working—we will do the rest of the mission on one TV I guess. Funny, they did not insulate it sufficiently. We had a plan to put it out the solar air lock for EVA but can't do it now.

Sue and Amy will be getting up soon over in London—we do not go over London, it above 50 degrees N latitude.

Europe is an awfully small part of the Earth—so is the U.S.. Jack has mentioned several times that each time he looks out the window it's either water or clouds or night.

The odds for man all beginning long ago back in the sea is reasonable just from a total volume of water available to total area of land available.

Jack has a small sty on his left eye, he wanted some "yellow mercury" but settled for Neosporin. Jack treated himself but Owen will examine it tomorrow—Paul Buchanan said for us to be extremely careful because that could be contagious. Perhaps a streptococcus of some type.

Our little Sony tape recorders are holding up so so. It is necessary to clean the rubber drive wheel about once per two days—lots of trouble.

Did Mk I exerciser TV today, also some Mk II and side horse swinging on the portable hand grips.

I keep my personal items in a locker above my head—it was across from me at eye height till I turned my bunk up side down, it has a hand hold with a thin rubber sheet covering it. The rubber sheet has a horizontal slit in it and serves to keep objects inside when my hand is inside searching for one of the 20 or so objects within. As I look inside I see chewing gum, stereo tapes,

Skylab sticker, my dental bridge, a string tether for books. Some tape drive wheel cleaning swabs, a couple of old teleprinter messages.

238

We are going to sleep just under one hour to the mission midpoint. Our science briefing today showed that we had made up the ATM observing time we missed early in the mission and predictions are for us to exceed even the 260 hr ATM sun viewing goal. We are ahead in corollary experiments. They are even having us try a new SO 63 using the AMS that was not to have been till SL 4.

Took my second shower, noticed that you could not hear much of the time less you shook your head because large amounts of water cling to your ear openings. I was the only one showering today.

We did urine refractory tests to determine specific gravity.

Exercised today although that is not my plan for day offs—not doing the exercise would be a nice reward but did not have time EVA day.

Think the reason the days pass so quickly up here is the variety of tasks we each do—we never do any one thing longer than 3 hours and the typical would be one hr or less—and you feel you're doing what you trained for for more than 2½ years. I spent much training time in the Navy for war and never went, thank God.

Talked to my mom and dad. Mom is so happy with Pepper and David's new baby, Brandi Frances, that she doesn't know what to do—if it had been a boy they were going to make the middle name Arnold. That might suggest Pepper's feelings toward her grandparents. Mom said Clay was leaving for school next week—had returned from California at 3 in the morning.

Am going thru all my music tapes to grade them pop or easy listening. And very good, good or so so. Seems I listen to the same 3 or 4 all the time.

Jack made an excellent observation when he said NASA should play down the spider after the initial release because it tended to detract from the more meaningful experiments we are doing up here. Will the tax payer say, now that I know what they are doing up there, I don't like my money going for that sort of thing.

Owen tried to do a science bubble experiment with cherry drink but didn't look too promising to me. He kept losing the drop of drink from the straw.

Jack suggested we put a pair of trousers in with our clothes in the CSM so we would have a full uniform on the ship—we will wear our hypertensive garment for entry—in fact we may wear it for the SPS burn since we found out Paul and Joe had a gray out during this burn.

We may want to give the entry guidance to the CM early if gray out occurs there too.

I have noticed if I do not force myself to drink then I will drink much less than on Earth and will dehydrate—I do not seem to automatically desire the proper quantity of water. Owen and I were talking about a possible advantage to the 18 days post flight urine and water measuring period—I suggested that when we get back we may not naturally readapt to one G and become dehydrated there. He does not agree at all.

Jack is playing his music in bed. Owen and I both agree on his music at least.

I slept 9 hours last night, a record for me in space.

239

Spent part of the morning composing a message to the dedication of the Lyndon Baines Johnson Space Center. Thanks to Owen's and Jack's suggestions, it turned out acceptable I think. It must have because Dr. Fletcher read only President Nixon's and ours.

When I used to go from compartment to compartment I would be lost when I got there—now I look ahead as I enter and do a quick roll to the "heads up" attitude for the space I'm entering.

Owen said his only regret was that he would never adapt to Zero G again—he thinks Pete is the only one besides ourselves that has ever done so—the weight loss and the poor appetite are the evidence. We could gain all sorts of weight if we wished.

My pockets are most useful—Lower left—all trash I pick up I empty from time to time. Jack just keeps it there till he throws the pants away. Right lower, tape and tools if needed. Upper left, timer and tools. Upper right sizzers and flashlight. The little pockets for pens and pencils, knife and sizzers. Far from perfect—flaps too short, needs snap & need more width.

I've noticed I enjoy the ATM time—The controls are an intellectual challenge and just interesting. Mostly I think it's because it's the only solitude that is available. No noise from anyone else—It's a time to be alone. In the sleep compartment someone is right next to you & it is not so private somehow—Here, seldom does anyone float by.

By the way, living in OG means you make no noise generally as you approach someone else, no footsteps, no ripples in the water, just silent swift movement. I've noticed that each of us tends to say something to another as we approach just to keep from surprising him—startling him. I noticed this silent movement first in skin diving—I could look around and suddenly there some 3–5 feet away would be a big fish—It all depended on the visibility of the water because much of the time one kept a good scan going.

The sty on Jack's left eye is going away—He found no "yellow mercury" but used Neosporin.

This message was sent to the ground by Alan for the dedication of LBJ Center.

Owen, Jack and I would like to send our best wishes to Mrs. Johnson, Governor Brisco, Dr. Fletcher, Dr. Kraft and all the other distinguished guests and officials at the dedication of the Lyndon Baines Johnson Space Center. The work in which we are engaged now in Skylab would not be possible had it not been for the strong support and leadership of Pres. Johnson in the Senate and the Presidency.

Our present preeminence in space is in no small way a result of his grasp of the fact that a progressive program of space exploration contributes significantly to the future well being of our nation and its people. We are proud to be representing nasa and the Lyndon Baines Johnson Space Center as we circle the Earth.

We believe the work we do here now is an extension of that which he chartered and championed.

240

So I feel on top of the world.

Ergometer broke, I fixed it then went back and overtorqued it (should have stopped at 30 in lb) and broke the bolt. Jack used the bone saw then his Swiss Army knife (wonder if the Swiss army really uses them) hack saw blade to fix it—I pedaled 6007 watt/min with 3 min at 300 watts. A new record both for time & for max power. We are whipping into good condition.

I got mad at Owen today—He has a habit of acting disgusted when I make a mistake—Today I had aligned the ATM wrong and he was nodding his head—I lost my temper and told him to quit. Maybe he will take the hint.

I asked the ground to let Owen fly M509 soon—He has not flown any simulator, it is safe & it would answer some important questions. Hope Jack won't mind.

Also asked the ground to invent a pointing device for locating ground sites we should photograph rather than just telling us which window—Hope they will because it would help. Most of the flyover time is spent verifying that you have the right place by looking at the maps, that other observations are compromised.

They wanted to know tonight if Owen would fly on our day off and I said sure—later I asked him and he said OK.

Did a TV tour with Jack today. He was shifting his eyes a lot at first and I told him—he never did so again, he likes TV and will be very good at it when we return. I try to teach them all I know—or think I know.

About every other night I get up because of unusual noises—mostly they are all thermal noises. The most unusual view occurred once as I left my bunk and peered up to the forward compartment. The ½ light from the lock revealed three white suited figures, arms outstretched, leaning several awkward ways—silent, large with white helmet stowed over them—one drying, the others waiting to dry. I was shook a little by the eerie sight so I went to the wardroom and looked out. The dark with white airglow layer and white clouds filed the lower right portion of the window like it was one foot away. It startled me even more.

241

Morale is high—work level is high. Last night after dinner Owen asked Jack if he (Jack) would like him (Owen) to take his late ATM pass. Jack said no he was looking forward to it—he wanted to find some more Ellerman bombs—(bright points in penumbra near sun spots) as he got some earlier—I interrupted and mentioned that the flight planners had voice uplinked a change in the morning, assigning me to the pass. Owen laughed—here we all are fighting for the last ATM pass.

Kidded Owen about wearing his M133 cap—said Jack and I better watch our PS & QS tomorrow, Owen will be in a bad, criticizing mood—he took the kidding well, hope it will have effect.

I have been decreasing my number of mistakes significantly by only doing one job at once—ATM is the one most enticing to doing special task as a set of instruments is operating. Invariably if I do, I do not get back in time or do not catch a simple error in the set up. I am waiting now doing S019 which requires little thought other than checking the exposure time every few seconds. Exposures are 270, 90 or 30 sec.

242

Had another run with T-20—suspect it was the last unsuited run. Worked somewhat better but the restraint straps were too time consuming to put on.

Noticed I have begun to think of places I like to go—spaghetti joint in LA, drive down the San Diego Freeway too. I can see it in my mind's eye. Our trip to Israel, I see the drive to the Galilee with Mathew Simon. Driving the Gulf Freeway in Houston, Christmas time in the Ft. Worth/Dallas area—shopping at Northwest Mall, driving home last Christmas Eve with gray skys, funny the places you remember, often the least memorable.

I declared "This is the captain speaking, this is proclaimed 'loud music day' aboard Skylab, only loud music will be played." I may have to wrestle Owen to the floor. I noticed Jack was playing loud Dixieland at the ATM panel—he sent some to the ground.

Owen was testing our Achilles tendon this morning—funny to hear Owen ask Jack "how about tapping my leg?" You must get used to touching one another in this job. Doctors do and probably some other professionals, but test pilots usually do not. Putting the sensors on the upper right and left buttocks (sacrum) is a difficult task for us—handling the fecal bundles before they are dried is also a bit new and not comfortable although completely natural and sanitary.

Owen has been unusually pleasant today—wonder if it was the kidding last night.

Jack and I did TV last night—finished up the tour of the lower compartments, Jack did an excellent job, soft-spoken, a few jokes sprinkled in here and there, shower bit with him inside was good. He had to do the ending twice but it turned out okay, I advised him to complete the show in the LBNP he did, it just wasn't right.

Been way ahead on my time line today—the flight planners gave me housekeeping time and a long lunch—they are really scratching their heads to think of useful things for us to do.

The ergometer must be easier to operate at zero G—perhaps because of no weight the legs only have to push pedal force 35# and not the weight of the body.

We did some activities Achilles tendon reflex TV today and later I accidentally put some 183 TV on the recorder over it. We repeated the test later.

Brought our 6 day food supply down—all helped put it in the pantry.

Be glad when EREP starts again so we have more good solid science to accomplish. We have been doing a lot of S/C tests and BMMD tests lately.

The condensate dump line plugged up last night—worked on it off and on all day—hot 35 psi water finally cleared the line—wonder if it will give more trouble, bet it will.

Exercised 10,041 watt/min today for a new world record for me—my normal is about 5,000.

Jack said Gratia said he was on TV last night with part of the tour. She liked it—Sue gets home from Europe, London, Braunwald (Switz.) late tonight. I'll call tomorrow.

Jack's boy (6) got on the phone and told Jack their dog was the father of a litter of puppies next door. Jack wondered how he knew and his son said they were black like his dog—Jack had a cat named Detail because as a kitten it slept on the warm car motor one too many times.

243

Good day—we all hustled all day long—Jack almost got a big flare. He was on the panel and the Be counter went up almost enough to have the size flare called for today (counter one, 4,150 counts). He had counter 2, 3,000 or so counts—when I arrived (had been exercising on the Mk I), he was slewed to a bright linear region or plage that was several times brighter than anything else on the display. It almost looked iridescent (in B&W) Jack could not make up his mind whether to go all the way or not—he elected a prudent course and went into a modified low film usage flare mode. Sure enough, it did not go any higher, he had calculated right. This is where man in the loop comes. Can use his judgment and commit resources as appears wise at the time.

Had a slight argument with Mission Control—we have for two days trying to make the waste management dump system work to empty the holding tank and water separators. They have had us do all sorts of procedures to clear it out, dump air, dump hot water at 35 psi-suddenly today Bruce McCandless calls up and yells, "CDR, that last dump you made may have backed up and cracked the separator plates, turn off both condensate valves, switch air flow to heat exchanger B on both Mole sieves and quit dumping." Well, I quit dumping several hours ago. Next site I told them we were operating this space station like an unmanned station, we could change probes bring up the old probe, look at it, plug it in and check the heaters, blow it out with high pressure nitrogen, they said they would take it under advisement. I also said you could not inspect the plates for cracking as they had a chamy cloth over, under and between the two plates (like a double decker sandwich).

Did my exercise while Owen did M092-93 and I was the observer—peddled 200 watts for 4 min then adjusted his setting, then 200 more for the next 4 minutes. While he did 93 I went upstairs and did Mk 1. There are many ways to get ahead on the time line.

The sun has lots of activity on it—it looks like a Christmas ball with many bright lites on it—strung across the equator mostly—wonder if could be like Owen's water bubble experiment the other day where as he spun the 1 ½ inch ball of water the air bubbles in it all lined up along the equator of the spin axis.

Owen and I were laughing about this night we both went up to see the lights of Enid—he talked of Mexico, Ft. Worth, Dallas, here comes Tulsa, Oklahoma City, look at St. Louis, Chicago—everything but Enid—Helen Mary called up there and tried to sooth the people—she gave Owen hell—I kept telling him to say something about Enid but guess he was too shy, they had a direct TV hook up, radio hook to us and all lights including the football field.

Found a stereo tape of mine in one of Pete's cassette holders. Must have been there since day one—knew some of my music was missing—found it because I was trying to mark the good/so so tapes. We have not played any of SL-4's music yet I may ask them if it's okay to do so.

We have been thinking about asking to be extended. We all feel good, obviously healthy. Don't exactly know how we can go about it ethically—there must be a proper way.

We are going to ask Dr. Buchanan (Paul) about Fred Haise.

Sue had a great time in London and Switzerland—She said Amy was so perfect. She is that anyway. I know the trip was good for both as 60 days is a long time to wait. She will not sit and wait but she will have to wait nevertheless. It felt good to hear her voice and know she is nearer.

Owen, Jack and I were laughing about the fact that we are on our 6th food bundle and have filled 6 fecal return containers. He thinks the whole flight might be a phony and we are up here just to process food—I said we could just invent a machine to do this and eject the waste for aircraft recovery. Think of the trouble it might save. Jack chimed in Hey, I'm going on strike, I won't eat. I'll just put my food directly in the fecal bags—yes, cans and all—Owen quipped don't forget to include the supplemental pills.

There are many islands in the world. Some small one lost way in the middle of a vast ocean. Some green, some brown, some sandy with light blue green water, the beach—some are just the circular edges of ancient volcanoes—who knows how they slipped beneath the surface.

244

Writing this on day 245. Yesterday we had our first EREP pass in a long time. It felt good to be back at it. We are good at EREP and it's a worth while experiment. I mentioned for the first time—since the timing seemed just right—that we wondered if they had heard anything about our staying an additional 10 days. It caught all by surprise, (except Jack and Owen, they were caught a little by surprise because I had intended to wait about 5 or so more days), the time just felt right.

We are all in such excellent health physically and mentally that this is the right way to do—it could be a great thing for NASA to keep us up here it wouldn't cost much and the sun is very active right now.

Jack, Owen and I are eating the overage food much more than any of us thought. Jack needs more calories, Owen more salt and potassium and I need more potassium also.

I sure feel different about our physical condition now than I did when we were losing leg circumference. The question was when would it stop.

245

Day off—Two EREP passes plus some good ATM—Owen caught a flare, Jack and Owen a couple of smaller flares. The sun is hotter right now than it's been in 14 years. We can even see 8 or so active regions on the XUV monitor without integration. Jack says it looks like a 3 day after Halloween pumpkin Jack o lantern.

Dr. Paul asked if we really wanted to go the extra 10 days and we said "You bet." He asked if we had talked to our wives and Owen said he did last night and she is 100% in favor, for as long as he can. Paul said he would just let that drop without comment. As predicted, Owen explained how she meant it.

Owen played his classical music all morning, so I would not put on my loud stuff. I'm enjoying the newly discovered music tape.

Came over San Francisco just before sunset—It's a beautiful place even from 270 miles. The city was clear but the fog was covering the bridge and just sticking into the bay. It will be a pretty night there tonight I bet.

Did not take a shower. Takes too long.

Heard Jack debrief M487—I need to get his comments (all the guys' comments)

246

Clouds are like snowflakes, no two alike—patterns are every kind imaginable and unimaginable—light furry white, furry grey.

Got my first flare today—A C6 in active region 12. I noticed it while doing some sun center work as an especially bright semicircular ring around a spot. There were 8 or so similar bright rings but this one became exceptionally bright both in Hydrogen-alpha and in the XUV. I debated with myself about stopping the scheduled ATM work and going over and concentrate on the possible flare. As of this time it had not reached full flare intensity and it is not possible to know whether it will just keep increasing in intensity or will level off then drop. As I elected to stop the experiments in progress and repoint I noticed about 5800 counts on our Be counter aperture 2—A true good flare would be 4150 counts aperture 1. It never got much higher. Owen hustled up at once to help—He noted 55 was not at sun center so we repointed it (It was 80 arc sec too low) We took pictures on all except 82A which is extremely tight on film at this point.

Paul had Joe Kerwin talk with us the other night. Joe had heard we had asked to stay and he indicated they had discussed it on day 22 or 24 and decided against it. He seemed to think they had made the proper decision. Joe indicated Pete, Paul & he were in preflight condition—the only real funny was the fact their red blood cell mass was down 15% or so and their bodies did not start making it up till about Day 17. Why it waited that long they do not know. I wondered if it were possible to affect the mechanism so that it stopped forever. Seems far fetched but a thought.

As an astronaut you become very health conscious—if we were not so healthy we could become hypochondriacs. I've worried about a rupture in the LBNP. My legs losing circumference, back strain on the exerciser, gaining too much weight, not having a good appetite in zero G, heart attacks. You name it—I worried about it—As Owen said—from a health viewpoint these may be the most important 2 months in our lives. He could be right with the changes going on and the remoteness of medical aid.

Owen stayed up late doing ATM because of the activity.

247

Jack told Story we had only 1 set of salt packets left—or 11 normal days worth.

Sounds like a monumental screwup to us. No one counted the salts as what everyone was using—I said it shot a hole in our staying extra days—Jack responded, yeah, all they need is an excuse to get us back on time. Besides if I don't stay longer, said Jack, how can I finish my BMMD work. It is a joke just how much BMMD work we have done over the last weeks. Repeatability tests, 50/150 gram weight change tests/insensible water loss tests (mounting food trays & calibrating).

Jack sure loves to talk on comm. and especially on TV—on EREP he chatters away constantly. Bet the guys on the ground enjoy his blow-by-blow descriptions. Jack started maneuver back to SI with too short a maneuver time in while talking.

Jack's extra food is eating into SL-4 supplies. It's hard to tell which his motive, to hold weight now or to eat up SL-4 food so they can't stay so long.

I mentioned to Owen that our attitude varies like the sun's activity—now it's way up because the sun's activity is up. Owen avowed that ours is up or way up. Always positive. I mentioned that the first few days it didn't seem way up. He allowed that. It wasn't down—sort of like rain on a camping trip—You just have to be patient, good times are ahead. He also allowed that that would be a quotable quote when we got back.

Forgot to mention Jack saying that if they extended us we could always do BMMD calibrations all day—Right, 1 on the BMMD, 1 on each SMMD, then rotate every hour.

In the middle of last night I heard a loud thump—It actually shook the vehicle. I got out of bed and looked at all the tank pressures in the cluster and even in the CSM. Nothing seen—I recorded the time 0725 and told Houston this morning. They called back and said they broke the data down at minute intervals and found nothing—they are now breaking it into ½ sec increments. Something happened, but what I don't know. Perhaps it was a sharper than usual thermal deformation possibly caused by lower beta angle & more rapid sunrise/sunsets. It may show up someday—I hope not entry day with it being a CSM pyro. Maybe we got a small meteoroid hit. Well, forget it, or at least, put it in the back of my mind.

248

I mentioned to Owen about it becoming colder now, especially noticeable in the early morning—You might freeze your ass off if you don't watch out. Owen said yes the FYAO (pronounced FAY-YO) number is going up—hope it never reached 1 because

Jack is pedaling the bike with his arms—good for shoulder and arms, he can do 125 watt/min for 5 minutes.

Owen just flew by with the evening teleprinter messages—We try to find a new record, our old record is from the dome hatch to the ceiling of the experiment compartment.

My pocket timer has been invaluable. It allows you to concentrate on two jobs at once without worry you will forget the first. I keep grey tape on both switches so they won't get accidentally knocked off while in my pocket.

I'm doing 2 16 min exposures on S019—forgot and left the wardroom window open for the first 30 sec on the first one.

Stowed the EREP film in CSMA9 locker complete with brackets—Easier than I thought. We are getting ready.

Owen was on the ATM almost all day doing JOP 12—Calibration rocket work to compare with a more recent sun sensor instrument to insure our instrument calibrations have not drifted.

EREP tape recorder easy to load, tape has not set and does not float off the reel & make a tangle.

Story Musgrave said there was a sound in the background like a roaring dragon—It was the sound of Jack pulling the MK I exerciser.

Jack's triangle shoes are wearing out—Hard work on the bike & MK II mostly. He is going to recommend SL4 bring up an extra pair.

Did TV of S019 set up—It was quick. Did not use the checklist much as it is a familiar one. Did it efficiently.

Took a soap (Nutregina) & rag bath today after work out—Do a soap one every other day—And a water one the other day. We are all clean—Body odor just is not present nor is a sticky feeling after exercise.

If one gets stranded in the middle of an open volume he could merely take off clothes & throw the opposite direction you would like to travel. With no clothes you could wait till the urge to urinate was upon you & jet over. Actually there is enough air flow to prevent any stagnation except against some other object—Sometimes though you will accidentally let loose or your triangle shoe will slip out and you will start a slow drift across the forward compartment—It's exasperating because there is nothing to do but wait.

Several things I have learned up here but the most valuable for any operation is "Do not try to do anything else while you operate ATM—You invariably make ATM mistakes."

Another is 2 to 3 minutes is too short a time to let your mind wander on another subject when you are within that time from a job that must be done then—such as a switch throw, photo exposure, etc.

Our condensate system vacuum leak has fixed itself—somehow when I connected it all back up after the dump probe changeout it did not leak. The ground thinks it's a fitting on the small condensate tank. Glad to see it go because the 1 to 1 ½ hours per night to dump was time consuming.

Changed my jacket today. Wasn't dirty but need the 2 patches, the flag and the name tag off them.

249

Good day. Did EREP. Got 3 Phoenix sites on one pass but did not find the 4th one (Sand Dunes National Monument) Probably had it in the VTS but could not see the dunes. It's hard to find in a short time. Had my chosen photos of the targets (NASA wants them called sites) all clipped or taped in close view—As I was waiting to start the EREP pass we had a 3 min maneuver time to Z-LV I did two chips of the ATM contingency plan for no EREP then powered it down for EREP. Bet that's a space first.

Grand Rapids blinked its lights for Jack Lousma tonight. He said some good words over the headset.

Our private medical conference which seems to drag a bit because we are all well with nothing to report. It's good to hear Paul because sometimes we get straighter information than from Capcoms.

My exploration map has not been of much use during the flight. I'll have to look at it more.—Our synoptic view just isn't as great as I had imagined. So many things seem commonplace now that prior to flight were unknowns, for example the physical condition we would be in by now or the final EVA. How we would get around inside the vehicle, how much work we could do, what our mental attitude would be.

Talked with Sue tonight—Amy had a sore throat—She, perfect person, knew how many days before I came home—what date today was, what date I would splash down—she is so competent. Sue seemed tired from her first few days of teaching school. It will do her good to earn the money to support the swimming pool. She is giving a party (going away) for the Scotts inviting all our group that can make it. Dick Gordon, who has the perfect job, the one Pete & I picked for him, and his wife Barbara may come from New Orleans.

Owen threw some peanuts from the experiment compartment floor all the way to the dome and I caught them in my mouth—may be a distance record for that sort of thing.

Fun to move objects like S019 around and just let them float.

Owen was playing some waltzes and remarking "There's music to waltz by, no better, music to float by"—Jack piped in with, "Hell, that's music to fart by—In fact all music is that way up here."

250

Owen got a X class flare first time manning the ATM panel this morning, we all hustled up there to help. It was well done. The big daddy flare we have been waiting for. All of us were laughing and cutting up. Owen had said yesterday he had used all his luck up. Guess he didn't or he's running on Jack's or mine.

3 of Jack's 4 EREP sites were overcast. We did TV prep prior to the pass.

Took apart the video tape recorder and removed 4 circuit boards, 63 screws did the job. No sign of circuit problems, burns, loose wires, etc.

Owen & Paul had it out on the exercise as Paul said last night Owen was slacking off. Jack was up at the ATM and was laughing and hollering as was I. We have been calling Owen slacker this evening.

Owen and I got 10 EREP film cassettes, 1 EREP tape, 3 Earth terrain camera mags and a S019 mag out of a plenum bag where the SL-2 crew had left it. Wonder if we could use it on our mission or on a mission extension.

Owen & I went up and got out some grape drink & lemonade & pears & peaches & pineapples from overage for all of us to eat. We probably won't eat it all but then will not have to keep going back.

Jack told us about boot camp. How he was told to get in line and be quiet. When he got to the airport. How the DI's treated everyone. How they handled the situation if your bunk was not made right or rifle was not clean or clothes not in place. Jack said it taught you to follow orders without asking questions among other things—Owen said he was not prepared to go thru that type of training at his age—one time maybe yes, but not with what his attitudes are now.

I am very happy with the way our crew is performing—We are doing the job without problems & without giving problems. In my view, it's a professional performance.

Pete was sitting by Bob Crippen tonight. He sent congratulations for passing him in space flight time—I told Bob to tell him the reason I said for him to never look back was that I was right on his heels. Pete said that's the name of the game. I then asked him to ask Pete to fly our entry timeline a couple of times when all the trajectory was finalized—He and Vance.

251

Made a decision for Owen & I to do the EVA. Talked it over with Jack before I asked Owen—Reason was that he would probably get another chance to fly & to EVA, but Owen would not. In my opinion Owen has made this space flight much more interesting than it could have been with 3 operational types.

Tried reading while exercising today. It was much better than just looking at the panel.

Jack mentioned he felt bad after the spin chair—He wonders if we are becoming sensitive again—I told him I had felt that way last time and believed it's because we have not been doing flips, somersaults, spins as much. We will start doing them the last 4 or 5 days.

Owen & Jack could not reach wives for private comm. because all the wives were going out to dinner. They will try tomorrow.

I timed a fastest time from wardroom table to ATM and got 11 sec. From CSM to wardroom 15 sec. Fastest.

Owen was measuring insensible weight loss during sleep again.

252

Day off. Good day—We worked at steady pace—1 EREP cancelled because of weather, the other very cloudy. Jack shot a site that they did not send up because they thought he would not be able to see it. Owen has been excited by seeing such fantastic aurora due to the big flares we have had recently. We found out in the science conference that the big X flares we had the other day peaked 3 times in the night before we saw it. We got it just after peak. Today on the ATM we got 3 flares—2 M1 class just at the last pass and Owen stayed up late to finish the synoptic the next rev. Jack earlier had an eruptive prominence that was large enough to see with the Hydrogen-alpha zoomed fully out (min magnification) You could not see it move but in a few minutes it would be in a different place. It was the only thing Jack said that he had seen move on the sun. It was bigger than any we had in the simulator.

Gratia told Jack not give up hope on a mission extension. Keep up a good attitude and in good health.

253

Looking back over the last few weeks I've had to laugh—we've been like hogs in slop—one day we had EREP, a flare, and aurora all in one pass. Once Jack saw both north and southern lights in the same day—he may be the only man in history to do so. I got the man in space record.

Tonight, I wonder what Dr. Buchanan is going to say. Today I rode the bike when Jack was doing M092—No real problem at all but the Earth doctors just don't think operationally—my guess is they want full attention to watching the subject. Owen thought it was bad form too because it was not called for in the experiment protocol.

Good EREP—did a dry lake near Boulder Colorado and also took data on a hilly site near by—These multiple sites are fun—Got to quit using the worlds targets & shoot on the air.

Did a TV pass over the Sahelian zone (a 500 mile strip just below the Sahara desert) running from the Atlantic to the Red Sea where a 4 to 5 year drought is causing hunger, disease, etc. We hope images from future space vehicles will help these situations—not make it rain but maybe aid in locating potential subsurface water areas, water runoff routes, and so forth. The state department had requested it and wanted a more "promising" speech but I thought best to play it straight.

I have been doing TV of our operations for SL4 guys—it hasn't had much class but I hope it will help them. Just showing on TV what things should help them get going faster and smoother up here.

Owen brought out his tape with Helen Mary's voice playing like she was up here—it started with her muttering over the comm box about the interaction between the comm boxes and causing a squeal. Then she said, "Hello, Houston, this is Skylab." Long pause. "Hello, Houston, this is Skylab." Now Bob Crippen, who was in on the joke says, "What's going on up there." Then Helen says, "Houston, is that you, Bob—(Owen had them made up for several of the Capcoms) I haven't talked to you in a long time." (pause) Bob says, "Where are you?" Helen replies up here in Skylab, the boys needed a hot lunch so I brought them one.

"Boy Bob, this forest fires are fantastic. Smoke as far as you can see, and the sunsets, they are magnificent."

Then in a low voice Helen continues "I better go now. I see the boys drifting up to the Command Module. They don't like me talking to the ground. Bye now."

It was effective, but the ground just changed the subject. It was kind of a letdown.

Owen has a tape of continual laughter, you know the kind, and one with party sounds. We are going to use the party one when we hear we are extended. We can tell them we are going.

On Sunday Jack always has the Capcom call his Sunday school class at the Harris County boys home and tell them he's thinking of them—He plays some of the religious music he has up at the ATM by himself.

254

This was a good day right up to the end. I had a M092/171 scheduled after dinner (supper) and at about 2 min from completion I had to punch out. I had a very warm tingly feeling in my arms and shoulders. Don't know whether it was too much hard driving today or just what—my urine output 2 days ago was larger by 100% my normal—It even beat Owen & Jack. It probably means something. Also I found a knot on the left back side of my neck just below the point where the neck joins the head.

Then while I was riding the bike Bob Crippen, the Capcom, said we had a teleprinter message and to check it—he obviously did not know what it said as it was from the Flight Director Don Puddy reading "Crip's birthday is today and we have a surprise for him. Maybe you could sing Happy Birthday from orbit. (Incidentally, our wives and kids were at MCC tonight.) We rounded up Owen's sound effects tape, found the party sounds and when he came up the next site we played the tape, told him we were having a party in his honor and sang Happy Birthday. Jack stood back and hesitated to sing for some reason. Crip was moved I could tell—they brought out a cake for him—He is one swell guy, and efficient too.

This was the last comm. pass tonight so he told us that he hated to be the bearer of bad news but our request to stay longer had been considered but that it was decided to hold to the present entry schedule of Day 60. We answered with a simple, "OK, thanks."

We talked of it the rest of the night—I ran around saying how great that was—now we could get home—Now we could get off the food—our Command Module would never last more than 60 days—Owen said, "He never thought we would be extended because there was no positive reason for doing so, ATM film used up, more EREP sites than ever thought possible, we're all healthy, all corollary experiments overkilled—to sum up—more risk with little to gain—we could not think of any directorate but our own who would support us. ATM wants us back for data to look at prior to SL4, EREP wants its data, medical wants our bodies. Jack was disappointed.

Got call from the ground wanting to know who had been riding the ergometer during Jack's M092/171—I said me. I knew Paul would ask about it later (by the way, this occurred yesterday) tonight Paul wondered if I thought I could monitor M092 from the bike—I said yes, but that I knew the medical directorate would not like it. I asked if he could ride a bicycle and carry on a conversation at the same time. He said he went over to the simulator and tried it & it seemed OK to him.

When you push off the floor to go to the dome, it's flatfooted and reminds me of the way Superman did it on TV.

In Apollo you go for just a visit or trip to zero G. In Skylab you live it.

255

Morale a little down today because of the mission duration decision. All of us moving a little slow—except in EREP I got my alternate site after looking for chlorophyll bloom in Chesapeake Bay then almost got another site but the gimbal down ran out. We know our sites and its fun to work on EREP—Jack & I both have 191 sites tomorrow—Jack has been down because he has had on nadir swaths lately.

I zeroed in on Seville, Spain, Marseilles, France, and Milan Italy this morning—Did not get Milan, too hazy, saw the airport though.

Got several changes to the entry check list—going to use the docked DAP while undocked to maneuver and to hold attitude—must still use the old undocked DAP for the burn—I enter it at 1 min to the burn.

I spent an hour or so reading the procedures. I will start to work and practice them in the CSM in the days to come—the other evening I spent an hour or so in the CSM touching each switch as I went thru the entry check lists. Nice to find out one does not forget too rapidly—I knew them all and felt at home—But somehow tense. I have noticed I get cold when I get tense—In an airplane in bad weather just prior to a landing approach I always get cold and have to turn up the heat—once in the approach I forget and warm up. It happens all the time, and I guess it is an evolutionary protective mechanism—blood internal to support life, external flow minimized so cuts would not bleed excessively.

Had to reduce Owen's schedule tomorrow—they had him doing post sleep, M110 (blood draw), hematology and urine specific gravity. Two hours assigned for a three hour job—My evening comm. with Sue is over the Vanguard—ship off some of these sites have good comm. Other just don't seem to have it. Guam's is one we have difficulty with.

The coolant leak in the CSM has dried up—we do not operate the suit circuit heat exchanger every 7 days any longer so the leak is probably in that valve.

256

Had 3 or 4 bumps on my head for the last couple of days—think it may be soap drying on my scalp because I have been washing it with soap a lot lately & the washrag rinse just doesn't seem to do the job. Will use the brush & just water for a few days. It is necessary to learn a new way of life up here in many ways not always preconceived. Here we live at zero G and in Apollo we just visit or take a trip at zero G.

Jack & I almost pulled a great trick on EREP. I had just taken the scheduled site (Mono Lake) and an extra site + granite outcrop (Walker Lake) and finished a nadir swath (Boy does Jack burn when he has those) and he suggested we swap for a minute and he would get his site up in South Dakota. He had mentioned this just prior to the pass but I didn't like the idea—Anyway as the time approached we swapped positions—I fumbled with the C&D pad but didn't miss a switch—Jack found the big dragon-shaped reservoir but his site located between the dragon's back legs was barely cover with clouds. We switched back—we lamented the weather the rest of the day because it was executed properly but could not be completed. We will try it again before coming home. Looking for the sites in the T-38's has paid off handsomely.

We then did T-20 suited—suiting up was long as was strapping in—we used the modified umbilical and found little tether dynamics—will be a good fix for SL4—Jack & I hope we do not do 509 again—we did not finish till about 8 o'clock (1300Z) so we then ate and put the equipment up & exercised and got in bed about 30 min late.

Houston said they were going to slip our circadian rhythm around by getting up 2 hrs earlier on entry-6 days and entry-8 days for a total 4 hrs. We didn't have to do it this way but if we can make it work. We thought one 4 hr earlier wake up would do the job better but it lost out—we can make this work though.

Owen commented on no time off & made up list for day off—This I transmitted to the ground along with a request for our day off (EREP & ATM) and get the trouble shooting done and all tests by entry -5 days. Our plan is to work entry whether they do or not and do other things after.

257

Eat, sleep, exercise and entry.

Jack was saying his memory was good but it just doesn't last long. Another of his favorites is "I have a photographic memory but it just hasn't developed."

Haven't felt too spunky the last few days. Owen mentioned it—I suggested it had something to do with my high urine volume 4 days ago—He thinks not, but that it is hard work, long hours—He may be right because we have hitting it steady for almost 2 months. I may go take a little nap prior to S019 & M092 today.

Had a hard time going to sleep—felt guilty emotionally. Intellectually I knew I should but emotionally I had difficulty.

The M092 was the best I've done. Guess the rest was good but probably the greatest reason for the difference this time and last was last I had gone all day on afterburner and the test was just prior to bed. I had essentially run out of steam. Overconfidence will get you if you don't watch out.

Should we fly the paper airplanes? NASA would love to show it on TV—but would it convey the mission the taxpayer is paying for? I do not know—Owen will have to decide.

258

One of the spiders died today—Anita, we think. The prime spider Arabella who is in the vial is OK—She had more time out of the vial. Got out earlier, and guess just remained healthy.

Tried to set up private comm. with Sue tonight—3 different sites—Either I could read them and they not me, or vice versa—or nobody read anybody—the ground always says to check our configuration. It hasn't been wrong in a month. Wonder if there is any follow up on this.

259

Day off. We did our usual 2 EREP & ATM plus not much else. We go to bed 2 hours early tonight to shift our circadian rhythm around—We did not want this but can live with it. I went to the CSM to get a Seconal to sleep on time.—Owen couldn't find the OWS Seconal—it was in some other drug cans that the ground had him move. Later he inventoried some drugs—This sort of thing always puts him in a bad mood. He gave the ground a partial hard time.

Pedaled the ergometer for 95 straight minutes, to establish a new world's record for pedaling non-stop around the world—and as Jack said, I did it without wheels too. Owen was interested and thought he might do it later in the week when our orbit had decayed and then beat my time by a second or less. Bruce McCandless pointed out that he must exceed by at least 5% to establish his claim.

I also took some movies of me exercising on the MKI, the MKII and the MKIII for use later in post flight pictures. Need to do others of acrobatics soon.

Owen did some good TV of how the TV close up lens could be used medically—He looked at Jack's eye, ear, nose, throat & teeth and discussed how the TV might be used by doctors to aid us in diagnosis and in treatment of problems we might have, say, an eye injury, a tooth extraction, suturing a wound or any number of things from a broken bone to skin rash. Owen has a mind that dwells on the scientific aspect of all that he does. He knows much about much—he is interested in all branches of sciences. He is a great back of the envelope calculator—able to reduce most problems to their simplest elements. He has done great school room TV demonstrations of zero G water, magnets, his spiders.

On the EREP C&D panel was

CDR / PLT DUMB SHIT
EREP CUE CARD
S192 Door—Open
S191 Door—Open
S190 Window—Open
Tape Recorder Power—On

260

Jack said he felt good today. I had noticed he was not as happy as usual the last four days. He says it's because of sleep lost.

Our circadian rhythm is in good shape today after the shift. The continuous day/night cycle makes it easer than on Earth—Also we are the only ones we see and we are all on the same schedule. Maybe that's why days off that we work are no sweat, everyone we see is working.

Found out today that we had 6 hrs from tunnel closeout to undock—then 1 hr 45 min from there to deorbit burn then 24 min to 400,000 feet. A nice slow timeline that will allow us to get set up, double/triple checked for our entry.—Maybe we should get up 2 hours later—well, we'll see.

ATM operations have become much simplified the last week—with all the solar activity the film is gone.—It was a peak on the sun and we were lucky to see it.

We have had the opportunity and were PREPARED—who could ask for more. They call it luck but PREPARATION/opportunity must both be present. I am very satisfied.

261

Owen & Jack were doing TV of paper airplane construction and flying. The trick is not to cause them to have lift or they will pull up into a loop—with more space they would continue in loop after loop. The designs were different than we've all made as kids—more folds in the nose and on the inside edge of the wings.

We interrupted our work to do some special TV—I took 2 of the M509/T-20 pressure bottles put a twin boom sunshield pole between them and taped that up. I then put some red tape and marked 500 on each. We now had a 1000-lb barbell. We showed Jack with Owen and I lifting the barbell on Jack. He grimaced till he was red as he lifted it up. We then moved it so it was to his front and he lifted it again as he came to full up he released his triangle shoe locks and kept going off the top of the camera. We then set up the ATM for 2-3 minutes and did the Bean push up—both hands first, then with Jack on my back, then also Owen on his back—then a one arm push up then the finale a no arm push up with all 3. The piece-de-resistance was a 3 man high with Owen at the bottom, me in the middle and Jack on the top. We had to mount the TV on the first floor and do our stand up on the forward compartment floor to fit all of us in the field of view. Owen was great. He wobbled around like we were toppling. We now must put it all on movies to use after we get back.—Funny, you never know what movies people will find funny—It gives a welcome relief from the science we do.

At lunch Jack was talking about "stretch gut"—wherein he postulates since our stomach has not been pulled down for 2 months by gravity the tendons that hold it up will be weak and when we get home to 1g, the stomach will drop and fall out over our pants. We laughed and further thought if that happened our testicles would have the same problem in the scrotum. "Stretch nuts" I guess. Jack said he finally found out why people had trouble sleeping in space—no gravity to hold the eyelids down when the body gets tired of doing it. Up they come and you wake up.

Just went to the CM to put our pants that we will wear on the ship on our couches so that as we come up with items we wanted to take home we could put them right in the pockets rather than have items hidden all over the s/c. We have been discussing bringing Vance, Don & Bill's patches (U.S.A. flag, NASA symbol) home for them but we wanted them to fly the maximum distance so should we leave them for SL4 to bring? We have been thinking what to bring to the troops we work with.

When Alexander the Great visited Diogenes and asked whether he could do anything for the famed teacher, Diogenes replied, "Only stand out of my light."

262

Woke up ~2 hrs early today. It's our last time shift to fit the post-landing medical checks. The doctors are on the boat right now, heaven knows doing what. Waiting for our landing in 6 more days. Owen told Paul to take along his Scopedex and to do head movements.

263

Out the wardroom window we saw a bright red light with a bright/dim period of 10 sec. It got brighter and drifted along with us for 20 min. or more. I said it was Mars but Jack & Owen said a satellite—it was because it also was moving relative to the stars. It may have been very near, it was the brightest object we've seen.

Also saw a laser beam from Goddard. It looked like a long green rod perhaps as long as your fingernail held at arm's length when viewed end on, that is 20 times longer than in length (which was parallel to the horizon) than in width. Tomorrow the ground will tell us that Goddard did not have our trajectory right and did not point at us—we may have seen the side view somehow. Owen said at the time a laser should appear only as a bright point of light and not a bar.

Entry -5 day. CSM checks went well—somehow I knew they would. We only look at the G&N and the real problem might be the RCS. Well, we'll know soon enough. There's no reason to believe anything's wrong with the two remaining quads A&C. Got to bed an hour and ½ late tonight because of late scheduled meals and not wanting to exercise prior to M092.—I talked with the ground about watching our meals and sleep periods and exercise periods and keeping them on time because of the physically demanding features of returning to the 1g on Earth next Tuesday. The days can't pass fast enough. We have done our job and are ready to get back. At least I am, I don't know about Jack but Owen would like to stay.

EREP went well till I got to talking too much then I turned off 190 (the camera) rather than 192 (the multispectral scanner). Wish we could come up with a big play on EREP like on ATM but we have not had the weather.

Took Seconal because I wanted to get to sleep. It worked.

264

Got up feeling a little high from the Seconal and went right to the ATM. Did OK, usually make mistakes early in the morning.

Thinking about EVA all day, did our pre-prep—installed new umbilicals without water. We will use O2 flow cooling for the first time. Dick Truly mentioned a 4 hour EVA limit before this but we should finish well before the time. I need to give my suit a good last minute check out. The drying and desiccants have helped my confidence with their integrity.

Had a TV press conference come off well—all of us had question asked us and we answered well I think. We are going to have many more of those the next few months—glad Jack and I are going to Russia in November. Wish Sue could go.

Found out where Elba and St. Helena Islands are—Dick Truly looked it up.

Helen Mary mentioned that the office's reorganization had begun already. 5 groups, don't know where I'll be. Glad to be backing the Russian flight.

Our last EREP today. U.S. & Sicily & Ethopia I saw an active volcano on Sicily and the Aswan Dam in Egypt. It is the most well defined man made object I've seen to date.

Bibliography

Books

Belew, Leland F. *Skylab: Our First Space Station*. Washington DC: NASA, 1977.

Belew, Leland F., and Ernst Stuhlinger. *Skylab: A Guidebook*. Washington DC: NASA History Office, 1973.

Bilstein, Roger E. *Stages to Saturn*. Washington DC: NASA, 1980.

BioTechnology, Inc. *Skylab Medical Experiments Altitude Test (SMEAT)*. Houston: NASA JSC, 1973.

Compton, William David, and Charles D. Benson. *Living and Working in Space: A History of Skylab*. Washington DC: NASA, 1983.

Cooper, Henry S. F., Jr. *A House in Space*. New York: Holt, Rinehart and Winston, 1976.

Cromie, William J. *Skylab*. New York: David McKay Company, 1976.

Eddy, John A. *A New Sun: The Solar Results from Skylab*. Washington DC: NASA, 1979.

Johnston, Richard S., and Lawrence F. Deitlein. *Biomedical Results from Skylab*. Washington DC: NASA, 1977.

Lyndon B. Johnson Space Center. *Skylab Explores the Earth*. Washington DC: NASA, 1977.

Mark, Hans, ed. *Encyclopedia of Space Science and Technology*. Hoboken NJ: Wiley-Interscience, 2003.

Newkirk, Roland W., and Ivan D. Ertel. *Skylab: A Chronology*. Washington DC: NASA, 1977.

Shayler, David J. *Skylab: America's Space Station*. Chichester, UK: Springer Praxis Books, 2001.

Summerlin, Lee B. *Skylab: Classroom in Space*. Washington DC: NASA, 1977.

Ward, Bob. *The Light Stuff.* Huntsville AL: Jester Books, 1982.

NASA Documents

NASA News Releases, 1972–1974.

Skylab Presentation Reference, Johnson Space Center, Revised May 1973.
Skylab 1/2 Technical Crew Debriefing. NASA Johnson Space Center, Crew Training and Simulation Division, June 30, 1973. JSC-08053.
Skylab Mission Report, First Visit. NASA Johnson Space Center, Mission Evaluation Team, August 1973. JSC 08414.
Skylab 1/2 Medical Debriefing. NASA Johnson Space Center, Medical Research and Operations Division, July 3, 1973. JSC 08054.
Skylab 1/2 Preliminary Biomedical Report, NASA Johnson Space Center, September 1973. JSC 08439.
Skylab 3 Preliminary Biomedical Report, NASA Johnson Space Center, February 1974. JSC 08668.
Skylab 1/4 Medical Debriefing. NASA Johnson Space Center, February 27, 1974. JSC 08811.
Skylab 4 Preliminary Biomedical Report, NASA Johnson Space Center, January 1975. JSC 08818.
Skylab 1/2 Flight Plan; as flown; 1973.
Skylab 1/2 Technical Air-to-Ground Transcription, Johnson Space Center, July 1973. JSC 08051.
Skylab 1/2 Onboard Voice Transcription, Command Module and OWS recorders, NASA Johnson Space Center, July 1973. JSC 08052.
Skylab Experiments. NASA Headquarters, Office of Manned Space Flight, August 1972.
USS *Ticonderoga* Cruise Book, 1973.
Skylab Astronyms, Philco-Ford Corporation, Houston, TX, 1973.
Numerous internal NASA memos and notes, on file, J. Kerwin.
NASA "Oral History" Interviews: Berry, Carr, Crippen, Garriott, Gibson, Harlan, Hutchinson, Kerwin, Kinzler, Lousma, Pogue, Thornton, Weitz.

Web sites

http://www.nasa.gov
http://www.astronautix.com
http://www.collectspace.com

Index

Page numbers in italics refer to illustrations.

In the Outward Odyssey: A People's History of Spaceflight series

Into That Silent Sea
Trailblazers of the Space Era, 1961–1965
Francis French and Colin Burgess
Foreword by Paul Haney

In the Shadow of the Moon
A Challenging Journey to Tranquility, 1965–1969
Francis French and Colin Burgess
Foreword by Walter Cunningham

To a Distant Day
The Rocket Pioneers
Chris Gainor
Foreword by Alfred Worden

Homesteading Space
The Skylab Story
David Hitt, Owen Garriott, and Joe Kerwin
Foreword by Homer Hickam

Ambassadors from Earth
Pioneering Explorations with Unmanned Spacecraft
Jay Gallentine

Footprints in the Dust
The Epic Voyages of Apollo, 1969–1975
Edited by Colin Burgess
Foreword by Richard F. Gordon

Realizing Tomorrow
The Path to Private Spaceflight
Chris Dubbs and Emeline Paat-Dahlstrom
Foreword by Charles D. Walker

To order or obtain more information on these or other University of Nebraska Press titles, visit www.nebraskapress.unl.edu.